Good Nutrition – Good Bees
ISBN: 978-1-912271-95-5
David Aston & Sally Bucknall

©2021 Northern Bee Books, Scout Bottom Farm, Mytholmroyd, Hebden Bridge, HX7 5JS
(UK). Tel: 01422 882751. www.northernbeebooks.co.uk

Book design by www.SiPat.co.uk

Good Nutrition
Good Bees

David Aston & Sally Bucknall

Contents

Preface   19

Introduction   21

1.   The British Isles – and their climates   23

2.   A short history of the honey bee in the British Isles   26

2.1   Early days in the evolution of honey bees 26
2.2   Before and after the last glacial period 27
2.3   The Dark European Honey Bee (*Apis mellifera mellifera*) 28
2.4   Honey hunters 29
2.5   Beekeeping in hives...30
2.6   As a symbol of love 32
2.7   The human need for sweetness...32
2.8   The human need for light...33
2.9   Art, literature and communication 33
2.10   Honey bees and religion..33
2.11   Beekeepers and their Saints 34
2.12   The honey bee, the hive and thriftiness 34
2.13   The concept of the honey bee community in religious and secular life.................34
2.14   The history of beekeeping in English gardens 35
2. 15 The Victorian beekeeping revolution in the British Isles...............................35
2.16   Hive products and their typical uses..37
       2.16.1   Historical uses of beeswax..37
       2.16.2   Medicinal honey...38
       2.16.3   Pollination...39

3.   Human relationships with the honey bee in the
        British Isles today   41

3.1   Taking stock 41
3.2   Beekeeping and gender roles 44
3.3   Changing weather and climate change 44
3.4   The human population in the British Isles 45
3.5   Changes in land use 45
3.6   Environmental aspects of agriculture and horticulture and beekeeping 46
3.7   Agriculture in the UK in 2019 50
3.8   Urban and suburban man 51

4.   Honey bees – domesticated or semi-domesticated?   53

5.   Beekeeping – applied honey bee biology   55

6.   Recent history and current status of beekeeping in
        England and Wales   57

6.1   Current status of the honey bee in the British Isles 63

7.   The differences between hunger, malnutrition and starvation   68

8.   The role of the beekeeper in understanding the honey bee
        colony and factors affecting the species' survival   71

9.   The honey bee colony structure – a superorganism –
        the fundamentals   74

10.   Wild honey bee colonies   77

10.1   Genetic adaptation to location 78
10.2   Colony density 78
10.3   Space for colonies to exist in 78
10.4   Forage diversity 78
10.5   Diet 79

11.   Honey bee colony life cycle   80

11.1   Overwinter survival 81
11.2   Nest homeostasis 82
11.3   Temperature and its effect on the colony 82
11.4   The Overwinter Cluster 83

12.	Optimum colony size	85

12.1	Size Matters	85
12.2	Lessons from bees in the wild	86
12.3	Colonies to be allowed to follow their natural cycle, including successful overwintering and reproduction (swarming)	87
12.4	Colonies to be managed to maximise honey production	88
12.5	Colonies to be managed for pollination services	89
12.6	Colonies to be managed for educational and instruction purposes	89
	12.6.1 Observation Hives	89
	12.6.2 Nucleus colonies	89
12.7	Colonies to be used for the breeding, raising, rearing and mating of queens	90
12.8	Non-grafting techniques	91
12.9	Grafting-based techniques	91
12.10	Mating hives	91
12.11	Breeder queen colonies	91
12.12	Drone-rearing colonies	91
12.13	Colonies to be managed for increasing colony numbers by 'splits'	92

13.	The composition of a honey bee colony	94

13.1	The Basics	94
13.2	The worker honey bee and her roles in the colony	95
13.3	Ageing, adaptation and changing roles in the colony	96
	13.3.1 Cell cleaners	97
	13.3.2 Nurses	97
	13.3.3 Middle-aged bees	98
	13.3.4 Foragers	98
13.4	The drone & his roles in the colony	99
13.5	The queen & her roles in the colony	100
	13.5.1 Eggs for colony development and reproduction	100
	13.5.2 Colony cohesion	101
	13.5.3 End of Life	101

14.	Some important aspects of insect and honey bee structures and physiology – with respect to nutrition	102

14.1	Hypopharyngeal glands	102
14.2	Mandibular glands	102
14.3	Worker jelly	103
14.4	Royal jelly	103
14.5	Digestion and excretion	103
14.6	Storage and utilisation of carbohydrates	104
14.7	Storage and utilisation of proteins and amino acids	104
14.8	Storage and utilisation of lipids	105
14.9	The Fat Body	105
	14.9.1 Functions	105
	14.9.2 Structure	106
	14.9.3 Fat Bodies and Varroa	107

15.   Honey bee pheromones and their role in nutrition   108

15.1   Some definitions...108
15.2   Queen pheromones and the queen signal 109
15.3   Worker honey bee pheromones...109
15.4   Tarsal glands 110
15.5   Pheromones, the Waggle Dance and foraging.............................110
15.6   Worker mandibular glands and 2- heptanone (2-HPT) 110
15.7   Brood pheromone (BP) 110

16.   Honey bee hormones and their role in nutrition   112

16.1   Introduction 112
16.2   Juvenile Hormone (JH) 112

17.   Key nutrients required for growth / reproduction & existence 114

17.1   Some basic considerations 114
17.2   Carbohydrates 114
        17.2.1   Nectar 115
                 17.2.1.1   Functions 115
                 17.2.1.2   Nectar composition 115
                 17.2.1.3   Types of nectar 117
                 17.2.1.4   Types of sugars and their importance to honey bees.............119
                 17.2.1.5   Sugars which are harmful to the honey bee 119
                 17.2.1.6   Examples of types of nectar and sugar values.........................120
        17.2.2   Honey 124
                 17.2.2.1   The transformation of nectar into honey..............................124
                 17.2.2.2   Physical transformation 125
                 17.2.2.3   Chemical and biochemical transformation............................125
                 17.2.2.4   Nectar consumption by the colony...................................126
                 17.2.2.5   Honey Granulation 126
                 17.2.2.6   Honey Fermentation 127
        17.2.3   Honeydew 127
                 17.2.3.1   What is Honeydew?..127
                 17.2.3.2   Origins of Honeydew 127
                 17.2.3.3   Plant sources..128
                 17.2.3.4   Composition of honeydew honey.......................................128
                 17.2.3.5   Use by honey bees...129
        17.2.4   Plant toxins in honey bee nutrition relevant to humans and honey bees129
17.3   Proteins and amino acids 129
        17.3.1   Proteins...129
        17.3.2   Amino acids 130
                 17.3.2.1   Essential Amino acids 130
        17.3.3   Pollen – more than just proteins and amino acids 131
                 17.3.3.1   What does pollen contain? 131
                 17.3.3.2   Colours of Pollen Loads....................................132
                 17.3.3.3   How honey bees use pollen 134
                 17.3.3.4   Nurse bees and their key role in protein availability.............134

17.3.3.5   Consequences for the colony when nurse bees do not have access to pollen 135

17.3.3.6   Regulation of pollen collection by the colony 135

17.3.4   Pollen quality and proteins 135

17.3.4.1   Protein Content and its determination 135

17.3.4.2   Nutritive Value expressed as Crude Protein Content 136

17.3.4.3   Crude Protein Content of some Plant Species 136

17.3.4.4   Pollen quality and the season 142

17.3.5   Bee bread 143

17.3.5.1   The importance of bee bread 143

17.3.5.2   Preparation of bee bread 143

17.3.5.3   Composition of bee bread 144

17.3.5.4   Nutritional value of bee bread in response to local land use and forage 145

17.3.6   Vitellogenin 149

17.4   Determining the nutritional status of a honey bee colony 150

17.5   Essential fatty acids 150

17.5.1   Source and Function 150

17.6   Lipids 151

17.6.1   Composition 151

17.6.2   Function 151

17.7   Sterols 152

17.8   Vitamins 152

17.8.1   Sources 152

17.8.2   Functions 153

17.9   Inorganic elements 154

17.9.1   Other sources of inorganic elements 154

17.9.1.1   Micro-organisms 154

17.9.1.2   Fungal spores 155

17.10   Water 155

17.10.1   Requirements of a colony 155

17.10.2   Functions of water in a colony 155

17.10.3   Quality and Availability 155

17.10.4   Provision of water 156

17.11   Dietary elements (minerals) and various other substances 156

17.11.1   Dietary Elements 156

17.11.2   Dietary elements in honey and honeydew 158

17.11.3   Dietary elements in pollen 158

17.11.4   Ballast Substances 159

17.11.5   Crude fibre in pollen 159

17.12   Beeswax 159

17.12.1   Composition 159

17.12.2   Functions of beeswax in the colony 160

17.13   Plant resins and propolis 160

17.13.1   Composition 160

17.13.2   Use of resins and propolis by honey bees 162

17.14   Venom production 162

18.   Metamorphosis and aspects of the physiological & nutritional requirements of each caste thoughout the individual's life cycle   164

18.1   Metamorphosis - definition 164
18.2   Caste determination 165
18.3   The worker 166
      18.3.1   Egg 166
      18.3.1   Larva 166
      18.3.3   Adult … various stages 168
18.4   The Drone 169
      18.4.1   Egg 169
      18.4.2   Larva 170
      18.4.3   Adult 170
18.5   The Queen 171
      18.5.1   Egg 171
      18.5.2   Larva 171
      18.5.3   Adult 172

19.   Gross nutritional requirements of the honey bee colony during its annual cycle: Carbohydrates   173

19.1   Factors affecting the productivity of the colony 173
19.2   Nectar consumption by a colony 173
19.3   Larval nutrition 174
19.4   Adult workers 174

20.   The gross nutritional requirements of the honey bee colony during its annual cycle: protein   175

20.1   Colony nutrition 175
20.2   Larval nutrition 175
20.3   Adult nutrition 176
20.4   Overwintering bees 177

21.   The honey bee gut (micro) biome – an introduction   178

22.   Bee health, vitality and nutrition   181

22.1   What is meant by health and vitality? 181
22.2   The honey bee immune system 182
22.3   Environmental stressors 183
22.4   Carbohydrates 184
22.5   Nutrition and Immunity 186
22.6   The Balance of Carbohydrates and Proteins 187
22.7   Pollen and Longevity 187
22.8   Propolis and honey bee health 187
22.9   Cannibalism 188

23.1   Honey bee Colony Number Fluctuations...189
23.2   Computer based Modelling 189
23.3   Beekeeper responsibilities..190
23.4   The Challenges to the Colony...191
23.5   Important threats to honey bee colonies..192
        23.5.1   American Foul Brood (*Paenibacillus larvae larvae*) (AFB) 193
        23.5.2   European Foul Brood (*Melissococcus plutonius*) (EFB)...........193
        23.5.3   Nosema (*Nosema ceranae* and *Nosema apis*)...........................193
        23.5.4   Amoeba (*Malpighamoeba mellificae*) 195
        23.5.5   Chalkbrood (*Ascosphaera apis*)...195
        23.5.6   Acarine (*Acarapsis woodi*) 195
        23.5.7   Tropilaelaps (*Tropilaelaps claraeae* and *T. mercedeseae*).........196
        23.5.8   Small Hive Beetle (*Aethina tumida*) 196
        23.5.9   Varroa (*Varroa destructor*)...196
        23.5.10   Greater Wax moth (*Galleria mellonella*) 199
        23.5.11   Lesser Wax moth (*Achroia galleria*) 200
        23.5.12   Mice 200
        23.5.13   Hornets..200
                  23.5.13.1 European Hornet (*Vespa crabro*) 200
                  23.5.13.2 Asian / Yellow- Legged Hornet
                  (*Vespa velutina var nigrithorax*) 200
        23.5.14   Wasps..201
                  23.5.14.1 Yellow Jacket / Common Wasp (*Vespula vulgaris*)................202
                  23.5.14.2 German Wasp (*Vespula germanica*) 202
                  23.5.14.3 Yellow jackets
                  (*Dolichovespula media, D. sylvestris, D. saxonica*) 202
23.6   Viruses found in honey bees..202
        23.6.1   Viruses and their symptoms found in honey bees in the UK.............202
        23.6.2   Deformed Wing Virus (DWV) 206
        23.6.3   Chronic Bee Paralysis Virus (CBPV)..206
        23.6.4   Black Queen Cell Virus (BQCV)..206
        23.6.5   Acute Bee Paralysis Virus (ABPV) 206
        23.6.6   Sacbrood virus (SBV)...206
        23.6.7   Can viruses be managed through beekeeping practices? 207
23.7   Other stressors...208
        23.7.1   Xenobiotics 208
        23.7.2   Pesticide residues in nutrients 208
        23.7.3   Veterinary Medicines for Honey Bees 209
        23.7.4   Effects of Antibiotics on the beneficial microflora of honey
                  bees and their Immune systems...210
        23.7.5   Transgenic products 211
        23.7.6   Genetically Modified Organisms (GMO's) and crops............211
        23.7.7   Minerals in Honeydew 212
23.8   Specific examples of nutritional impacts and stressors.......................212
        23.8.1   Pollen and honey bee health 212
        23.8.2   Bee bread and fungicides and chalkbrood symptoms............212
        23.8.3   Essential amino acids and their role in survival of
                  infection of Deformed Wing Virus (DWV)................................212

23.8.4   Nectar and honey bee health...213
23.8.5   Non-Infectious diarrhoea or dysentery 213
23.8.5   Propolis and honey bee health...214

24.   Phenology & Pollination   215

24.1   Phenology – definition 215
    24.1.1   Recent findings in the British Isles.....................................215
24.2   Pollination – definition and introduction 217
24.3   Crop pollination by honey bees...220
24.4   Historical aspects of beekeeping, crop pollination and honey production 225

25.   Plant and flower structure   228

25.1   The availability of pollen – timing of pollen presentation...............228
    25.1.1   Flower Opening and Closure...228
    25.1.2   Presentation of the pollen and its attachment to the honey bee 232
25.2   The availability of nectar and nectaries 233
    25.2.1   Floral and extrafloral nectaries 233
    25.2.2   Nectar as an attractant 233
    25.2.3   Nectar as a protectant 234
    25.2.4   The availability of nectar 234
    25.2.5   Types of Floral Nectaries...237
        25.2.5.1   Sepal nectaries 237
        25.2.5.2   Petal nectaries 237
        25.2.5.3   Staminal nectaries...238
        25.2.5.4   Carpel or Gynoecial nectaries..........................238
        25.2.5.5   Receptacle Nectaries.......................................238
        25.2.5.6   Axial nectaries..238
        25.2.5.7   Extrafloral nectaries..238

26.   What we understand honey bees know about flowers   240

26.1   What do honey bees know about flowers? 240
26.2   Honey bee vision 241
26.3   Optic flow...244
26.4   Use of landmarks for Navigation and Homing............................244
26.5   Foraging and Flower Detection..244
    26.5.1   Foraging behaviour in relation to pollen quality 244

27.   The hoarding instinct   246

28.   Honey bee foraging   247

28.1   Scout bees and foraging...247
28.2   Flower constancy...248
28.3   Forager worker honey bee ways of working different types of flowers 248
28.4   Some facts about foraging 249
28.5   Basic extrapolations re nectar...250

28.6   Basic extrapolations re pollen...250
28.7   Pollen gathering by honey bees 250
28.8   Communication dances and signals 251
28.9   Food odour density...253
28.10 Optimisation of colony foraging efficiency and effectiveness 253
28.11 Foraging for water..255
28.12 Foraging for plant resins 256

29.   Landscape, land use and forage   257

29.1   Types of landscape and land-use patterns in the British Isles................257
29.2   Changes in forage with especial reference to honey bees as reported
       by beekeepers..259
29.3   Most useful nectar-yielding plants in England compared with Scotland, Ireland,
       and Wales. 260

30.   The plant palette in the garden and pollinators   267

30.1   Where to Begin?...267
30.2   Garden plants rich in nectar and / or pollen which attract 'pollinating insects' over
       the course of a year in a garden. 268

31.   Hedgerows and boundaries and screening   273

31.1   The importance of hedgerows, boundaries and screening 273
31.2   Plants species commonly found by hedgerows, on ditch banks and
       banked walls visited by honey bees 274

32.   Roadside Verge Management   277

32.1   An important resource 277
32.2   Top 10 plant species for roadside verges 278

33.   Gardeners and their grass ('lawns')   279

33.1   Useful flowering plant species which can be grown in grass and
       the effects of different management techniques 279

34.   Trees and shrubs   281

34.1   Trees matter 281
34.2   Tree and shrub species visited by honey bees 283

35.   British floral sources considered to be of importance
       to the honey bee and their flowering periods   292

35.1   Early records 292
35.2   The situation today...295

36.   Some likely effects of climate change on species change
       and flowering times in plants visited by honey bees   305

37.   Archaeotypes, neophytes and invasive non-native species   307

37.1   Some plant introductions into the British Isles 307
37.2   Archeophytes...310
37.3   Neophytes 310
37.4   Invasive non-native species..310

38.   Land management and its impact on forage availability
       for honey bees throughout the year   312

38.1   Current situation..312
38.2   Agro-forestry..313
38.3   Managing farmed landscapes 314
38.4   Managing urban and suburban landscape areas 316
38.5   Managing the Semi-natural landscape 317

39.   The feeding of honey bee colonies – options available   318

39.1   Initial consideration 318
       39.1.1   Why feed?...318
       39.1.2   Terms used 319
       39.1.3   HMF (5-Hydroxymethylfurfural) 320
39.2   Honey as a feed...320
39.3   Honeydew honey – as a feed...322
39.4   Carbohydrates - the honey bees' energy source 322
       39.4.1   Why feed sugar as a supplementary feed?.......................323
       39.4.2   Main reasons for feeding sugar...323
       39.4.3   General Precautions in feeding sugar to colonies.........323
       39.4.4   When to feed the sugar? 324
       39.4.5   What form should the sugar be in? 325
       39.4.6   Dry Sugar 325
       39.4.7   Sugar syrup 326
       39.4.8   Syrup for stimulating a colony...327
       39.4.9   Thick syrup suitable for processing by bees into overwinter stores.......328
       39.4.10 Emergency feeding to recover starving bees...................329
       39.4.11 Preventing the Fermentation of sugar syrup in feeders.........................329
       39.4.12 Feeding in the comb...330

40.   Types of sugar feeders   331

40.1   Small quantity feeders 331
       40.1.1   Jar Attachment feeders 331
       40.1.2   Wooden entrance feeders 331
       40.1.3   Plastic entrance feeders 332
       40.1.4   Frame feeders (also called division board feeders) 332
       40.1.5   Tin feeders...332

40.2   Larger quantity feeders..332
    40.2.1   'English feeder' / tray feeder..332
    40.2.2   Rapid feeder 333
    40.2.3   Ashworth feeder 333
    40.2.4   Contact feeders 333
40.3   Good apiary hygiene when feeding syrup...333

41. Sugars in the form of fondant and candy   335

41.1   Fondant 335
41.2   Candy 335
    41.2.1   Hard Candy..335
    41.2.2   Queen Candy 336
41.3   Syrup..336
    41.3.1   Invert sugar (acid-inverted sugar) syrup 336
    41.3.2   HFCS (High Fructose Corn Syrup) - HFS (High Fructose Syrup)
            also called Isoglucose 336
41.4   Other potential sources of sugars..337
41.5   Sugars poisonous to honey bees 337

42. Protein - through the feeding of pollen supplements or substitutes   339

42.1   The need to Feed Protein..339
42.2   Whether to feed 340
    42.2.1   Colony size and its nutritional status.................................340
    42.2.2   When conditions are unsuitable for bees to fly and forage 340
    42.2.3   Weak colonies 340
    42.2.4   Timing and availability of surrounding forage 340
    42.2.5   Pollen-deficient 'nectar' flows..340
    42.2.6   Pollen sources of doubtful quality.....................................340
    42.2.7   Pollen clogged combs...341
42.3   Quality of the protein supplement to be fed 341
42.4   Factors affecting the consumption of pollen substitutes and supplements 341
42.5   Supplementation by British Isles beekeepers.....................................342

43. Pollen / protein supplements with / without fondant   344

43.1   Physical forms in which protein supplements can be given to
    honey bee colonies..344
    43.1.1   Pollen / Bee bread..344
    43.1.2   Dry Feed...344
    43.1.3   Patties...344
    43.1.4   Recipe for a pollen supplement (bulk)................................345
    43.1.5   Recipe for pollen substitute (smaller quantities) 345
    43.1.6   Recipe for high protein and high carbohydrate pollen substitute 345
43.2   Commercial pollen substitute...345
43.3   Candy Board Cake 345
43.4   Liquid Amino Acid and Protein Supplements.....................................346

44. Other Supplementation   347

44.1 Vitamin supplementation 347
44.2 Propolis supplementation 347
44.3 No additional feeding..347

45. Current supplementary feeding practices in England and Wales   349

45.1 Proportion (%) of beekeepers in England and Wales reporting feeding
their honey bee colonies...349
45.2 Use of different types of bee feeds in England and Wales....................350

46. Beekeeping strategy considerations for meeting colony nutritional needs throughout the year   352

46.1 The Seasons..352
46.2 Useful skills..353
46.3 Preparations for any Eventuality 353
46.4 Assessing Stored Carbohydrates..354
46.5 Assessing protein / bee bread reserves..357
46.6 Considering what resources will become available in the future 357
46.7 The importance of Varroa management and awareness of
Deformed Wing Virus (DWV)..358
46.8 Acoustic Hive Monitoring...359

47. Overwinter colony survival   360

48. During the winter until early spring   365

49. The spring expansion   367

50. Early and mid summer   372

51. Late summer / autumn   374

Protein intake 374
Nectar intake and processing...374
In summary 375

52. For colony consolidation before onset of winter   376

53. When drawing foundation   379

54.	When queen rearing	380
55.	Preparing colonies for pollination services	383
56.	Feeding nuclei, small swarms, queen mating and small colonies	384
57.	Emergency feeding	385
58.	Hive hygiene	386
59.	So what is in store for our honey bees in the future?	389
60.	Conservation agriculture	393
61.	Implications for decision makers	395
62.	In conclusion	397
	References	398
	Annex I	415

Some families of plants used by honey bees, with notes on their nectaries 415

Annex II	419

Recommended wildflower and grass seed mixtures to benefit bees on
field margins or grassland sites of specific soil types 419
Cultivation notes 419
Neutral / loamy soils 420
Calcareous soils 421
Acidic / sandy soils 422
Wet soils 423

Contents

Preface

Within the British Isles can be found a diverse flora which is determined to a large extent by the environment and the climate experienced across them.

Beekeeping and the ability of the honey bee to survive in these Isles is a fascinating subject.

Both of us have long been interested with the landscapes and the habitats of the British Isles and in which we expect the honey bee to survive.

The ability of the honey bee to survive and more importantly to thrive in the British Isles is a reflection of the state of the environment and the current problems being experienced by honey bee colonies suggest that all is not well and needs to be addressed.

This book is intended to encourage and inform those interested in the welfare and fate of the honey bee and about the needs of the honey bee with particular reference to honey bee nutrition.

The implications and practicalities of addressing these needs are profound, but are indicative of the changes which 21st century humans will have to make if there is to be a future for the human species and the honey bee on Planet Earth.

In preparing this book we have tried to offer a source which is very amenable to 'dipping into' to finding and reading something of interest. The interconnectedness of so many of the subjects we are intent on covering make writing a book to read as a novel nigh on impossible. There is a price to be paid for this and that is repetition, but we hope that the way in which we have tried to portray the information will encourage and stimulate readers to discover this high degree of interconnectedness and the interrelatedness of so many of the topics.

We hope that you will enjoy reading the book and use it as a source of reference, a starting point to pursue topics in more detail and a springboard to promote ideas and values which will be of use to humans and will benefit the honey bees of the British Isles, an insect species with whom humans have had a long and beneficial relationship and which enrich the flora and fauna on which we all depend.

Wressle

East Riding of Yorkshire

2021

Introduction

Many people are fascinated by honey bees and this reflects the long relationship that *Homo sapiens* has had with them. In particular the benefits mutually gained from the honey bee (*Apis mellifera*) and the opportunities that man's activities have opened up to the honey bee.

In recent years the fate of honey bees and pollinators in general have been the subject of much concern, wringing of hands, research, teaching, education and extension work by beekeeping associations, government agencies, schools, charities of all kinds all aimed at meeting the demands to try to fulfil all sorts of objectives.

The importance of pollinators cannot be overstated, and this is recognised in many initiatives both nationally and internationally and these include identifying policies, practices and their implementation to redress the currently declining fortunes of pollinator species of all kinds, including the honey bee.

It is true to say that through the reporting in the media and the great interest shown in schools that there is a greater awareness of the importance of the role of pollinators in helping *Homo sapiens* produce enough food in the form of crops, fruit and seeds through pollination and other delights produced by the honey bee such as honey and beeswax.

'Bees need food and shelter' is the strapline for the current UK Insect Pollinator Initiative and in common with many other species of bees and pollinators in the British Isles the honey bee also has these needs.

Most honey bees living in the UK are managed by beekeepers but there are some wild and feral colonies often living in their natural locations of cavities within the trunks of old trees, as well as in roof spaces and in chimneys.

This book has been primarily written as an introduction to the nutrition and nutritional management of honey bees and which can be 'managed' through practices such as beekeeping.

It has been written not just for beekeepers to help them in their craft, but also to also interest and inform those who have an inquiring mind and would like to know more about the honey bee and what can be done to create and sustain an environment of greater diversity for themselves as well as the bees. It is also hoped that those who are responsible for management decisions which could impact the honey bee, both positively and negatively, will also find the information contained in this book helpful.

Whilst this book concentrates on the honey bee (*Apis mellifera*) in the British Isles it is also relevant to many of the other pollinator species upon which we depend.

Because of their economic importance and that they are able to be studied there is an abundance of research and information reported in the literature on the honey bee whether in the form of research papers, books and interpretations of research for practical application, in extension and training purposes and often readily accessible on the internet.

We have only been able to sample some of the key literature, however we have tried to cite those documents which are often review papers thereby introducing the interested reader to other, more detailed sources of information and data. We have tried as much as possible to use open access publications readily accessible on the internet to anyone. Regrettably, much research is never made more widely known and readily accessible outside the immediate circle of the researchers working in a specific area and getting access can be an expensive and time-consuming process, which is concerning because much of the research conducted is financed from the public purse and should be more readily available.

This book sets out to encourage us to reappraise and re-evaluate the role of the honey bee and the recognition of its place and relationship with humans in our society, including their environment and survival. More importantly how humans can work symbiotically with honey bees.

As we shall see understanding and meeting the nutritional needs of the honey bee are, as in any organism, key to their survival and future.

1. The British Isles – and their climates

The British Isles are a group of islands in the Atlantic Ocean off the coast of continental Europe. They consist of Ireland, Great Britain, the Isle of Man, Shetland, Orkney and thousands of smaller islands. The British Isles include the Republic of Ireland, the United Kingdom of Great Britain and Northern Ireland which includes the nations of England, Scotland, Wales and Northern Ireland.

In this book the use of the words British Isles will refer to the geographical description and not the political one. The distribution and diversity of plants and animals in all their forms are determined primarily by the climate and the geology throughout the British Isles.

Generally, they have been described as having a mild temperate oceanic climate, moist and changeable with abundant rainfall and a lack of extremes of temperature.

Assuming adequate amounts and distribution of rainfall throughout the growing season temperature is the biggest factor that affects the distribution of plants throughout the British Isles and this of course impacts on the potential forage and nutrition which is available to honey bees.

Some plant species benefit from higher summer temperatures which may affect the ability of the species to set viable seed, as is the case in the dwarf thistle *(Cirsium acaule)* and the small-leaved lime *(Tilia cordata);* both species being visited by honey bees.

The distribution of other plant species may be influenced by the winter temperature.

Both Holly *(Ilex aquifolium)* and Ivy *(Hedera helix)* benefit from mild winters and are valuable sources of forage, both nectar and pollen

The average range of temperature between summer and winter is around 10-12^{0}C. Lowland sea-level mean July temperatures in Britain and Ireland range from about 17.5^{0}C around the

Thames Estuary to about 12^{0}C in the Shetlands. The mean January temperature ranges from 3^{0}C or less in northeast England and much of Scotland to over 7^{0}C in small areas of Cornwall and southwest Ireland (Proctor 2013).

Temperature is affected by latitude, altitude, slope and aspect.

Wind affects plant growth and also determines the days and the conditions suitable for when honey bees are able and willing to forage. It can also induce water stress for plants under dry conditions and in recent years in conjunction with high temperature has affected the flowering of plants and crops (e.g., brambles and field beans) and the amounts of nectar available.

Rainfall is extremely variable around the British Isles, in terms of events, their duration and quantity. Recent years have seen an increasing tendency for periods of greater rainfall intensity and quantity especially across the western and north-western parts of the British Isles. Whilst in eastern and south eastern parts of England rainfall and water shortages are becoming more frequent. Both phenomena being the result of changing weather patterns and the effects of the shortages being exacerbated by the domestic and industrial demands of the human population and the crops and livestock grown to satisfy its needs. Rainfall can dilute and even wash out nectar secreted in open style flowers reducing the value of the flower as a source of forage to the honey bee.

Proctor (2013) describes the British Isles (in this case excluding Ireland) as having two 'natural areas' these being roughly separated by a line from the mouth of the River Exe in the southwest to the mouth of the River Tees in the north east. To the south and east of this line the countryside is based on relatively easily eroded Mesozoic and later rocks (frequently calcareous) which have and continue to weather to give good depths of soil parent materials, fertile soils and farming makes good use of them. There are few summits more than 300m and the annual rainfall is typically below 1000mm. to the north and west of this line the rocks are hard, the topography of steep relief and high summits and the soils are commonly thin and infertile. Annual rainfall varies widely but is typically more than 1000mm. Here the farming is largely livestock based (typically sheep) with areas of forestry and rough grazing land. Within this large area of the northwest British Isles is a second dividing line based on the Highland boundary fault.

The topography of Ireland has been likened to that of a saucer with a centre of low-lying and flat expanse of lowland plain based on Carboniferous limestone (typically covered in drift and peat) surrounded by discontinuous groups of mountains comprised of hard, acidic rocks.

These general comments on 'natural areas' become much more blurred when looking at the finer detail of the landscape and the current patterns of land use some based on history and practice and others which have become possible through changing economics and the application of many kinds of technologies which have enabled previously problematic land to be taken in hand and exploited.

The geology and soils derived from the underlying rocks as old as 2600 million years as in the north west of Scotland as well as the impacts of events such as Ice Ages, the formation of sedimentary rocks, massive translocations and redistribution of rocks caused by ice sheets, the effects of rivers, flooding and so on forming the more recent soils mean that the British Isles have varied soils and a diverse pattern of vegetation and land management and use which, as we shall see, determines the opportunities for honey bees to survive and thrive.

Proctor (2013) contains a very readable and enjoyable account of the effects of glaciation and the changes in vegetation and then into the historical period and the arrival of man and his profound impact on the landscape and the vegetation of the British Isles.

2. A short history of the honey bee in the British Isles

2.1   Early days in the evolution of honey bees

The early evolution of the honey bee took place many miles away from the current location of the British Isles and many years ago.

Eva Crane in her book ***The World History of Beekeeping and Honey Hunting*** (1999) Duckworth gives a good introduction into the ancestry of honey-storing insects and the evolution of the genus *Bombus* and *Apis*.

Smith (2020) writes of the biogeography of honey bees. The first honey-storing honey bees appeared in the Tertiary (the geological period from 66 million to 2.7 million years ago).

The term Tertiary is now considered obsolete, but it is still often referred to. It has been replaced by the Cenozoic Era (66 million years to the present) contained within which is the Quaternary Period (2.6 million years to the present).

The advanced cavity-nesting species which build a nest containing a several parallel combs and now recognised as *Apis cerana* (Asian Honey Bee) and *Apis mellifera* (Western Honey Bee) appeared in the late Pliocene (3.264 to 3.025 million years before present). This was a warm period in the Pleistocene Ice Age with changes in the ice sheets covering the Earth.

The origin of the evolution of *A. mellifera* and its divergence from *A. cerana* is still the subject of debate and investigation but it is thought to have taken place around 26 mya (range 23.5-15.6 mya) (million years ago).

It was considered that the ancestors of the modern *A. mellifera* originated in Asia or the Middle East. However, *A. mellifera* may have an African origin and not European or Asian as previously thought. It may even have taken place elsewhere in Europe (or further east) before *A. mellifera* appeared in Africa. Honey bee genome sequencing studies have confirmed the antiquity of honey bees in western and northern Europe. Today the native range of *A. mellifera* can be summarised as most of Africa and Western and central Europe. (Bonoan and Starks 2020).

The development of the subspecies and races of *A. mellifera* was further influenced by the effects of geographical isolation caused by barriers of sea, desert and high mountain ranges, as well as the effects of the changing climate during the Pleistocene period.

The species' original range of Europe, western Asia and Africa is considered to have 30 subspecies within which are ecotypes. that is populations which are highly adapted to their local conditions, e.g., the timing of colony expansion to exploit the local forage availability.

2.2   Before and after the last glacial period

The area now known as the British Isles formed the western edge of the European continent landmass permeated by large rivers deltas and so if honey bees were present in what is now the British Isles they would be from similar stock to those which existed in north and western Europe, (Carreck 2018).

Around 22,000 years BP (Before Present) the last glaciation reached its maximum extent and ice reached the northern parts of the British Isles. Then around 11,700 years ago the ice retreated resulting in increased routes by which plant and animal life could move from one hitherto isolated area to another. Then as the ice sheets began to melt the sea level rose cutting off areas of land and islands, the land bridges were inundated, and plant and animal populations became isolated and so evolutionary divergence began to occur.

About 8000 years ago a major change took place in the vegetation in Britain and Ireland and the neighbouring parts of western Europe. This was the sudden increase in the pollen record of alder and the beginning of ombrogenous bog (a bog whose surface is raised above the level of the surrounding landscape and only receives water from precipitation) growth, indicating a warmer more oceanic influenced climate. The tree species representation changed in time and there was regional variation in geology and climate. Small-leaved lime (visited by honey bees) was widely to be found in lowland England right up to the Scottish border and there was a big expansion of holly and ivy, especially in western Britain and Ireland. (Proctor 2013).

It has been suggested that this woodland and forest cover was more of a wood – pasture patchwork whose patterns changed as trees aged and died and herbivores consumed the grasses and other plants which grew in the openings, but there is little evidence in the pollen record to corroborate this.

The separation of the island of Ireland from Britain occurred about 8,000 BC with the separation of Britain from mainland Europe in 5500 BC, although there are examples of plant species spreading across from Britain to Ireland until at least 5500 BC.

And so, we come to the time from when until now it is believed the honey bee became part of the fauna in the British Isles. There may well have been periods of time during the various cycles of the climate warming and cooling and warming when the vegetation supported an insect flora which included the honey bee but as mentioned already the pollen record available is not very extensive for non-tree species and the chances of fossilised remains of honey bees (for example preserved in amber) being formed in the British Isles and then being found are very slim.

2.3   The Dark European Honey Bee (*Apis mellifera mellifera*)

Apis mellifera mellifera is the only honey bee species considered to be native to the British Isles. Its original range stretched from west-central Russia through northern Europe and probably down to the Iberian Peninsula. Traditionally they were called the German Dark Bee, or the Black German Bee but today they are more likely to be called after the geographical / political region in which they are kept e.g., the Black British Bee, the Native Irish Honey Bee, the Cornish Black Bee or the Nordic Brown Bee (with or without the word 'native added).

Honey bee colonies were to be found associated with the cool temperate deciduous (broad-leaved) forests. Long-lived trees such as oaks provided cavities in which the colonies could exist and grow and survive the cold winters characteristic of areas in which the deciduous forests grew.

At the end of the Neolithic period (2,000-1,000 BC) the climate in Europe was mild, there was a rich forest vegetation in which the limes were an important component and a rich source of nectar and thence honey. Additionally, in open areas throughout the forest there were other species of plants which were good sources of nectar and pollen.

The presence of hazel in the flora of a location in time has been taken as an indicator that honey bees could survive at that time and place and on this basis they may have become established in Britain by 5500 BC.

The earliest evidence of man in Ireland (as the ice sheets retreated at the end of the last glaciation) has been set at around 7000 BC well before the Neolithic period in the British Isles (between 4000-3000 BC).

Neolithic pottery found in England dating from 3000-2650 BC contained substances which were of a honey bee source, namely beeswax, glucose and resin and separately in a second sherd with deposits which were typical of residues from honeydew type of honey.

Colonies of the temperate-zone *Apis mellifera* which evolved and developed in Europe were able to survive the winter by establishing nests in sheltered and insulated hollows inside a tree. During the winter period no brood is produced, when the colony temperature falls below 35⁰C, which is the temperature required for brood rearing. This conservation of energy especially at a time of low outside temperatures and lack of forage (nectar and pollen) coupled with the ability of the colonies to establish reserves of stored honey and be quiescent in a cluster over the winter period meant they could survive in many parts of the northern temperate zone. This ability and need to survive cold winter periods are still important factors in the distribution of honey bees today on the northern and western fringes of Europe. As we shall see much of our beekeeping season planning and practice are geared towards ensuring our colonies go into winter in secure and weatherproof hives, in healthy colonies of adequate size and which are well- provided with stores of food, ideally including capped honey.

The recommencement of brood rearing usually takes place in late winter depending on the season and the geographical location of the colonies. The colonies then expand and through the swarming process new queens and new colonies are produced and the continuity of the species is assured.

Human beings have had a very long relationship with species of the genus *Apis* (honey bees). Humans have harvested or robbed honey and wax (possibly also including larvae and pupae) from colonies of the genus *Apis* since prehistoric times and these activities have been shown in prehistoric cave and rock paintings dating from 4,000 to 6,000 BCE (Before Current / Common / Christian Era before the year 1).

In this book we are primarily concerned with the relationship of *Homo sapiens* and the honey bee, *Apis mellifera,* in the British Isles. Crane (1999) contains extensive information on the history of beekeeping in the British Isles and readers are encouraged to consult her book and build a mental picture of the increasingly reliance on the interdependent relationship of the honey bee and humans.

Carreck (2008) concluded that taken together the biological, written sources and archaeological records prove beyond all reasonable doubt that the honey bee (*Apis mellifera mellifera*) was present in Britain several millennia before either the Roman or the Norman invasions.

It is now believed that this bee still survives as *Apis mellifera mellifera* in various places around the British Isles and there is an increasing interest in recovering and discovering the properties and characteristics of this subspecies, and in spite of the fact there has been a history post around 1859 of the importations of other subspecies of *Apis mellifera* e.g. *A.m. ligustica* from mainland Europe which has interbred with the *A. mellifera mellifera*. Other importations have occurred including Carniolan and Caucasian bees including the creation and importation of hybrid-based bee types collectively known as Buckfast. The original cross was between the Italian Honey bee (*Apis mellifera ligustica*) and *Apis mellifera mellifera*.

It should be noted that a hybrid is the resulting offspring of a *cross* of two genetically dissimilar parents especially offspring produced from breeding between different breeds, strains, varieties, races, species or genera. We often see the term mongrel used to refer to honey bees. A mongrel is the resulting offspring of a cross between varieties or races of a species. This term is usually used to refer to dogs of mixed or undetermined breeds but is used for other species and often with a derogatory overtone.

Wilson (2004) in her book **The Hive** gives a very readable account of the evolution of science, religion, politics and social history exploring the impact of the honey bee on food and human ritual. Her bibliography contains reference to other titles which cover the many facets of the relationship of man with the honey bee.

In this book whose central theme is bee nutrition we have set out to gently remind, or perhaps even inform, our readers of this longstanding and on-going relationship with the honey bee and through using this information lay out the foundations for ways in which we can ensure their (and our) ongoing survival and future.

2.4   Honey hunters

The practice of honey robbing continues to this day in many parts of the world and uses essentially the same techniques as those depicted in Palaeolithic rock art using long ladders, ropes to access colonies of bees often living on cliff edges, using poles and baskets to dislodge and collect the honeycomb.

The honey robbing of bee trees took place in Europe up to the Middle Ages. Colonies were located by capturing small numbers of bees and watching where they flew after they were released. The bees would then fly in a 'beeline' to their nest and its position could be more precisely determined by using triangulation techniques. Variations in the use of this technique are still used today in research studies and is described in Seeley's book *Following the Wild Bees* (2016).

Honey hunters / gatherers looked for mature trees which were likely to have cavities in which there might be colonies of wild honey bees nesting. Whole or part honeycomb harvesting was carried out in early autumn and honey and comb composed of beeswax removed for human use. Honey gatherers in medieval Europe developed techniques which started the change from hunting / robbery to forms of activities (husbandry) which would aid both the honey bees and the gatherer. For example, the cutting of small entrance holes to help the bees access the cavity and the fitting of small wooden doors to enable easier access for the gatherer to the honey combs inside the cavity. Tree trunks with nests were often marked with symbols identifying the person who claimed ownership. The form of beekeeping was termed forest beekeeping. These techniques were probably also used in many parts of the British Isles as well, although there are no records for Scotland where perhaps much of the country was too cold for the survival of honey bees in the wild.

In the large mature forests of Europe honey bees began to be kept in 'hives' made from logs containing nests cut from the tree which were then located at ground level directly in contact with the ground or on pieces of rock or wood. This tradition does not appear to have been practised much in the British Isles as the earliest hives in England seem to have been made of pliable willow (osier) or hazel wands woven around a conical circle of stakes joined at the top.

The number of large suitable trees such as those found on the European mainland were not so readily available further west of the deciduous forest zone and this probably encouraged the development and use of such containers.

2.5   Beekeeping in hives

Hive beekeeping was probably introduced in the east of England from continental Europe and the practice spread westwards. Beekeeping in the British Isles was influenced by cultural, geographical, economical and resource factors, such as the available of cavities in large old oak trees. Such influences came from areas west of the Forest Zone, namely Central Europe, France and the Low Countries. In the case of Ireland, Celtic speaking peoples who arrived in Ireland and Britain between 1000 BC and 500 BC probably brought knowledge of beekeeping with them which included the use of wicker skeps (also known as an 'alveary'). An initial wave of settlement took place in Ireland and then Scotland, followed by a second wave who reached England and spread to Cornwall and Wales.

The earliest known written evidence about hive beekeeping in England dates from c. 705 and refers to hives made of wicker (osier – *Salix viminalis*, (Crane and Walker 1999).

This structure was then coated or 'cloomed' with mixtures of wet cow or oxen dung and ashes or gravelly soil smeared between the spaces and the stems of the framework to give a smooth surface which hardened and dried to become a waterproof covering.

In other parts of the country where there were cereal growing and grassland areas, straw, reeds and sedges were made into coils of ropes which were coiled, secured with stripped blackberry

/ bramble stems inserting the binding using a bone or metal awl and made into dome-shaped basket or bushel shapes containers called skeps (lip-work skep). These were then adorned with hackles made from reeds which were formed into a shape to fit over the top of the skep thereby improving its resistance to the weather. The skeps were placed on a stone, a wooden base, plinth and often set under a cover or into bee boles (recesses), or other structures to provide protection from the weather. These special structures were constructed in stone, brick or cob (lean-to) shelters, bee houses or recesses in walls. Crane and Walker (2000) describe wall recesses for beehives across the British Isles. A register of these structures in can be found on the website https://www.beeboles.org.uk.

Alston (1987) describes the English methods of making coil worked skeps.

Many Roman citizens in Britain were bee owners, but the tasks of working (managing) the bees were carried out by a British slave who was known as a *mellarius*.

There are many stories and legends from the Anglo-Saxon period abound concerning the consumption of mead. The word 'honeymoon' is derived from the ancient custom of presenting newlyweds with mead to drink on each of the first 30 days after their marriage.

The *Rectitudines singularum personarum* is the only known systematic account, probably put together in Bath Abbey of the rights and duties of workers and tenants on an English estate prior to the Norman Conquest. It was translated into Latin c. 1100. It includes beekeeping. In Anglo-Saxon society the person who attended the estate's hives was called a *beo-ceorl* (bee churl) was a free man with the same social standing as the swineherd, these being the lowest rank of freemen. As a vassal and bound to his lord his bees and the land they were on belonged to the lord of the manor or the king if they were recorded in the Domesday Book in the king's royal demesnes. He held them by virtue of his job description and reverted back to the lord on his death. If the hives were gafol-hives i.e., those on which rent was to be paid, in the 11th century this was payable at a rate of five sesters (a variable measure of volume) of honey.

Honey is referred to numerous times in the Domesday Book, suggesting that beekeeping continued to flourish under the Normans in the 11th century. References to mead are largely to be found in Anglo-Saxon records. There is little mention of beeswax which it is presumed was paid directly to the Church to satisfy its needs.

Throughout the preceding centuries beeswax had been used to make models, seals, writing tablets, cosmetics, medicinal products and candles.

The Domesday Book which was a detailed inventory of the characteristics, assets and property rights produced for William I after his conquest of 1066 refers to *mellitarii* (singular *mellitarius)*, translated as 'honeyers,' who it is suggested were men who checked on nests of wild bees and collected wild honey in the woods and forests. The person who looked after the bees was also referred to *custos apium* (guardian of the bees).

Crane and Walker (1999) contains some details of the referencing in the Domesday Book on the extent of beekeeping and honey transfers in England, but it contains little information about beekeeping practices.

The honey and wax obtained from the Royal hunting forests was the property of the King. The rights of others to honey and wax were set out in the laws contained in the Carta Forestae in 1217 and later in 1225.

The *grete herbal* published in London in 1526 distinguished between 'tame' honey from hives and 'wild' honey from the woods.

Crane and Walker (1999) detail information on Early English Beekeeping citing evidence found in local records up to the end of the Norman period.

In a later publication, Walker and Crane (2001) describe the changes taking place in English society from c 1200-1850 with respects to honey bees, beekeeping and honey bee products. These are illustrated by reference to purchases and sales, wills, bequests and as gifts of beekeeping- related items.

Continuing the story, Fraser's book **The History of Beekeeping in Britain** (1958) as its title suggest describes the early history of beekeeping and its progression up to about 1900.

Honey bees and their products have been of significant social and economic importance and inevitably become the subject of human law and customs and as a currency used to pay taxes and ransoms. They have even been used in warfare hurling colonies of bees in skeps onto the enemy.

Walker and Crane (2001) describe the changes taking place in English society from c 1200-1850 with respects to honey bees, beekeeping and honey bee products. These are illustrated by reference to purchases and sales, wills, bequests and as gifts of beekeeping- related items.

But what underpinned this long association of humans with the honey bee and what has led to this closeness?

Apiculture was a prominent part of the Minoan culture around Crete (*c.*3000 BC – *c.* 1450 BC) and their representations of honey bee-related imagery were absorbed into the later Egyptian, Greek and Roman cultures and further developed.

In Egypt the bee was a symbol of royalty and power.

This is all well- described in Crane's book **The Archaeology of Beekeeping** (1983). A more recent, very readable and well-illustrated book is the **Collins Beekeeper's Bible** (2010).

2.6   As a symbol of love

As a symbol of love, the bee was associated with Cupid during the Renaissance. Paintings of bees were depicted stinging Cupid, the god of love and desire. Cupid is often depicted dipping into honey, without thoughts to potential danger of protective bees. The lesson being 'Beware of love's stings.

2.7   The human need for sweetness

Humans like sweet-tasting things and for centuries honey was the main source of sweetness, a high calorie containing food, however sweetness was not always a common characteristic desired in food. It could also be stored.

Honey retained its pre-eminent position until it was superseded by the spread of knowledge, cultivation and the extraction of sugar from sugar cane (*Saccharum officinarum*). Crane (1999) describes how in 1250 honey was one-fiftieth the value of sugar, but by 1987 honey was said to cost nearly seven times as much as sugar. A survey of BBKA members conducted in November 2020 showed that the average price for one pound of honey was £5.86. Some 34.7% of the members sold their honey at between £4.01- £5.00, 6% between £5.01-£5.50 and 30.3% over £6.00 per Ib.

Mead has also become a drink which today has a dedicated but small following. Its popularity reduced and ii was replaced by various forms of beers including the historically more recent discovery of hops which improve beer taste and longevity. The popularity of beers and lagers in turn is reduced with increasing access to low- priced wines from around the world. Even more recently the development of specialist gins.

2.8   The human need for light

Light was a precious luxury to have, and candles made from beeswax burned with a much brighter, smoke-free flames and were less smelly compared to the smokiness of cheaper tallow (hard animal fat) based candles and tapers.

2.9   Art, literature and communication

Honey bees, beeswax and honey were key aspects of many civilisations and whilst fascinating it is too big a topic for our book and we invite the reader to look at some of the suggested sources.

2.10   Honey bees and religion

Bees and honey are referred to many times in the Bible (Judaism and Christianity).

In Christianity the bee has been seen as a symbol of Jesus Christ's attitudes, with honey representing his sweet and gentle nature and the bee sting to justice and the Cross.

Beeswax candles played an important part in the rites and rituals of the Christian Church. Large numbers were used in the mass and in funerals.

For Muslims there are references to bees and honey found in the Quran with 'The Bee' being considered a revelation of God.

Some of the ancient scripts in Hinduism refer to pollen and honey with honey being one of the ingredients of Panchamrit ('the five Nectars') which also include milk, sugar, ghee and buttermilk.

In Buddhism honey and honeycomb plays an important role in the festival of Madhu Purnima.

Honey bees (not just *Apis mellifera*), honey and drinks made from honey together with bees wax have been a significant part of the religious life of many civilisations going back thousands of years.

These relationships and importance have been described in Ransome's book ***The Sacred Bee in Ancient Times and Folklore.***

She describes many historical aspects in societies of the British Isles including myths and legends.

2.11   Beekeepers and their Saints

Beekeepers have their patron saints, and some are listed below.

Ambrose (also known as Ambrosius, or Ambrose Aurelius) is a patron saint of beekeepers, beggars, learners and Milan. He lived from 340 to 397 AD.

Saint Valentine is also a patron saint charged with ensuring the sweetness of honey and the protection of beekeepers. He died around AD 270 and has been a patron saint since 496 AD.

Saint Gobnait (also known as St Abigail) is a patroness of bees and beekeepers and died in the 6th century.

Saint Bernard of Clairvaux, a French Abbot, is also a patron of beekeepers and candlemakers. He died in 1153.

2.12   The honey bee, the hive and thriftiness

The bee and the hive have long been symbols of industry (as in 'a hive of industry'), regeneration, wisdom and obedience.

Today, we use the expression 'a hive of industry' to describe a place where people are working very hard, or where there is a lot of activity taking place.

2.13   The concept of the honey bee community in religious and secular life

Throughout history there have been members of society who felt a special affinity with honey bees and saw a link between their own way of life or what their ways should be with that of their understanding of the honey bee society as they understood it at the time. This was very much involved with the understanding of the relationships and functions of the different types of bee found in a colony and the perceived relationships of the queen (long thought to be a king) with other colony members.

Honey bees signified immortality and resurrection.

Political theorists such as Aristotle Plato, Virgil, Shakespeare, Marx and Tolstoy used their perceptions of the honey bee colony (community) as a model of human society in their works.

This in turn played into the political and religious aspects of society. The writings of Aristotle remained unchallenged for many centuries and were only displaced as science, the scientific methods and the understanding of honey bee biology developed.

This is described more fully in Crane (1999).

Honey bees appear to tolerate human beings in the sense that they can live or be kept in relatively close proximity with humans, together with evolutionary adaptations they had undergone meant they could be 'domesticated' and managed to the benefit of humans. For example, honey bees establish nests in cavities and humans can aid this by establishing such protected cavities in the form of hives. Honey bee swarming and colony reproductive behaviour meant they were relatively easy to capture, transfer to another cavity (hive) and allowed to develop. Such hives could be moved (within certain limits) from one location to another thus further increasing the closeness of the relationship between humans and the honey bee.

2.14   The history of beekeeping in English gardens

Throughout the greater period of time that beekeeping has been practiced it has mainly been on a small scale with most of the of colonies being managed in small numbers by individual beekeepers. The places where they were kept were referred to as bee gardens. The term 'apiary' (from the Latin *apiarum)* was first used in 1650.

Walker and Crane (2000) describe the history of beekeeping in English gardens and readers interested in this aspect of beekeeping are encouraged to read the descriptions to gain a broader appreciation of this fascinating aspect of the relationship of the honey bee and human society in England through the ages.

In the eighteenth and early nineteenth centuries books concerning beekeeping dealt less with garden aspects and focussed much more on the developments which had been taking place with the development of new hive designs and methods and techniques for keeping honey bees.

2. 15   The Victorian beekeeping revolution in the British Isles

The scientific basis of beekeeping began to be established in the late 1500s and early 1600s as part of the whole world of natural philosophy (the precursor to natural science) coupled with the development of scientific instruments which aided observation, anatomical studies and increasing knowledge in chemical analysis.

Lawes' book (2011) *The Victorian Beekeeping Revolution* describes the rapid changes in beekeeping and knowledge of the biology of the honey bee which took place during the Victorian period, the legacy of which is still very much influences the craft of beekeeping today in Britain.

It was in the fall of 1851 that the Rev. L.L Langstroth recognised the significance of providing a space of about three-eighths of an inch made available by reducing the size of the frame and any points of contact of the frames with the hive body. He described the movable frames and the fact this this resulted in the bees not attempting to attach the frame to the sides or the bottom board of the hive. In 1853 he obtained a US Patent No. 9300 in the United States titled 'Beehive' which claimed five inventions the most significant of which for beekeeping was this space and followed this by a beekeeping manual in 1852 *On the hive and the honey-bee* and in 1859 *A practical treatise on the hive and the honey -bee*. Langstroth did not invent the term 'bee space' nor did he detail its significance. The term seems to have been introduced by James Heddon in 1885 where gave fuller detail of what bee- space meant in terms space, location, the use of propolis and the production of brace comb and **he** referred to Langstroth as the inventor. It wasn't until the 1887 edition of The *ABC of Bee Culture* by Root in the US and then in 1888 in England that the term *bee-space* enters beekeeping literature, although mention was made in a paper by T. W. Woodbury in 1867. It was Woodbury who introduced Langstroth's movable frames hives into England, in particular Devon.

Today we recognise the term bee-space as a gap whose size if larger than 3/8 in or 9 mm would be filled by the bees with comb and if less than ¼ in or 6 mm would be filled by the bees with propolis and that led to the development of a completely movable and removable frame / comb which could be used in a box- shaped container to mimic the parallel combs found in wild bee colonies, but which could be easily removed for inspection and relocation.

There are a number of books which chronicle the development of beekeeping techniques and the increasing knowledge and understanding of the natural history of honey bees up to modern times. Such a vast literature cannot be done justice in a book such as this and interested readers are encouraged sample the literature. Such books describe the evolution from a swarm- based system of beekeeping utilising skeps to the evolution to and introduction of the movable frame system of beekeeping which is the most commonly used colony management system of today.

Not only is the ability of being able to remove frames and inspect them for the state of the colony, its demographics and the presence of brood related and other diseases but to be able the assess the quantity, state and form of the food reserves held in the comb. Such abilities enable the beekeeper to adjust their management strategies according to the status of the food reserves, the colony and assess this in relation to potential foraging opportunities.

The use of the moveable frame also encouraged swarm management through use of prevention and management control techniques, rather than the key aim of the promotion of swarming as utilized in skep beekeeping.

Good places to start appreciating the significance of the moveable frame hive are Crane (1999) and Kritsky (2010).

The practical consequence of the introduction of the moveable frame hive was that larger colonies could be developed which could give increased crops of honey and in turn this meant a greater interest in the availability of nectar and nectar yielding plants.

There are many fascinating books describing beekeeping at various times since the end of the Victorian period, e.g., Herrod-Hempsall (1930 and 1937). Many titles can still be acquired in second- hand bookshops and via online booksellers. Many copies pass from beekeeper to beekeeper and there is something to savour when you have acquired a book which you know has been handled and used by a beekeeping predecessor and made even more special when the book is signed by the author and / or the name of the beekeeper written in it. If you are member of a beekeeping association, it is highly likely your association will run a library. Do use it!

There was much discussion and competition between the merits and demerits of straw versus wood and skeps versus boxes. Skep beekeeping was practiced for centuries and then alongside the development of various designs of hives.

Skep beekeeping waned in the British Isles after World War 1 when a government restocking scheme only gave a subsidy for bees kept in boxes and not skeps. However, skeps continued to be made for keeping bees and today are mainly being used in the collection of swarms. It is one of life's beekeeping pleasures to lower a swarm into an upturned skep, invert the skep over a crown board, insert a small wooden block to raise the rim of the skep and watch the rest of the swarm follow under the rim of the skep to be with the queen inside.

It is however easy to get carried away with the ingenuity of man creating hives of various patterns, designs and complexity together with the tolerance (to some extent) to manage and manipulate the honey bee to our wishes. The recent interest in the British Isles in the simple top bar styles of hive is of note.

We must not forget that honey bees are cavity nesters, which utilise cavities having certain desirable qualities, namely:

- The volume is around 40 L (range 20-80).
- It has been previously occupied by bees.
- It has a south facing entrance which is 1-5 m off the ground.
- The size of the entrance hole is 12.5 cm2.

2.16 Hive products and their typical uses

2.16.1 Historical uses of beeswax

Beeswax has been used by human societies for a long time and these were described in Cowan's book **Wax Craft** published in 1908 and Crane (1999).

In the British Isles historical uses of beeswax have included its use in

- Torches / candles
- Images and sculptures
- Wax models of deities
- Death masks
- Embalming the dead
- Seals
- Medicinal uses
- Bronze casting
- Encaustic painting

Writing tablets in the form of a book with raised borders so when the message was written and the book closed, the incised wax surfaces were protected from friction and damage. The closed books were sealed with wax and an impression made on the seal of a device using a ring of the writer. These tablets would be recycled by smoothing out the writing made with the use of a stylus. Examples of these tablets with their message still decipherable have been found in the excavations associated with the settlements built by the Romans to service the Roman frontier which was Hadrian's Wall in today's Northumberland's Borderlands.

Beeswax was a hugely important commodity and there is still a guild in London called the Worshipful Company of Wax Chandlers (www.waxchandlers.org.uk)

This list illustrates the many ways in which honey and wax and other hive products have been used in society and so the population of the British Isles came into contact with and were aware of the honey bees.

In more recent times the honey bee was an important part of the 'war effort' not only in terms of producing a sweet substitute for sucrose sugar but also used in the manufacture, maintenance and storage of war materials.

This second table on this subject using information detailed in Krell (1996) lists current uses of hive products. It can be seen they still figure significantly in the modern economy and satisfying the needs of today' society, even though it is highly unlikely that today the majority of the population are aware of many of them.

Hive product	Typical uses
Honey	As a food / a food ingredient / an ingredient in medicine-like products / Products of honey fermentation
Pollen	As medicine / as food / in cosmetics / For pollination
Wax	In beekeeping / for candle-making / for metal castings and modelling / in cosmetics Food processing / Industrial technology / Textiles / Varnishes and polishes / Printing / Medicine
Propolis	In cosmetics / In medicine / Food technology
Royal jelly	Dietary supplement / As ingredient in food products / an ingredient in medicine-like products / In cosmetics
Cosmetics	Lotions / ointments / creams /shampoos/ soaps /toothpaste / mouth rinses / deodorants/ facial masks/ make-up/ lipsticks / perfumes

2.16.2   Medicinal honey

Prior to the modern era (early- to mid-20th century and the advent of antibiotics) honey had a long history of medicinal use but it became generally, from a medical point of view, to be considered as a worthless, but harmless, substance. The ever- increasing rate of resistance development by microbes to antibiotics in medicinal, agricultural, veterinary and other end uses coupled with the lack of the development of new drugs are causes for concern. This has prompted a re-evaluation of the antimicrobial and wound-healing properties of honey. It appears unlikely that microbes could develop resistance to medical honey products because of the various properties which make it antimicrobial.

Whilst this short section is primarily considering the relationship of honey and the human species we should appreciate its properties and importance in relation to honey bee health and honey bee colony nutrition. It should also inform us of the need in our beekeeping activities not to consider that sugar solution is equivalent to the honey which the beekeeper may have deprived the colony of during honey harvesting activities.

The taste, colour and medicinal properties of honey vary in accordance with the range and relative proportions of nectar collected from forage plants. All honeys have some level of antimicrobial activity, some (e.g., Manuka honey made from nectar collected from tea trees (*Leptospermum* spp.) are well-studied, but there is a wide variation in activity.

Medicinal honey (prepared as a wound dressing product and registered as a wound care product) is particularly effective in wound-healing and in applications such as:

- Burns
- Chronic wounds
- Ulcers and pressure sores
- Gingivitis
- Eczema
- Infected surgical wounds and superbug-infected wounds including MRSA (Methicillin-resistant *Staphylococcus aureus*)

2.16.3   Pollination

During prehistory and for most of the historical period, it was assumed that sufficient pollinating insects were generally available to pollinate crops grown by humans and, with a few exceptions, no special attention was paid to the process of pollination.

An understanding of pollination in general and of the part played by bees was gradually achieved between the 1670s and the 1880s. In 1750 Arthur Dobbs described the role of honey bees in collecting pollen and moving it from plant to plant, thereby bringing about fertilisation and successful fruiting of crops. In 1793 the German naturalist C.K. Sprengel recognised that the function of flowers was to attract insects and that nature favoured cross-pollination for hybrid vigour and recognised the role that bees (and other pollinators) played in achieving this. Charles Darwin also studied the co-evolution of insect pollinators and their plants as well as performing his own experiments on cross-pollination publishing his work about 100 years later (Brian and Crane 1959). It was included in his book *The Origin of Species*. Darwin showed that bees were necessary for the fertilization of clover and also the vital importance of bumble bees for the production of red clover seed. Incidentally he also studied comb building by honey bees and how it was done.

Proctor, Yeo and Lack's book *The Natural History of Pollination* (1996) contains an excellent opening chapter on the study of pollination: a short history.

Although social bees (not just honey bees) have been valued for their products since early times, their greatest significance as a world resource is in the pollination of plants.

In the 1930 beekeepers in Denmark were able to demand payment for the pollination services provided by their honey bee colonies and by the 1940's this practice was also adopted in a number of other countries.

This also takes place today and many beekeeping enterprises are set up to provide pollination services to the agricultural and horticultural industries.

The honey bee *(Apis mellifera)* is by far the most important managed pollinator and is especially effective on many crops grown as monocultures. The part played by *A. mellifera* in the pollination of native wild plants is more complex, especially where in the world this bee species is not considered native and may compete with native pollinators for food and nest sites. There are those in the conservation movement and even in the scientific community who promote the interests of bumble bees and butterflies and who are quite antagonistic towards honey bees *(Apis mellifera)*. In some ways such people ride on the back of the relationship and interest of the public towards honey bees in attempting to divert resources towards their own interests.

Now, in the 21st Century, it is important that we reflect on and cherish the relationship that humans have had with honey bees over many centuries. It is a relationship quite unlike any other and the reader is encouraged to read more widely and to draw on the breadth and depth of this relationship and the extent to which it has played a part in our social, political and economic history. Then to begin the process of looking at our relationship in the years to come especially during the predicted effects of climate change on biological systems, rising population and the effects of economic and infrastructural developments on the habitats that are currently found in the British Isles.

3. Human relationships with the honey bee in the British Isles today

3.1   Taking stock

At this point it is worth reflecting on a number of terms which are widely, but often, inaccurately used, in particular the use of the word ecosystem.

Natural is the term used to describe structures and processes that have not been substantially altered by human intervention, or which persists despite human intervention.

Nature in the broadest sense, is the natural, physical, or material world or universe, It is often used to describes wildlife and geology, and is analogous with the more strictly defined concept of ecosystems.

An **Ecosystem** describes all the organisms living in a community and the abiotic (non-living) factors with which they interact.

Honey bees have been admired for their perceived virtuous behaviour – orderly, industrious, cooperative, productive and have been held up as models for human society

It is important to reflect on our recent history in the British Isles and to recognise that whilst there have been many improvements to our material wealth, healthiness and perhaps quality of life it has come and continues to come at a price. This part of the book is not a lament or harkening back to 'the good old days', they probably never existed. Rather to look forward and plan to creating a place where there is a healthy and environmentally sustainable use of the resources we have, to enable ourselves to be surrounded and to be part of a richer, more biodiverse and yes, more beautiful environment.

Even without taking into account the economic value of hive products such as honey, bees wax, royal jelly and propolis because of its pollinator activities the honey bee is the third most important 'domesticated' animal in Europe; third only to cattle and pigs and ahead of poultry.

In the British Isles we live in an intensively managed and in many places overcrowded landscape such that there are few, if any places, which have not been influenced by the hand of man over many years. This may be in the form of more intensive cultivation, agriculture and an ever-increasing urban landscape and the loss of land to infrastructural developments.

The countries of the British Isles have a long history in the global movement of goods, plants, animals and people, including the importation of honey bees and in the past the taking of colonies of honey bees to countries which were being colonised and absorbed into the British Empire. In recent decades globalisation of people through rapid and cheap travel and the movement of manufactured goods and agricultural products has increased dramatically and one of the undesirable impacts of its increase has been the increasing speed at which pests, pathogens and diseases have spread around the world. The British Isles have not been immune to this. In beekeeping such an example can be seen in the parasitic mite Varroa destructor which was first identified in the UK in 1992 in Devon. Its effects on the honey bee colony in relation to nutrition will be described later.

Beekeepers, the bee inspectorate, plant health and other authorities continue to be on a high alert for the spread of the Asian Hornet (*Vespula velutina*) and the Small Hive Beetle (*Aethina tumida*) from mainland Europe or elsewhere into the British Isles. If either, or both, of these introductions were to occur and planned contingency measures proved inadequate to eradicate the incursions this could lead to severe damage to the honey bee population and probably many other pollinator species as well.

 In 2010 a publication by Maclean **Silent Summer** became available and it was clearly intended to be a link with the seminal work of Rachel Carson's **Silent Spring** (1965) . Silent Summer was an in-depth audit of wildlife in the British Isles over the past 50 years and predicted future developments. In effect it was a 'Domesday Book of Wildlife'. It was a large publication and in 2015 a '**A Less Green and Pleasant Land**' was published by Maclean which presented the topic in a more concise and accessible format. No doubt in coming years there will be many more such reports prepared but will the recommendations from them be transformed into decisions and then implemented?

As we shall see in this book the future well-being of the honey bee is inextricably linked to that of the state of the environment as a whole and it is no overstatement to say that man is very reliant on the honey bee for pollination. Hive products such as honey, beeswax, and propolis and all of the derivative products which utilise these materials are found widely in our society.

Biodiversity is also reliant on adequate pollination activities by other pollinator species such as bumble bees, butterflies, hoverflies and moths to name but a few, however both their absolute numbers and number of species have been in continuing decline.

Today the World's plant and animal species are faced with the 6th Great Extinction Event which is considered to be substantially man-made due to the ever-increasing impacts of the demand for food, natural resources, living space and commercial and industrial development associated with human activity.

There are few wildlife species whose relationship with man is akin to that of the honey bee. On the one hand the majority of honey bees in the British Isles are kept and managed by beekeepers and are reliant on the help of the beekeeper to control the incidence of the parasitic mite *Varroa* and associated viruses, such as Deformed Wing Virus, but there are an increasing number of reports of the ability of some colonies of honey bees to have a greater tolerance or ability to control the varroa parasite themselves. Perhaps there are evolutionary processes taking place in the honey bee species and *Varroa* itself and their interrelationships.

There are feral honey bee colonies existing throughout the British Isles and their extent, distribution, origin and health status are of increasing interest.

Browne et al. (2020) reported on a study of unmanaged, free-living colonies in Ireland including some colonies which have persisted naturally and unaided over multiple years. They published molecular data from mitochondrial, microsatellite and SNPs which demonstrated that the free-living population of colonies they studied largely comprised pure *Apis mellifera mellifera*.

Sadly, it appears that today in many parts of the British Isles without the intervention of the beekeeper the fate of many honey bee colonies would be for them to succumb to the risk of starvation due to the lack of forage as well the potential for weakened colonies to become infected with a range of pathogens and harassed by pests.

A further consideration is the pursuit of beekeeping practices which have to be radically re-appraised to ensure the continuation of honey bees and from a selfish perspective for humans to benefit from their presence and activities in our world today.

Legally they are considered to be livestock and this simple definition has meant significant impacts with respect to the attitudes of governments and regulatory bodies to their management and control (both positive and negative). In addition, because honey is a foodstuff this in turn has meant regulatory control and oversight.

Consequently, the teaching and mentoring of beekeepers in particular from Victorian times to the present day has focused on honey bee colonies being seen as economic units and techniques and practices developed to maximise honey production in particular. The perhaps more essential role of honey bees in their pollination activities only being recognised in specific areas of horticulture and agriculture and not the wider environment and society.

This mindset will have to change as the continuing demise of non-Apis pollinators because of a range of factors cannot easily be reversed. Hope must lie in the messages setting out the needs of pollinator species not only reaching policy makers but to all parts of society. In the British Isles the honey bee, (*Apis mellifera*), is the main pollinator species which has the ability to pollinate a wide variety of plants and the only one which can be managed in a way that could ensure there is sufficient pollination capacity to meet the human need to be able to feed itself and sustain the biodiversity upon which humans depend.

Today's highly visual culture and imagery together with the capabilities of social media are being harnessed to publicise and highlight the plight of 'bees' and other species of pollinating insect. Regrettably, and for various mixed motives, 'bees' are totally misrepresented and given cartoon appearance and persona. There is a mixing up of bumblebees with honey bees and wasps. Honey bees are made to look like large furry, large round eyes, striped black and yellow objects which one would normally consider a possible literal cartoon representation of the bumblebee. Wasps are portrayed as bad and to be eliminated. Some supporters, advocates and defenders of the cause for protection of bumblebee species are happy to attribute honey bee characteristics to the bumblebee in particular in fund-raising activities and thereby causing further confusion and misrepresentation and misinformation.

There is an urgent need for accurate representation and objectivity in communicating the correct features of all types of pollinating insects so that everyone can understand their true nature. Cartoon representations and films with human social norms made in the name of seeking to attract the attention of small children are actually contributing to the lack of understanding of these creatures and their crucial role in human survival and the natural world.

3.2   Beekeeping and gender roles

The role of gender in UK society in the 20th and 21st centuries has had and continues to have a high profile and beekeeping is no exception.

Historical records (Crane 1999) describe the beekeeping activities of women and these are further described in Walker and Crane (2000). They write that from the late sixteenth century some English authors assumed that country*women* would care for both the garden and the bees.

Since the development and spread of beekeeping practices in the 1850s which utilised the movable frame and the demise of skep beekeeping together with the creation of Beekeeper Associations and beekeeping publications women have done much to practice, promote and enjoy the craft. After the First World War with the loss of so many men it was more common for women to be beekeepers than men and this situation continued after the Second World War. Then the increasing recognition and status of women in society lead to more women wanting to continue to work after their invaluable and crucial role in the war effort. The trend towards equal status with men continued apace and is still unfinished business. In more recent years there has been an increasing interest and number of women beekeepers and in some beekeeping associations they are in the majority. Many of these women have busy careers as well as having households to run and families to raise.

Wooden hives containing honey are heavy to move and this has encouraged women to work co-operatively with other beekeepers to share the load in some of the beekeeping operations involving moving and lifting. There is scope for other changes to be made and adopted by the craft.

3.3   Changing weather and climate change

The Earth has experienced periods of climate change before and in forms very much more severe than the apparently apocalyptic predictions in today's period of climate change. For beekeepers we think terms such as changes in weather patterns are much more tangible at the local level and enable a more focussed appreciation of the threats (and opportunities) that changing weather patterns may bring. From my own experience of keeping honey bees continuously in one geographical location (my home apiary is in an arable area of the East Riding of Yorkshire) for 38 years I have seen the significant impacts on the crops grown and forage available, agricultural intensification, destruction of habitats and associated forage potential and the challenges of rearing adequately mated queens and bringing new colonies into useful pollination capability and honey production for colony survival and some surplus for me to enjoy.

Queen production, successful mating and introduction and period of active brood-rearing are phenological events (phenology is the study of the timing of biological events) and all can be impacted by weather.

Honey bees, unlike bumblebees, can overwinter as colonies through their ability to store their energy source as honey and a protein reserve so crucial at the start of the breeding season in their bodies as storage proteins during a period when there is little pollen (protein) source or nectar available.

Current predictions of the effects of climate change for the British Isles indicate a shift to warmer more unsettled conditions. Warmer conditions may mean more forage is available but there is a risk that these conditions may favour the survival and transmission of pests and disease. Unsettled conditions may mean less forage and foraging. Adverse weather (cool, wet and windy) may stop foraging activities leaving colonies nutritionally stressed and potentially more exposed to pest and diseases such as *Nosema* spp. and the Foul Broods (American and European).

Long term observations made on first flowering dates of some plant species have shown many species to be flowering earlier in the season. The seasonal distinctions we all believe exist are now much less pronounced. In years when the weather is suitable for honey bees to fly and forage such earlier flowering can be a bonus but it can also be an opportunity lost if the flowers are finished in periods of weather when the bees cannot forage or the weather conditions do not favour the secretion of nectar or the availability of the pollen after the anthers have split. Honey bees are very generalist in the range of plant species they can access for nectar and pollen, however they cannot access some flowers because of the size of their mouthparts and the relatively short length of their proboscis, for *A. mellifera mellifera*.

The weather which dictates the plants and crops which can be grown in an area and honey bee colony development, reproduction and nutritional status is intimately linked.

3.4 The human population in the British Isles

The UK population in 2020 of 67,886,011 humans inhabits around 242, 900 km² and averages out at about 279 people per km², but this is not distributed equally throughout the country, with the highest density being in England.

Ireland (the Republic) has a population of 4,937,786 occupying an area of 70,273 km² with a density of 70 / km².

(Source: https://worldpopulationreview.com/country-rankings accessed 01/10/2020).

3.5 Changes in land use

Over the past 150-200 years land use changes have transformed the Earth's land surface. Approximately 60% has been altered primarily by agriculture, forestry, livestock grazing, mining and quarrying and a smaller amount (2-3%) by urban development and transportation corridors.

Major changes have taken place since the 1930's in the landscape of the British Isles in both the rural and urban environments.

Rural environments have experienced agricultural intensification and there has been a near trebling of the land area growing insect-pollinated crops which now cover about ⅕ of UK cropland.

The mechanization of arable farming and improved transport to markets has meant larger fields and a loss of boundary features such as banks and hedgerows. This increasing uniformity of the rural landscape has significantly reduced pollinator forage diversity and availability.

Urban and suburban environments and infrastructure developments continue to expand at a fast rate.

These effects have impacted all insect pollinators, including the honey bee, but the effects are mitigated to some extent by the beekeeper's husbandry of the honey bee.

3.6   Environmental aspects of agriculture and horticulture and beekeeping

Appendix 1 of Sims' book *Sixty Years with Bees* (1997) refers to the keynote address which he gave to the BBKA (British Beekeepers Association) Spring Convention in 1983 and describes his reflections on changes in farming practice in England and Wales and their implications for beekeeping. He kept his own bees from 1926 and his experiences included running about 40 colonies on farms in Cambridgeshire, Essex, Kent, Devon and Northumberland. He spoke about sheep grazing in the orchards of the Weald of Kent and wild white clover grown for seed. In the 1930's the sheep were replaced by gang mowers, clover seed crops finished and the loss of cherry orchards. In Suffolk in the 1930s former agricultural land had been abandoned and there were field boundaries with hawthorn 15 years old with much bramble and in the former fields, charlock and thistles. Honey bees thrived and were very productive.

In between the years 1932 and 1982 there was a significant increase in the proportion of land used for arable farming with the need to grow our own food in the Second World War being a key driving force behind this. Pasture decreased except in the west (Devon and Wales) despite areas which were changed during wartime to arable farming reverting back to pasture. In these areas and in the northern parts there was an increase in the number of cattle and sheep on the land.

These trends continued with intensification of arable farming in eastern England whilst Wales, the north, and west remained largely pastoral, but with higher stocking rates of cattle and sheep.

Orchards with large fruit trees have been replaced by fruit grown on dwarfing stocks which are easier to manage, control pests and diseases and pick the crop but the large areas of grass under the trees have not been re-instated.

Richard Mabey in his book titled *Weeds* (2010) begins with the words 'Plants become weeds when they obstruct our plans, or our tidy maps of the world'. The best-known and simplest definition of a weed is that it is 'a plant in the wrong place'. In other words, growing where you would prefer other plants to grow, or sometimes as perhaps is the case in our increasing urbanisation and installation of hard surfaces to accommodate the ever- increasing possession of multiple cars per household and actually have no plants at all.

He writes that the criteria for weediness can change dramatically with time and begs the question of what is the 'right place' for a plant.

The sanctimonious attitudes of many gardeners and organisations which purport to encourage the correct choice and cultivation of plant species claiming their benefits to nature and the environment would do well to read Mabey's book and learn of the history and incredibly important social and historical events which took place over thousands of years resulting in the increase in the number of plant species (many of which we now persecute as weeds) that form the flora of the British Isles.

Hopefully this book will encourage the reader to re-appraise their views on definitions such as weeds and the consequences and perceptions of these definitions and reconsider them in the light of the needs and survival of the honey bee (and other pollinators of course) and not just we humans.

During the past century the control of weeds, pests and diseases of farm crops by spraying with pesticides and the cutting and harvesting of forage crops at an earlier stage of its development have reduced the variety and quantity of forage available to honey bees.

In the UK the treated area of arable crops (number of hectares x number of applications) has remained relatively stable since 2008, whilst the total weight of pesticide applied has shown an overall decline. (National Statistics 2020). Sources of information on the products used can be found in the Pesticide Usage Surveys carried out from 1990 until today.

Flowers from clover, lucerne and sanfoin have been lost as sources of forage for bees in forage crops.

Fields growing cereals such as wheat and barley had charlock growing in them and then later in the year the flowering of thistles which gave good honey crops but have now been eliminated by spraying.

The widespread increase in the growing of oilseed rape which gave a honey crop in an arable landscape in many areas of the British Isles has changed in recent years because restrictions on the use of certain pesticides (neonicotinoids in particular) and increased resistance to pyrethroids to help in the control of cabbage stem flea beetle damage have meant that the areas of wheat, barley and oats have increased.

For some years there was an increase in the amount of borage (*Borago officinalis*, also known as starflower) grown for its oil which is highly valued for its high gamma linoleic acid (GLA) content. Good honey crops were harvested from this crop, but it is less widely grown today but there some encouraging signs of a revival in its popularity.

Today, about 70% of our land area in the British Isles is under some form of agriculture.

However most agricultural land in the British Isles, whether it is a monoculture or gazed pastures) is a hostile environment for the honey bee (and other pollinators) due to the lack of nutrients and the use of pesticides, including herbicides, which remove potential forage sources. As we shall see honey bees requires a variety and continuity of supply for food sources to be able to survive and thrive, as well as satisfy the desire of the beekeeper and society at large to produce some delicious honey.

Estimates state that half the nation's farmland needs to be transformed into woodlands and natural habitats to help fight the climate crisis and to help and restore wildlife.

The increasing acceptance that people in rich nations need to eat less meat and this will improve their health will impact on the numbers of livestock required to be raised. Most livestock production in the UK is unprofitable without public subsidy. Livestock production also results in environmental damage such as greenhouse gas emissions and water pollution and such effects are the subject of research and development taking place in livestock production techniques as ways to mitigate them are investigated. Uplands and upland pasture farmland cannot usually be used to grow arable crops and with the proposed changes in the subsidy payments made to farmers away from production to 'public goods' or environmental benefitting activities it is argued it could be better to store carbon, water management to reduce flooding, rewilding and make the land available for people to improve their health and well-being and this could be achieved by the growing of trees. Growing trees removes carbon dioxide, a major driver of climate change, from the atmosphere, however there is now a debate as to whether this is entirely true.

Consequently, farmers could find themselves being paid for growing and managing trees and thus removing carbon dioxide from the atmosphere. This increasing focus on trees is of potential importance to beekeepers and later in this book a section has been devoted to trees and their potential role as a forage resource and other benefits to the honey bee.

This is for the future; what is the position we find ourselves in now? Recognizing and accepting where we are and what we humans as a species, and all of us as individuals with a collective responsibility, have done to the environment is key to helping us to decide what positive actions need to be taken, even if it is only to ensure the ongoing survival of the human species, let alone the other life forms which co-inhabit the earth with us.

Once common hay meadows, pastures and thinly grazed pastureland and their associated flowers are much reduced in number, reputedly a 97% decline since 1945.

Some were key forage plants for honey bees, including Charlock, Corn mint and Cornflower. Corn cockle was also common.

The perceived cultivation techniques, weed control, use of herbicides, insecticides and other pesticides necessary for modern agriculture in the UK began to be challenged and affected by the adoption of the Common Agriculture policy (CAP) by the EU in June 2003. This introduced a Single Farm Payment for direct subsidy to farmers. In return for this payment of a set amount per hectare of land under cultivation farmers were required to comply with the careful use of pesticides and fertilisers and 'set-aside' 8% of productive land for possible benefit to wildlife. In 1996 the UK had already introduced a Countryside Stewardship Scheme, and this was superseded from 2004 by Environmental Stewardship. There were other initiatives including a scheme for Environmentally Sensitive Areas (ESAs) also developed.

This book is not the place for a detailed description of these schemes as they were and are continually evolving in response to political initiatives driven from the European Union (EU) (following Revision of the CAP, Common Agricultural Policy in 2013); national perspectives (as a consequence of the Brexit saga); the UK no longer being a member of the European Union (EU); international actions on the environment; the future trading and regulatory relationships between the countries of the world.

An important point to note is that these schemes are intended, targeted and designed to encourage differing sorts of habitats and encourage biodiversity. Such biodiversity would include pollinating insects and honey bees would be one of the pollinator species which could benefit from such schemes.

In addition, these schemes begin to address the concerns of forage availability on a landscape scale and bearing in mind that some 70% of the British Isles is under some form of agricultural activity the potential benefits are of a scale which will help in the survival of many species, including honey bees.

Not only is the planting of trees (solitary and in woodlands), shrubs, hedges and meadows being encouraged but the all-important way in which they are to be managed is being addressed. It is a sad sight to see so many of our hedgerows decimated by mechanical flayers each year and seeing the result of these activities in the loss of a suitable place for birds to nest and shelter and for other small animals to live with the obliteration of the all-important berries and fruits which are lost to the overwintering birds and animals. Good and well-maintained hedgerows not only enhance landscape and land values but also yield a huge return in biodiversity. Pollinating insects benefit from the nectar and pollen from hedgerows managed to promote flowering and then other species benefit from the fruits, nuts and berries, including humans and their interest in flavouring gins, brandies and cordials with sloes and other hedgerow resources.

There are other important habitats which occupy a significant land area and these are the moors kept and managed for grouse, partridge, deer shooting and other sports. These areas include the North York Moors, Exmoor, Northumberland in England together with the huge areas in Scotland and elsewhere in the British Isles. These moors are usually comprised of ling and various other heather and species (such as *Calluna vulgaris* and *Erica* spp.) and provide a significant monoculture of flowering in the late summer. At one time a very important event in the beekeeper's calendar for those beekeepers who had access to the ability to transport hives for 'heather going'. Lowland heath occurs in a few areas. To get a flavour and sense of heather going, Ian Copinger's ***Anthology of Works on Heather Honey*** containing articles written by famous and respected beekeepers on their experiences of going to the heather is a delightful read and one can almost conjure up and smell the beautiful distinct aroma from heather comb and its honey.

As we shall see the quantities and variety of forage required by a honey bee colony are substantial and forage is required to be available for a significant proportion of the year. It is only when considered on a landscape scale will the potential to understand the nutritional needs of honey bees (and other pollinators) be quantified and planned for.

At the end of the book we will take a further look at how government at the national and international level are planning to address environmental issues utilising the principles developed in the evolution of the concept of 'natural capital'.

But let us take a snapshot of agriculture in the UK in 2019. These are statistics published each year by the Office for National Statistics on behalf of Ministers to fulfil the requirements of the 1993 Agriculture Act. They contain data and information issued by Defra and the corresponding organisations in the devolved administrations of the UK.

3.7   Agriculture in the UK in 2019

We have abstracted some of the information from the report that has a bearing on the potential forage available to honey bees (and other pollinating insects).

Some 72% of the total area of the UK is 'utilised agricultural area', amounting to 17,532,000 ha, in addition to which there is 1,197,000 ha of common rough grazing. The total agricultural area is 18,849,000 ha.

In 2019 the total 'croppable' area was composed of:

Arable	4551,000 ha
Cereals	3,211,000
Oilseeds / linseed and borage	547,000
Potatoes	144,000
Other	649,000
Horticultural	163,000
Temporary grassland < 5 years old	1,193,000
Total permanent grassland	10,193,000
Other land on agricultural holdings	1,328,000 (of which 1,033,000 is woodland)

In terms of specific crops, the areas (ha) for relevant crops were:

Oilseed rape	530,000 ha
Peas (not visited by honey bees) for harvesting dry and field beans	178,000
Maize	228,000
Orchard fruits	24,000
Soft fruits including grapes	11,000
Outdoor plants and flowers	11,000
Glasshouse crops	3,000

In 2019 because of the threat from the cabbage stem flea beetle the area sown for oilseed rape was reduced by 9.2% from the previous year to 530,000 ha., the lowest area since 2002. The crop value was £586 million.

Field Beans were grown on 137,000 ha with a crop value of £104 million.

The report also contains statistics on the environmental performance of agriculture, and it was notable that the Farmland Bird Species Index for all farmland bird species in 2018 was less than half of the 1970 levels. The availability for many such birds of feeding and sheltering resources having been lost due to agricultural practices and in turn this also means the loss of forage to honey bees.

3.8 Urban and suburban man

It is now official! More people now live in towns, urban and suburban areas than live in the 'countryside'. Sadly, the understandable wish for families to move into the countryside and escape the urban environment very often results in the incomers destroying the very thing they believed they wanted to live in and experience. Too often the impact of urban values can be seen in rural properties. Close-mown lawns, extreme tidiness, the profligate use of herbicides and insecticides, highly trimmed hedges, gang mown grass verges (the ride -on tractor mower, chainsaw and grass and bush strimmer) all a rite of passage. Then there is the decking, paving, and the hardstanding for the increasing number of cars and SUVs (Sports Utility Vehicles) now spilling out onto village verges and streets together with the security lighting and electric gates all impact detrimentally on the environment. We should all learn to recognise that we are part of the environment and the earth's ecosystems and not see 'nature' as something detached and to be exploited or destroyed without due consideration of the consequences. The inability of the political classes and our increasingly urban based attitudes and cultures to recognise this will lead to problems and these will be compounded by the effects of climate change and changing weather patterns. The apparent acceptance that it is socially a sign of success to have multiple children and the subsequent cries for help (and assumed rights) when family welfare and stability change or are threatened are signs of a society and as a species losing its relationship with biological reality.

Many people have always been aware, and many others are now being persuaded by the facts and by pressure from the next generations to take positive action in these matters and it is through these pressures and the increasing interest in growing plants and fresh food for our table, together with gardening on a large or small scale that cumulatively in an urban or suburban environment will benefit pollinating insects including honey bees. The increasing diversity of our diets in part reflects our social history and cultural changes but it also suggests there is an increasing awareness of our relationship with the food we eat and how and where it is produced. Food security should be uppermost in our minds. We import large quantities of fruit and vegetables from Europe and Africa, however in order to reduce our dependency on such imports by growing more home-grown food could impact adversely on wildlife. On a global scale the perceived responsibility to 'feed the hungry' of the world and produce more and more food cheaper and effectively could in direct contradiction to local biodiversity and low-impact agriculture.

The Lawton report entitled *Making space for nature*, published in late 2010, highlighted the benefits of what are called 'ecosystem services'. In other words, the way in which nature and its wildlife contributes to our health, our pleasure and the welfare of our agriculture, horticulture and forestry. Personally, I do not like the term 'ecosystem services' because they can be interpreted as those being of use and primarily of benefit to humans and hence exploitable without due regard for the consequences. This is the wrong dynamic for the relationship of humans and their surroundings and has done much to contribute to the current perception of the relationship of humans and their environment encapsulated in the Biblical statements that man shall have dominion over the fish of the sea, over the birds of the air, and over the cattle, over all the earth and over every creeping thing that creeps on the earth (Genesis 1:26). Looking positively at the proposals of the Lawton Review one of which is that we should move away from an approach in which we try to hang onto what remains, to one of large-scale habitat restoration and re-creation.

4. Honey bees – domesticated or semi-domesticated?

The question is often posed are the honey bees which are kept by beekeepers **domesticated** or not?

Crane (1984) defined 'domestication' in terms of honey bees as those living in man-made hives and defined 'wild' colonies as those living in natural sites (or adventitiously) in man-made structures not intended to house bees, such as roof spaces in buildings.

Domestication is the process by which humans select and deliberately breed wild species to obtain cultivated variants that will thrive in man- controlled environments and that produce things of benefit to humans.

The honey bee is often listed as a domesticated species. However honey bee colonies do not require daily care. Their relationship with humans is different to that which humans have with chickens, ducks, cattle, horses and other farm animals.

With all these species it is humans who largely determine and select for the desired traits, and the ability to live in man-made environments, in other words human selection. In the honey bee the selection pressures are much more natural in origin and bees will survive in the natural environment and without human assistance..

Having said that at present honey bees are vulnerable to the pests, such as Varroa and associated viruses, such as Deformed Wing Virus and require beekeeper intervention to control the varroa population. However, there is some anecdotal evidence of some indications of increasing resistance to or tolerance of varroa by honey bee colonies over and above their hygienic behaviour. It is often said that without beekeepers unmanaged colonies face an uphill struggle to survive, including the colony succumbing to the parasitic mite *Varroa* and its associated viruses. Good beekeepers usually dealt with varroa in managed colonies by a pro-active active varroa control management programme.

A swarm of honey bees which evades capture and collection by a beekeeper adds to the natural or feral population of the area. Its subsequent survival depends on many factors, including the ability to find a nest site, build a brood nest and be able to enlarge the nest to store honey for use in periods of dearth, including overwintering and also its vulnerability to varroa infestation.

Virgin queens mate with drones they encounter on their mating flights. The drones will be from a variety of locations and colonies with varying genetic traits and not under the direct control of the beekeeper.

The recognition of bee space and the introduction of movable frame hives enabled techniques to be developed in which swarming could be managed and colonies of honey bees with desirable characteristics replicated to some extent.

It is only in the last one hundred years that beekeepers could, through the use artificial insemination techniques select to some extent the genetic stock (drones and queens) and obtain bees with desirable characteristics such as high honey production and low colony defensiveness (humans still refer to bee behaviour as being aggressive towards them).

Yet in spite of the many breeding programmes which have been run to improve disease, increase production Seeley's opinion (Seeley 2019) is that 'there is no evidence that artificial selection has altered in any general way the behaviour of honey bees'.

Having said all this there is no doubt that beekeepers through technological adaptations and greater understanding of honey bee biology can decide and manage where and how many honey bees live (i.e. colonies in hives). These hives are located in and moved to places of man's choice and management.

Honey bees which escape from the beekeeper through swarming and relocate in trees or cavities in buildings show they can exist independently as long as they are not killed by the *Varroa* and its associated viruses. Such colonies can be replaced through the more frequent swarming and creation of colonies in the wild means that there is more material for natural selection to work on and a greater chance for resistance and coping mechanisms for colonies to develop.

We should consider honey bees as being semi-domesticated. In this way we recognise their relationship with us and their place in the wider environment. We should also consider more widely those aspects of their behaviour which we should take greater account in the way we manage our colonies and the provisions as beekeepers we make for them and as members of human society who rely heavily on the pollination activities of the honey bee for crops and other biodiversity.

5. Beekeeping – applied honey bee biology

Beekeeping (apiculture) is the care, maintenance and management of strong healthy honey bee colonies usually kept in hives designed for the convenience of the operator (beekeepers also known as apiarists) and the removal from the hives (and subsequent processing) of the products for which the colonies are kept. These can be honey, beeswax, young queens, package bees, pollen, royal jelly, propolis, bee venom, bee brood (used as food). It can be practiced as a craft and an occupational and economic enterprise.

Man (*Homo sapiens*) has had a long relationship with the honey bee and this has been well described in Crane's books.

For the purposes of this book, we restrict our description of the relationship of humans with honey bees to the British Isles, (Crane 1980 and 1999).

The evolution and development of beekeeping in the British Isles changed significantly with the introduction of the movable frame hive and the recognition of 'bee space'.

Honey bee colonies could be kept and managed and the colony divided up into discreet areas such as brood area and honey supers and thus become vulnerable to analysis and manipulation according to human logic but perhaps with little knowledge of the natural history and biology of the species. Aspects of today's beekeeping practices and beekeeper attitudes are legacies of this relationship of humans with honey bees in the modern era.

We refer to the 'art of beekeeping'. Today skilful and competent beekeepers have to have a thorough knowledge about the biology of the honey bee, the interrelationships of plants and honey bees as well as be competent in other skills such as observation, management and organisation as well as a number of related practical skills. And then there is the science and understanding of honey bees and beekeeping including their pests and diseases knowledge of which is extensive as the honey bee is probably one of the most widely studied insect species. In effect beekeeping is applied honey bee biology.

Crane (1999) wrote that certain characters of bees were important to all peoples that harvested honey, whatever their harvesting technique: opportunistic hunting, owning and tending natural nests, traditional hive beekeeping or more modern 'rational' hive beekeeping. Such characters included:

- Good colony survival through dearth periods.
- Good colony survival through an active season in which honey flows were poor.
- Resistance to injury from diseases and pests.
- Storage of much honey in the nest.
- Tolerance of handling, and low tendency to sting when disturbed.
- Bees easily pacified by smoke.

It is always interesting to ask aspiring and established beekeepers why they became beekeepers and increasingly an important motivation was because of environmental concerns as well as the social aspects and mental well-being derived from keeping honey bees; and not simply honey bees being seen as an economic unit producing honey and wax.

Whitaker in his book, *The Ethics of Beekeeping* (2018) invites his readers to consider some of the ethical issues which beekeepers should consider in their interactions with the bees they have undertaken to keep. It is an excellent and stimulating read.

6. Recent history and current status of beekeeping in England and Wales

The keeping of honey bees has been widely practiced for a long time and there were often multiple beekeepers in the local community perhaps only keeping a few hives or working enough colonies to derive a whole or significant contribution to a household's income.

The number of hives kept can often reflect social status. (Walker and Crane 2000). The small garden of a labourer or cottager might have contained a few hives. Such gardens might also have a wall with a few bee recesses built into them. Larger country estates have been found with as many as forty-six wall recesses for hives and also included in such garden structures such as alcoves and bee houses. There does not seem to have been substantial change in the number of hives kept as determined from contemporary documents from English counties during the period c.1550 – 1750. For example, 2.0 for a labourer, 3.6 for a husbandman or yeoman and 4.4 for a gentleman. More hives were kept in areas where the climate was good and yields were high, a situation still reflected today.

Today honey bee colonies are kept by beekeepers in varying numbers. Many beekeepers are considered amateurs or hobbyist beekeepers and keep relatively few honey bee colonies.

However, this label of hobbyist belies the fact that many such beekeepers are extremely competent and knowledgeable beekeepers.

There are beekeepers who derive their income in whole or in part from their beekeeping activities and these are generally referred to as bee farmers.

In the post First World War era beekeeping practices, forage availability and honey production and the situation of beekeeping and beekeepers are well described in Manley (1936) and Herrod-Hempsall (1937).

There are publications which describe beekeeping post Second World War when the number of people keeping honey bees probably reached a peak when sugar was rationed. The number of honey bee colonies kept in the UK has declined since the end of the Second World War due to the end of sugar rationing, increasing urbanisation with post war reconstruction and the expansion of public and social housing and reduced levels of public interest in beekeeping and involvement in producing honey.

Currently there is no compulsory requirement for the beekeeper to officially register their interests however the National Bee Unit operates a voluntary register as well as providing information, guidance and support in particular concerning honey bee health (pests, diseases, nutrition, training and on-line learning).

In the years 1984 to 1986 David Little of the British Beekeepers' Association (BBKA) organised a survey of its member beekeepers asking about numbers of colonies they managed, their years of experience and the aspects of their beekeeping practices concerning forage availability, honey production and supplying honey bee colonies for pollination services, (Carreck 1994).

In more recent years the BBKA has carried out other surveys.

In 2009 the Government, in co-operation with beekeeping associations and representatives, launched the Healthy Bee Plan initiative which was to last 10 years until 2019. This was successfully completed.

In July 2013 Defra published a Review of Policy and Evidence on Bees and Other Pollinators: their value and health in England.

A Review of the Healthy Bees Plan (published in November 2020) as part of preparatory work to inform the development and establishment for the next Healthy Bee Plan which is scheduled to run until 2030.

In the first decade of 2000 honey bee colony losses over winter were very high which were believed to be partly due to the impact of the varroa mite and its associated viruses and other well-known problems such as inadequate food reserves and failing queens. A summary of the findings of these surveys in terms of colony losses and factors which may have contributed to their demise from the winter of 2007-2008 up to the winter 2019 – 2020 will be described in the following section. The surveys also requested information on numbers of colonies managed by BBKA members and their number of years of experience. In more recent years, the BBKA has also conducted an Annual Honey Survey asking members to respond on their honey harvest and other forage related questions, including whether they practised migratory beekeeping and moved their colonies to potential crop sites.

Year	Average and Median No of colonies / Beekeeper	%age of Beekeepers / average no of colonies	Years of beekeeping experience	%age of Beekeepers / no of years
1985	< 4	62	< 2	17
	>10	10	>20	22
	>100	1		

Data taken from Carreck (1994)

The BBKA November 2020 Honey Survey revealed the following profile of the respondents to their number of years of beekeeping experience.

Number of Years	%age of Respondents
1-2	22.38
3-5	25.87
6-10	24.13
11-14	7.8
15-20	6.56
21-25	1.38
26-30	2.33
31-35	1.90
36-40	2.04
>40	3.64
Don't know	Balance

In Appendix 2 of the Healthy Bees Plan Review there is a series of tables which summarise the data obtained through the annual husbandry review carried out by the National Bee Unit from 2009-2018. The following table concerns the changes which have taken place in the numbers of colonies managed by beekeepers (whether or not they were members of beekeeping associations) at the April and October time points for each year and who responded to the survey over the years.

Year	Number of Responses	Date	Mean No. of colonies	Standard Error of Mean	Median
2009	1658		N/A		
2010	1915	Apr-10	10.59	1.28	2
		Oct-10	12.62	1.53	2
2011	1226	Apr-11	4.81	0.56	2
		Oct-11	5.98	0.62	3
2012	958	Apr-12	4.16	0.24	2
		Oct-12	4.77	0.26	3
2013	1248	Apr-13	4.54	0.31	2
		Oct-13	5.13	0.34	3
2014	1572	Apr-14	4.65	0.32	2
		Oct-14	10.57	5.14	3
2015	1254	Apr-15	4.94	0.34	3
		Oct-15	5.57	0.35	3
2016	1166	Apr-16	4.50	0.19	3
		Oct-16	4.75	0.21	3
2017	1054	Apr-17	4.15	0.20	3
		Oct-17	4.63	0.23	3
2018	1201	Apr-18	4.78	0.30	3
		Oct-18	5.41	0.34	3

Data abstracted from Healthy Bee Plan Review (2020)

Explanatory Notes:
The mean is the total number of hives divided by the number of respondents.
The standard error of the mean is an indication of how accurate the mean is compared to the true population.
The median is the middle number between the highest and lowest responses.

The next two tables contain data from the November 2020 BBKA Honey Survey and the 2019-2020 Overwinter Survival Survey.

Number of Productive Hives	%age of Respondents
1-2	43.88
3-4	22.45
5-10	18.80
11-15	3.72
16-20	0.51
21+	2.83

The mean was 4.81 colonies.

Number of Colonies with > 5 frames	%age of Respondents
0	1.38
1	20.09
2	21.01
3	14.85
4	10.75
5	7.12
6	5.01
7	3.87
8	2.76
9	2.42
10	2.22
>10	7.96
Don't know	0.57

The mean was 6.72 and the median 3.

In the BBKA November 2020 Honey Survey respondents were asked to quantify the average number of pounds of honey harvested for their productive colonies.

Average LB of honey / productive colony	%age of Respondents
1-5	11.30
6-10	11.44
11-20	19.61
21-30	12.90
31-50	15.74
51-100	11.81
>100	1.31

It is interesting to review these data in the context of whether those responding to the bee husbandry questionnaire were members of a beekeeping association, bearing in mind the survey covered England and Wales.

Year	BBKA and other	BBKA only	Other only	None reported
2010	1.7	78.1	9.3	10.8
2011	8.2	62.2	11.5	11.8
2012	9.8	60.6	11.1	18.5
2013	9.9	65.6	11	13.5
2014	11.7	62.7	11.3	14.4
2015	11	62.9	11.4	14.7
2016	9.8	63.5	10.8	16
2017	10.6	63.3	9.8	16.3
2018	10	64.9	11.2	14

Source Healthy Bees Plan Review (2020)

Carreck (1994) observed that in 1985, beekeeping was mainly an activity confined to those who keep honey bees as a hobby, for interest or for domestic honey production. There was wide geographical distribution of colonies and variation in the experience of beekeepers.

In 1991 beekeeper numbers published by the Ministry of Agriculture were 28,947 managing 152,216 colonies

In 2020 most honey bee colonies in the British Isles are managed by beekeepers (rather than being wild / feral). The Hive Count survey made by the National Bee Unit of the Animal and Plant Agency (APHA) estimated there were around 264,000 colonies of honey bees managed by hobbyist and commercial beekeepers in the UK in 2019. In Scotland there are around 2400 hobbyist beekeepers working between 1-12 hives and about 25 commercial bee farmers (no numbers of hives seem to have been published). In Ireland there are estimated to be around 24,000 hives.

Recent years have seen a huge increase in interest in the honey bee and the wish to take up beekeeping. Many beekeeping associations have risen to this challenge and have provided 'taster days', and training courses introducing prospective beekeepers to the craft and either helping them on their way to becoming practicing beekeepers through mentoring schemes, or whilst satisfying the interest and curiosity tactfully reveal that the person or their circumstances may not be conducive to beekeeping. It is essential that such people do not feel themselves to be failures because whilst they may not be able (for various reasons) to become practical beekeepers they are often effective and passionate ambassadors for the craft.

Beekeepers are encouraged to engage in training and education and there is a range of courses, including on-line learning and with them qualifications which can be gained through study, assessment and demonstration of their beekeeping knowledge, skills and competence.

6.1   Current status of the honey bee in the British Isles

The longevity of honey bee colonies of wild (native) populations or colonies lost from the control of the beekeeper which become feral colonies is poorly understood in the British Isles and the subject is attracting increased research interest. Whether feral or native they may struggle to survive because of their vulnerability to the impacts of pests and diseases (some such as varroa (*Varroa destructor)* of relatively recent appearance in the British Isles and its associated viruses and not benefitting from the husbandry of the beekeeper.

It is generally considered that there are few truly wild (native) colonies of honey bees in the British Isles but attempts continue to try and quantify the numerical and health status of honey bee colonies in the British Isles. Many colonies found in tree cavities or other non-hive locations are most likely to be feral, in other words colonies which have escaped the control of the beekeeper or had originated from managed honey bees which have set up nests in the wider environment for example in tree cavities or under roofs in the built environment.

The Western Honey Bee *(Apis mellifera)* has a number of subspecies one of which is *Apis mellifera mellifera.* There is a very dark race of this subspecies often called the 'black bee'. This may be a descendant of an old native honey bee and they seem to be more winter hardy, less likely to swarm and to show much more activity in wet and cool summer weather than does the 'normal' spectrum of honey bees kept elsewhere in the British Isles

Beowulf Cooper in his book *The Honeybees of the British Isles* (1986) provided a description of 'native' bees and their characteristics.

Research in the pursuit of such bees has long been a key part of the work of Bee Improvement and Bee Breeders' Association (https://bibba.com). These 'black' bees have dark pigmentation on the dorsal parts of their abdomen and other distinctive features and behaviours considered to be characteristic of these bees. The book argues that these characteristics are the result of natural selection pressures which benefit natural honey bee colonies and could be of direct benefit in colonies managed by a beekeeper. This book is well worth re-reading or even reading for the first time as it describes and interprets the expression of these characteristics in relation to the climate, geography and the topography of the British Isles. It is interesting to relook at these in the light of today's developments and the recognition of the significance of climate change and the characteristics of honey bees that may make them fitter to survive the stressors of today.

Over the course of 8 years the Annual Bee Husbandry survey has asked respondents to report on the race(s) of bees they were keeping and the proportion (%) reporting which race. These are shown in the following table.

Race of Bee	2011	2012	2013	2014	2015	2016	2017	2018
A.m. carnica	2.6	1.4	2.4	2.7	1.6	1.6	2	1.6
A.m. ligustica	0.5	0.3	0.7	0.7	0.2	0.5	0.8	0.4
A.m. mellifera	17.6	15.4	17.2	21.3	17.2	16.2	15.2	17.8
A.m. macedonia	0.2	0.2	0.1	0.1	0	0	0	0
Buckfast	6.5	7.8	10.4	8.8	9.6	125	13.1	14.3
Hybrid	20.6	19.8	18.9	22.5	21.8	23.6	23.7	23.8
Other	3.7	2	2.9	2.4	3.6	2.7	3.6	3.7
Don't know	48.3	53	47.4	42.1	45.9	42.9	41.6	38.3

The plight of the honey bee and indeed other pollinator species has been the subject of much study and speculation for the causes of the decline rife throughout the world. Very often the public is fed a confused diet of mixed facts and ideas relating to honey bees and other pollinating insects such as bumblebees. Honey bee colony numbers are said to have been in steep decline with a whole range of possible explanations for the cause of their demise. In realit,y the picture is not quite so simple. Honey bee colony survival should not be confused with that of the fate of many insect pollinators such as bumblebees.

Managed honey bees do benefit from the assistance of beekeepers to improve their chances of combatting pests such as Varroa and associated pathogens such as Deformed Wing Virus. Such assistance cannot be given to bumblebee species whose biology and relationship with man is different to that of the honey bee.

In addition, the reproductive potential of honey bees each year assisted by good beekeeping practices can help mitigate colony losses and so the pattern of changes in colony numbers in the British Isles when graphed is much more of a sawtooth in nature throughout the time when honey bees are active.

Honey bee colony survival may not truly reflect whether or not the health status of honey bee species as a whole is improving, stable or declining in the British Isles. It all depends on the question you ask.

Since the winter of 2007/8, the British Beekeepers Association (BBKA) has conducted an annual overwinter survival survey of the experiences of its members and tried to quantify the survival (or losses) of colonies over the period from the 1stOctober to 1st April in the following year. These surveys started in a time when honey bee colony losses were high and there were a number of explanations put forward on potential contributory causes of the colony losses. Successive press releases on the yearly results on the explanations put forward to explain the changing survival figures very often highlighted starvation both in terms of lack of food but also 'isolation starvation' where colonies starve because of loss of contact of the cluster with the food reserves, even in colonies which still had significant food reserves in them.

In the 2019-2020 BBKA Overwinter Survival Survey, 8.46% of the respondents reported starvation as a cause of colony loss and 6.81% reported colonies being lost because of isolation starvation.

Problems with granulated honey stores were cited by 6.12% of respondents.

Both these explanations point to poor beekeeping practices which can be improved by beekeeper awareness and remediation techniques.

Year	% Survival	% Loss
2007-2008	69.5	30.5
2008-2009	81.3	18.7
2009-2010	82.3	17.7
2010-2011	86.4	13.6
2011-2012	83.8	16.2
2012-2013	66.2	33.8
2013-2014	90.4	9.6
2014-2015	85.5	14.5
2015-2016	83.2	16.8
2016-2017	86.7	13.3
2017-2018	75.0	25.0
2018-2019	91.0	9.0
2019-2020	82.7	17.3

Source: BBKA News July 2019 (updated)

Poor weather conditions are often given as a cause of colony loss and this can be manifested through several ways. Weather can affect the growth and flowering of forage plants and even if the forage plants thrive and produce pollen and nectar the weather conditions at that time of maximum availability may be unsuitable for honey bee foraging. Low temperatures and high wind speeds can affect the bees' willingness to go out to forage and in any event these conditions themselves may lead to the lower availability of nectar and pollen to the foraging bee. The state of the colonies at the time of such forage availability may have been such that they we unable to benefit from it.

A number of pests, parasites and environmental stressors have also been linked to honey bee colony losses. It is now recognised that sublethal effects of stressors on honey bee colonies have consequences for the colony and that any factor that compromises a bees' abilities to be able to forage effectively or carry their other tasks to serve the colony can result in a colony's decline. (Barron 2015).

The same can be said for diseases in that they do not need to kill individual bees to kill a colony because if the disease sufficiently compromises the functioning of a colony this can result in colony decline and death.

In the British Isles honey bee colony survival has tended to improve in recent years and there are probably several contributory factors. Aspiring beekeepers are now more likely to want to study, train and study the craft and implement their learning into their beekeeping practices. Beekeeping associations at the local, county and national levels have extensive education and training programmes and there is a comprehensive structure of beekeeping qualifications aimed at encouraging beekeepers to improve and demonstrate their skills, knowledge and competence.

The Healthy Bees Plan Review (2020) contains a table showing the proportion (% age) of respondents to the Annual Husbandry Survey carried out by the National Bee Unit holding one or more beekeeping qualifications.

Year	2 or more	3 or more	4 or more	5 or more	6 or more	7 or more	8 or more	9 or more
2012	12.35%	4.94%	2.06%	1.23%	0.41%	0.41%	0%	0%
2013	12.98	4.97	2.76	1.66	1.38	0.83	0.55	0.28
2014	9	2.6	0.8	0.2	0	0	0	0
2015	7.51	2.35	1.17	0.47	0.23	0	0	0
2016	11.46	4.3	1.91	1.19	0.95	0.48	0	0
2017	14.89	6.38	3.19	2.66	1.6	1.33	0.27	0
2018	11.09	5.31	3.93	2.77	1.85	0.69	0	0

Access to information and help on the internet has revolutionised the ways in which we can learn more about the honey bee, including beekeeping. This process of continual improvement is very much part of today's workplace (e.g., Continuing Professional Development in the workplace and on-line learning in schools and tertiary education) in modern day society and comes second nature to those who are keen on managing their honey bees in a healthy and sustainable way.

There is no doubt that honey bee nutrition and feeding are key factors in the survival of honey bees through periods of forage unavailability throughout the year.

Can beekeepers learn to be able to anticipate and remedy shortages of forage? Achieving this state does require effort and the willingness of the beekeeper to think laterally and widely. Beekeeping has traditionally been taught formerly in a very mechanistic way (thinking of living things as if they are machines and units of production) and the techniques and approaches taught have been designed to make a honey bee colony conform to the norms and wishes of the beekeeper. This is changing and there is now a much greater willingness to consider the understanding and awareness of the natural history of the honey bee colony and how techniques used in beekeeping should satisfy the natural instincts of the honey bee as well as seeing if they meet the wishes of the beekeeper.

To reinforce this approach, we now refer to healthy honey bee management and integrated bee health management with the emphasis on disease prevention rather than the traditional approaches of diseases control through identification / diagnosis and its subsequent treatment, (Aston and Bucknall 2010).

An aspect of this involves the recognition of the importance of nutrition and the keeping of colonies in a healthy state, in preference to resorting to treating honey bee colonies when they display symptoms. In order to achieve and sustain this situation means that beekeepers and interested parties need to look much more widely and consider the broader needs and means required to achieve a sustainable, abundant food availability for honey bees and alter land management and beekeeping practices accordingly.

7. The differences between hunger, malnutrition and starvation

In the context of this book nutrition is a term used to encompass all the processes by which the honey bee both as an individual and as a colony acquires and utilises food substances to enable development, function and survival. It concerns physiological and behavioural processes involved in the acquisition of nutrients and other chemicals required for energy, growth, tissue maintenance and reproduction.

Honey bee nutrition has been extensively reviewed, e.g. Brodschneider and Crailsheim (2010). This review was updated by Wright et al in (2018).

In 2018 Zbigniew Lipiński published his very comprehensive book on *Honey Bee Nutrition and Feeding*.

Lipinski (2018) makes the following differentiations:

Hunger is the term generally applied to a temporary lack of food and can occur at any time in the year and anything that is suffering hunger signals that it needs to eat. In the honey bee there are several behaviours associated with hunger:

- An increase in the propensity to sting.
- A sudden reduction in brood rearing.
- The lack of nectar (nectar dearth) can cause competition between colonies and this can result in robbing which induces aggressive behaviour in both the bees which are the robbers and the bees of the colony which is being robbed.

Hunger may arise because the colony density is too high for the forage resources available.

Malnutrition can be defined as less than optimum nutrition.

Colonies which are being used for the pollination of crops which may be deficient in pollen or nectar, an important nutritional component or have low nutritive value can result in impaired immune systems and increased susceptibility to pesticides.

Starvation is a severe deficiency in the calorific energy intake needed to maintain honey bee life and is the most severe form of malnutrition. Dead bees and larvae can often be seen in front of the hive entrance (assuming the blue-tits and mice have not eaten them; in which case the beekeeper probably remains unaware of what is actually going on inside the hive).

Starving honey bee larvae signal their food needs to adult workers using a volatile pheromone E-β-ocimene, which is a component of the brood pheromone, which attracts adult worker bees to the larva. (Xu et al. 2016).

Even low levels of stress can have consequences on both honey bee foraging behaviour and as a result on the nutritional status and demographic balance of the population of honey bees in a colony.

In this book 'demographic' is used as a term to express the composition, size, sex ratio, age structure, fecundity and mortality of the group of individuals of the same species interacting within the same space (e.g., a colony of honey bees in a beehive or nest cavity) and how it changes as a result of the interactions of the external environment influencing the characteristics of the individual honey bees in the colony.

For a colony, food collection, processing, assimilation and excretion are decentralized processes achieved collectively by different individuals each with divergent nutritional needs.

In the honey bee colony there are nutritional cascades:

- Foragers collect nectar and pollen from flowers that they select based on learning and memory.
- On returning to the colony foraging bees indicate the location of forage resources by various dances e.g., the waggle dance.
- They share samples of the nectar resource held in their crops with nest mates. through trophallaxis to give information on the quality of the nectar available at their chosen location.
- They directly offload pollen directly into cells.
- Workers ingest and process the food resources (nectar and pollen) relocating it in the hive.
- Storage and the state of the resources gives direct feedback to the foraging bees and indicates the colony's need for nectar and / or pollen.
- Nurse bees feed the larvae, the queen, drones, foraging age worker bees and the workers of other ages.
- The larvae release brood pheromones which indicate the amount and type of food given by the nurses and collected by foragers that gives feedback to the foragers.
- Workers clean the cells and the complete hive.

As will be seen honey bee nutrition is multifaceted but there are several key aspects which we should always have at the back of our minds, namely:

- The needs of the individual in its development and in fulfilling its role in its contribution to the life cycle of the honey bee colony.
- The needs of individual bees will vary in accordance with the demands of the colony throughout the year from when they become part of the colony to the time that they die. Spring and summer bees may live around 30 days whilst those who will constitute the overwintering colony (referred to as overwintering bees) may last for several months.
- There is a seasonal factor depending on the phase in the colony life cycle:
  - Colony growth and reproduction in the spring and the summer
  - Reduction in brood production later in the summer and the autumn
  - Overwintering

This seasonality in the life cycle of the colony and the individual honey bees is found in the different nutritional requirements of bees at the time and not just a reflection of the forage that is available at that time.

8. The role of the beekeeper in understanding the honey bee colony and factors affecting the species' survival

Honey bees are essentially vegetarian requiring pollen, nectar, and water to provide the proteins, lipids, carbohydrates, vitamins and minerals to meet the needs of the colony. The beekeeper's understanding of these needs is fundamental to becoming a competent beekeeper, able to provide for the needs of their bees and to achieve a sense of satisfaction and pleasure in their relationship with their honey bees.

It seems there are few truly wild honey bee colonies in the British Isles and many colonies which are found outside hives, e.g. in tree trunk cavities or in house roofs are feral and are derived from colonies which have been lost to a beekeeper. The survival of such colonies rarely exceeds a few years and although there are reports of colonies inhabiting a particular space for many years probably reflects the fact that colonies die out, the wax combs are destroyed by wax moths and the cavities are re-occupied with a new colony of bees, most likely a swarm.

In spite of the assertions that there are colonies of bees that are resistant to or can cope, through traits like hygienic behaviour with agencies such as Varroa these are still the exception and not the rule. It is still the case that the long-term survival of the greater numbers of honey bee colonies is dependent on the activities of the beekeeper.

Increasingly experiences are borne out by surveys that show the most significant cause of honey bee colony losses is starvation or through conditions, such as poor colony development and poor overwintering which can all result from poor nutrition.

It has long been recognised that the weather is probably the most factor in honey production both for the colony and the ability of the beekeeper to harvest what is considered to be the surplus amount which can be made available from the colony. Unfavourable weather affects the development and flowering of plants with consequential effects on the production of pollen and nectar. It also effects the potential for the colonies to be able to forage successfully, thrive and importantly achieve sufficient storage of honey reserves to enable the colony to survive overwinter and during periods of dearth in forage availability at other times of the year.

Beekeepers must now think much more widely about the environment in which their bees are expected to, at a minimum exist and ideally to be healthy and thrive.

Such considerations will inform beekeeper strategies and enable the beekeeper to be more realistic about the expectations they may have about their colonies and the scale of the beekeeping operation which the geographical area can sustain.

The continuing significant losses of colonies, premature failure of queens and the demand for colonies from the increasing interest in beekeeping has put pressure on the supply of colonies and replacement queens often giving little choice in their origin, source and availability. More and more colonies are being produced from an ever-decreasing genetic pool.

Having acknowledged such problems there is a need to be more knowledgeable and aware of the resources that are available to us.

An understanding of the sources, origins and characteristics of the colonies of bees which the beekeeper uses or if a new beekeeper is first introduced to is very important and as we shall see the relationship of these characteristics to the local situations in where the colonies are established will determine whether the bees thrive, merely survive with unsustainable beekeeper support, are lost and die out.

Managing colonies which are good tempered, by which we mean do not show heightened defensive behaviour (which we take as aggressive behaviour towards us) makes for pleasurable beekeeping and mean reduced potential problems with the public.

The pleasure of a purchase of a pure strain of Italian bee which is usually docile can soon become less so when the colony reproduces and the new virgin queens each of which will mate with several drones (around 8-10, sometimes more) which she will have encountered when flying through clouds of drones (referred to as drone congregations) which have gathered from other colonies often of varying origin, often hybrids. Offspring from such mated queens can become bad tempered, difficult to manage and handle. They can also become 'followers' and follow and harass anyone walking away from the area of the hive often for as far as 100 metres from the hive.

So-called Italian bees (*Apis mellifera ligustica*) are adapted to living in a warmer climate than that found in the British Isles especially the long summers found in Italy. Italian bees do not store large quantities of stores for overwintering and instead are likely to consume substantial amounts of any reserves and convert it into bees and so in the climate in the British Isles they are likely to run short of reserves over the winter period or in times when there is a dearth of forage through inclement weather or just lack of forage. They need close observation to ensure they do not starve. The rapid population growth in the colonies initiates the swarming impulse which can result in many swarms, a reduction in the honey crop because the resulting colonies have too small foraging age bees in their populations and an increased chance of losing valuable queens of known origin.

In comparison advocates of the British Dark or European Dark Bee (*Apis mellifera mellifera*) today which is considered to be the original native honey bee have identified a number of characteristics which suggest that this form of honey bee has evolved and adapted to survive the climatic conditions which exist in the British Isles. Such characteristics include:

- Excellent overwinter survival capability.
- Low tendency to swarm.
- High longevity of workers and queens.
- Collects significant amounts of pollen.
- Will forage in poor weather.
- The adult bee is very hairy thereby reducing heat loss and can maintain the winter cluster temperature consuming less stores compared to other bees.
- The colonies naturally replace their queens about every three years.
- Its brood cycle and colony life cycle each year are suited to the British Isles climate.
- Exhibits good defensive behaviour against predators such as wasps and hornets (for example The European Hornet (*Vespa crabro*).

Hence there is an increasing interest in the Dark Bee and its potential in the coming years, in a changing environment.

There are opposing aspects of wild/feral and managed honey bees. Wild / feral colonies are often portrayed as being reservoirs of diseases and pests and should be brought under active beekeeping management wherever possible. This issue is discussed in Pirk et al. (2017). They reason that interactions between wild / feral colonies and managed honey bee colony populations can be both beneficial or deleterious. In order to survive newly introduced pests and pathogens wild / feral populations have to adapt and increase their resilience. These natural selection processes may not be present or as selective when considering managed honey bee colonies. A high ratio of wild to managed colonies could mean that the beneficial adaptations developed in the wild populations filter through to the managed honey bee population.

A note of caution…….

Whilst it is a good thing that there continues to be interest in people wanting to take up beekeeping there are risks associated with too many beekeepers and their hives populating an area too densely with colonies for the locally available forage to sustain. Malnourished colonies are more disease prone and are unable to reach their potential resulting in disappointed beekeepers and the population of honey bee colonies containing bees which at best are merely surviving and not thriving.

9. The honey bee colony structure — a superorganism — the fundamentals

Many insects overwinter as eggs or in a state of diapause (a period of suspended or arrested development e.g., as a pupa during a stage in an insect's life cycle usually triggered by the environment) in order to survive recurring periods of adverse environmental conditions, such as winter.

Eusocial insect species in the insect order Hymenoptera include bees (honey bees, bumble-bees, orchid bees, stingless bees), wasps, hornets and ants.

Advanced eusocial insects, such as honey bees, typically:

- Show continued care of their young.
- Have cooperative brood care.
- Have reproductive division of labour.
- Have colonies with at least two adult generations.
- The egg-layers are morphologically differentiated.

See Crozier and Pamilo (1996).

Honey bees are described as being holometabolous, that is to say progressing through four life stages, namely egg, larva, pupa and adult and can be said to have a complete metamorphosis.

Colonies of the honey bee, *(Apis mellifera),* are 'active' throughout the year and have an annual cycle. In spring there is brood rearing, colony growth and colony reproduction and replication by swarming. Brood rearing and population growth continue through the summer period and then with the onset of autumn when egg-laying and consequently brood rearing decline the bees accumulate resources in their nests (combs) for use over the winter period when they are confined to their hives for extended periods of time.

Tautz in his book *The Buzz about Bees* (2008) describes in detail the concept of the superorganism and its applicability to the honey bee.

Its relevance to the content of this book is to understand and appreciate the fact that the honey bee colony is not only in control of its internal environment, but through its genetic structures under optimal conditions be potentially immortal. It is in a continual state of responding to feedback mechanisms and colony behaviour is adjusted homeostatically. Homeostasis being any self-regulating process in which biological processes tend to maintain stability, whilst adjusting to conditions that are optimal for survival. In this way the colony acting as a superorganism.

The superorganism acts as the vehicle for the genome (the genetic material containing the genetic material of the organism and contains DNA) to enable the genome to be successful and able to compete and thrive in an everchanging world. In effect the queen and the drones are the key individuals in the colony because they enable the genome to proliferate and compete, hopefully successfully and adapt to an ever-changing environment.

The queens and the drones (which are the reproductive members of the honey bee colony) are relatively few in number and their function is to facilitate the direct transmissions of the genes. The worker bees provide all the support mechanisms which are required to enable the colony genome to be sustained and successful.

The honey bee colony is a super-organism and from a nutritional aspect there are three levels of nutrition…colony, adult and larval. The individuals are part of a colony which is a superorganism whose individuals through their co-ordinated physiology and behaviours share the characteristics of an organism. Let us look at these levels more closely.

Poor colony pollen stores may result in inadequately fed larvae, or the possibility of not all larvae being successfully reared to becoming adults.

The quality and the quantity of such adult bees may be poor; and this can affect the nutritional status of the colony and in turn this affects the subsequent brood rearing capability and ultimately the success of the colony.

Malnutrition can occur at any level. Larvae reared during a period of shortage of an essential nutrient, such as an amino acid, may develop into short-lived adults or adults with impaired brood rearing, and foraging abilities and compromised immune systems. It seems little is known about sublethal larval malnutrition or sublethal adult nutrition or the occurrence of both conditions in a honey bee colony.

The nutritional levels in a colony are influenced by the numerous adult – brood interactions and the feeding (trophallaxis) contacts between adults, and adults and the brood.

Trophallaxis describes the social transfer of food, partly in a directed manner and partly generating a 'common stomach' that enables all bees to obtain knowledge of the nutritional status of the colony.

Adult and larval development and sustenance rely on the state of the colony stores. Adult bees may adapt their foraging or brood-caring strategies according to the respective need for and the supply of proteins and carbohydrates.

Honey flows from one bee to another throughout the colony by mouth-mouth transfer. This ensures that each member of the colony receives energy and other resources.

When considered in combination with other stressors, such as low temperature during larval development or parasitism or sublethal pesticide effects, the potential stress and consequent impacts on a colony's health condition and survivability could be considerable.

The following diagram illustrates the relationships of colony, adult and larval interactions through the process of trophallaxis which enables all the bees in the colony to access knowledge of the nutritional state of the colony.

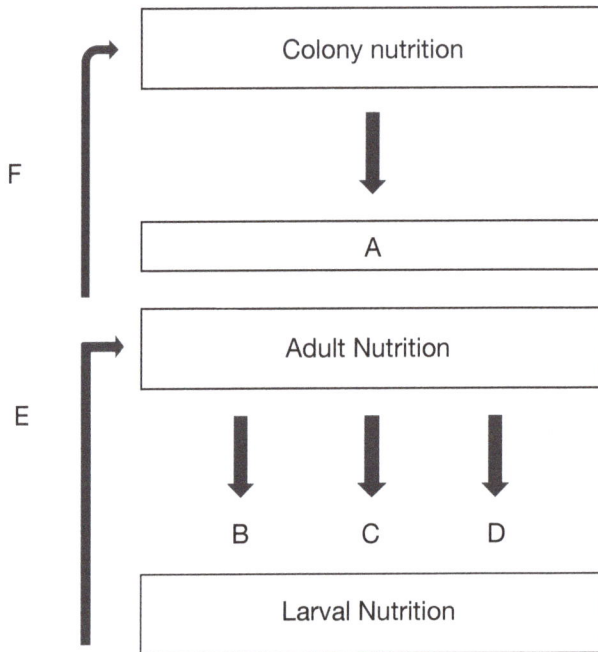

A -indicates the dependence of adults on colony food reserves

B – investment in larval quality

C- regulation of larval number

D- Cannibalism

E- Impact of larval nutrition on next adult generation

F – Impact of Adults on colony nutrition

(Adapted from Brodschneider and Crailsheim (2010)

This constant flow of honey via the bee's mouths is also important for the communication of information about the state of health and fertility of the colony queen.

10. Wild honey bee colonies

There is an increasing interest in the existence of wild honey bee colonies in the British Isles and their ability to survive.

In Europe honey bee colonies have been found especially in the following trees (Crane 1999)

Common Name	Botanical Name	Common Name	Botanical Name
Alder	*Alnus glutinosa*	Limes	*Tilia* spp.
Ash	*Fraxinus excelsa*	Maples	*Acer* spp.
Beech	*Fagus sylvatica*	Oaks	*Quercus robur* *Q. petraea*
Birches	*Betula* spp	Willows	*Salix* spp.
Elms	*Ulmus* spp		

Observations and research by Seeley and others over the years have shown that honey bee colonies in the wild state and living in forests and woodlands commonly establish colonies in cavities in particular in old tree trunks which have the following characteristics

- The volume is around 40 L.
- It has been previously occupied by bees.
- It has a south facing entrance which is 1-5 m off the ground.
- The size of the entrance hole is 12.5 cm2.

Loftus et al. (2016) discusses the importance of small nests and frequent swarming in wild colonies.

Seeley's book, ***The Lives of Bees. The Untold Story of the Honey Bee in the Wild***. (2019) should be read by all who are interested in the honey bee. It gives guidance on how beekeepers should approach their management of colonies and understand the often very divergent way that natural selection pressures through evolution have resulted in the honey bees of today, in contrast to the ways in which beekeepers attempt to make managed honey bee colonies bend to their will and expectations.

We have taken some of the key points of this book in particular with respect to honey bee nutrition and modern beekeeping and included them in this book. They are adapted from Table 11.1 of Seeley (2019) and selected with respect to nutritional aspects of the colony.

Recognition of the colony's needs will inform the beekeeper on how best to set out and manage their beekeeping activities.

The definition of 'wild' used in this instance is the occurring, growing or living in a natural state, not domesticated, cultivated or tamed.

10.1   Genetic adaptation to location

Wild colonies are genetically adapted to their location whereas unless reared locally from stocks of known pedigree many colonies worked by beekeepers are headed by queens bred from limited genetic strains, often in other countries and imported or are strains of bee not suited for the locality.

10.2   Colony density

It is usual beekeeping practice to keep multiple colonies of honey bees in apiaries, mainly it has to be said for the convenience of the beekeepers. This closeness of a lot of colonies causes greater inter colony competition for forage, robbing, and parasite and pathogen transmission.

Wild colonies are found distributed amongst woods and forests. Seeley (2019) summarises studies which have been made to estimate the density of wild honey bee colonies in the USA and Europe (Germany and Poland). He writes that the colony densities range from 1-3 colonies per sq.km (2.6 – 7.8 colonies per sq. mile). In other words, they are very low.

Keeping managed honey bee colonies in higher densities, such as in apiaries or in suburban gardens, has a potential impact on their health and nutrition.

10.3   Space for colonies to exist in

Wild colonies tend to occupy smaller cavities than those contained within hives. There is a stronger natural selection for strong healthy colonies in wild bees. Bees kept in hives are often not space-limited and in fact 'non-swarminess' is often selected for by the beekeeper. Large, non- swarming colonies can act as a large breeding pool for parasites such as *Varroa*.

10.4   Forage diversity

Wild colonies are found in locations where there is often a wide range of floral sources to forage. Managed colonies on the other hand are regularly kept or moved to crops such as oilseed rape *(Brassica napus* ssp. *napus)* which are monofloral and may have no additional floral sources nearby to forage to give the floral diversity and continuity of supply which has been found to be necessary to give the colony the range and quantities of nutrients, e.g., essential amino acids it requires. Crops are also likely to be sprayed with pesticides including insecticides, fungicides and herbicides.

10.5   Diet

Natural sources of forage are available to non-managed bees and usually have the range of floral sources required. Managed colonies may also be fed an artificial diet, e.g., with protein supplements and pollen substitutes. A lack of adequate diet, or feeding an artificial diet, stresses colonies and the worker bees have reduced longevity, a shorter period of their lives as foragers and begin foraging at an earlier age.

11. Honey bee colony life cycle

The honey bee colony is an everchanging entity in which the genetical composition and the resulting fitness of the colony to survive in today's conditions is determined by the queen and the drones. Both castes producing sex gametes, male and female respectively. These genetically based properties are expressed by the interaction of the worker caste with the queen and the drones and the abilities of the workers to raise the offspring and carry out the activities which will feed, sustain and defend the everchanging colony as it grows and produces new colonies through the phenomenon of the swarming process and new queens mating with drones from different colonies.

The abundance of honey bees is very much determined by the seasonal, annual or longer-term climates and these usually determine the peak flowering periods of potential sources of forage. The arrival of spring, the progression to summer and then to autumn promote the emergence of adult bees. Day length is also a fundamental determinant of bee activity.

Because of the multiple mating behaviour of the *A. mellifera* queen, colonies of *A. mellifera* contain within them genetic variety which enables the colony and, in the number of colonies in population terms, the species to adapt to changing environmental conditions thereby ensuring the survival of the species.

The strength and survival of a honey bee colony is dependent on having a fertile laying queen and sufficient numbers of worker bees to perform a whole range of activities required to ensure colony integrity including collecting, processing and storing food, rearing brood, and defending the colony.

Worker honey bees segregate tasks on an age-related basis with the young adult worker bees specialising in brood rearing roles and older adults defending the colony and in foraging. This system enhances colony efficiency by delaying the time before beginning to engage in the highest risk activities (i.e., outside the hive) after they have made their contribution to productivity of the colony in terms of the colony population.

Slight changes in brood-rearing or the longevity of the worker bees has significant effects on the survival of the colony. Factors which affect the population size of the colony include the rate at which worker brood become adult bees and their longevity are crucial.

If the adult population is large and there is good forage and space in the brood nest, then more brood can be reared thus continually increasing the size of the colony.

Larger colony sizes are usually associated with the need for the colony to reproduce, namely swarm, to be competitive and to survive.

Small colonies rear smaller numbers of brood and may be more vulnerable to death from starvation, the impacts of disease and pathogens and overwintering loss.

The western honey bee *(Apis mellifera)* has a number of subspecies, races and strains having developed a wide range of variation in response and adaption to local conditions. Modern beekeeping and the tendency for honey bee colonies and queens to be reared and exported / imported or moved around the country if raised in the British Isles have reduced this diversity and compromised the local adaptations developed by honey bee colonies over many years.

The graph charts the numbers of adult and brood stage honey bees through the year, assuming the colony does not swarm. A critical point to be aware of is the period, March- May, when the population of brood can exceed those of the adults and this has significant impacts on the ability of the colony to secure sufficient forage, especially protein, to feed the developing brood and the young adults.

11.1   Overwinter survival

The ability of the Western honey bee to be able to survive as a colony of several thousand bees overwinter has resulted from a number of morphological, physiological and behavioural adaptations. These changes and adaptations developed in response to the pressures and the opportunities the honey bee experienced during the course of its evolution. In spite of these adaptations the ability of the colony to survive overwinter successfully depends on the conditions the colony actually experiences and as we discuss the nutrition of the honey bee it will become obvious how important the nutritional and health status of the colony is to enable the colony to successfully overwinter.

Lipinski (2018) contains a table of data on Polish bees concerning the Winter Survival of Worker bees in relation to the time when the adult honey bee worker emerged as an adult.

Time of Emergence as an Adult	Percentage of the worker bees surviving the Winter
Up to 26th July	39.2%
27th July to 26th August	82.5
27th August to 7th September	88.4
8th September to 13th October	70.0

These data are interesting because they demonstrate the importance of the worker bees emerging in late August and their role in maintaining the colony over the winter.

11.2   Nest homeostasis

Nest homeostasis is the term used to describe the maintenance of ordered and stable conditions with respect to the nests of honey bees. Aspects of nest homeostasis include the regulation of temperature, humidity, oxygen and carbon dioxide levels, but they also include the maintenance of the dynamic order within the colony such as the properties of the nest and the spatial organisation of the adults and the developing brood within it.

Homeostasis therefore maintains the conditions under which tasks are done, the metabolism of the bees, how the colony's bees function and the larvae grow under conditions that are favourable to the honey bee, but not to parasites and diseases.

11.3   Temperature and its effect on the colony

Honey bee colonies can maintain the temperature of their brood nests within a narrow range (34-36^{0}C) over a broad range of external temperatures and at a relative humidity of 60%.

Honey bees can uncouple their wings and have the ability to activate their thoracic flight muscles when the external temperature drops and set them trembling to generate heat which is used to keep their own bodies and the brood at the required temperature levels. Individual unoccupied (in the sense there is no brood present) cells can be seen dotted throughout a comb of sealed brood and these cells are temporally inhabited on a rotational basis with worker bees (called heater bees) which conduct thermogenesis. The heat generated travels through the walls of the cells into nearby cells which contain developing brood.

The energy supplies of the heater bees soon become exhausted and have to be replenished. Within the colony there are worker bees which act as 'refueler' or honey dispensers and these are continually servicing the heater bees in their cells. Heater bees are completely exhausted after 30 minutes. Refueler bees pinpoint sources of heat on the brood combs and there the feeding process is triggered by the bee in need of nourishment by making contact with their antennae (Rasille et al. 2008).

Drones older than 2 days can also actively generate heat mainly when the ambient temperature is low, but the temperature-related energy metabolism in drones is not fully understood. See later comments on drones being ectothermic.

The worker honey bees will also shiver to generate heat to maintain the temperature of the winter cluster.

The glycogen stored in the flight muscles will only sustain flight or shivering for 10 to 20 minutes after which trehalose is taken from the haemolymph including sugars held in the crop and used as the primary energy source.

Each worker appears to respond according to her own internal threshold, and these may be modulated by reinforcement learning. The cumulative effect of this is to closely regulate the temperature through a graded colony-level response. Reinforcement learning is an adaptive process whereby the bee utilizes its previous experience to improve the outcomes of further choices.

The hatchability of eggs destined to become female (worker and queen) larvae is affected by the temperature it experiences in the cell and in the case of *Apis mellifera* the optimal temperature is about 35° C and the duration is between 72-76 hours at this temperature.

In the case of the unfertilized eggs which are destined to become drone honey bees the time is increased to 75-79 hours at 35°C.

Below 30°C less than 1% of the eggs will hatch and are taken out of the cells by the worker bees.

The location of the colony and the ambient temperature effect the temperature the colony experiences inside the hive and can induce behavioural changes which may impact on the nutritional status of the colony through the rate of food consumption and the status of its food reserves. The changes which take place in the colony are shown in the following table.

Key temperatures inside the hive °c	Effects on the colony
8	Bees become chilled and fall off the cluster and die
14	Clustering takes place
32-36	Brood survival
35 +/-0.5	Maintenance of brood temperature
35-37	Secretion of wax
47	Combs collapse

At temperatures from 5-20° C, drones are ectothermic (they cannot regulate their body temperature which oscillates with the changes in the ambient temperature), whereas workers were endothermic (they can generate heat to maintain their body temperature above the temperature of their surroundings). Between 25-35° C both the drones and the workers are endothermic.

The chill coma temperature (when the bee becomes motionless) is around 14° C for drones and lower at 11° C in workers.

11.4   The Overwinter Cluster

A colony will spend much of its time over the cold winter months and during periods of cold weather as a cluster of bees covering the surfaces of the frames of comb. The bees regulate the temperature conditions in and around the cluster, but they do not attempt to heat the whole volume of the space in the nest site / hive outside the periphery of the cluster. The regulation of the temperature of the cluster is vitally important if the temperature conditions are not to fall outside the levels referred to above.

The bees have different functions depending on their location in the cluster at any one time. The bees in the interior of the cluster produce heat by the micro-shivering of their flight muscles located in the thorax without moving their wings. If brood is still present the local temperature around the brood area is maintained at a temperature in the range of 32-36^{0}C. If the colony is brood less the bees maintain a lower temperature of the cluster, but not below 18^{0}C which is the lowest temperature for them to generate heat. The amount of heat produced is equivalent to that coming from a small light bulb with a brightness of 20- 40 watts.

Towards the outside of the cluster there are several layers of bees on close contact with each other forming an insulating shell around the heat-producing bees. The bees in the shell layers of the cluster orient their heads to point inwards towards the centre of the cluster such that those bees forming the outer layer of the shell have their abdomens exposed to the cold air surrounding the cluster.

Honey bees are covered in branched (plumose) hairs which trap the air and retain the bee's body heat and when the bees are tightly packed as they are in the shell of the cluster they also help retain heat and keep the cluster warm

 The heat-producing bees in the centre of the cluster generate enough heat to maintain the temperature in the outer layers of the shell above 10^{0}C. At temperatures less than this the bees will chill, fall from the cluster to die on the floor of the hive.

12. Optimum colony size

12.1 Size Matters

Beekeepers are often taught to 'intervene' perhaps more accurately described as 'interfere' to help honey bee colonies achieve situations which the beekeeper believes the honey bee colony should reach at particular points in the beekeeping season.

Clearly beekeeping techniques and practices should always be aimed at preventing problems and help the colony in the event of harm or disease threatening the colony, but often unintentional effects of beekeeping practices can arise.

Throughout the honey bee colony's natural yearly life cycle the beekeeper often intercedes and interacts with a colony to enable the resources of the colony to be utilised as demanded by the beekeeper. It is important that the beekeeper must always understand and respect the colony drivers which enable a colony (or its genome) to persist and the honey bees species to survive and thrive.

Good practice beekeeping should mean the beekeeper has a strategy or some idea of how they intend to work with the colonies they manage during the course of a season; be it the timing of colony development to concentrate on honey production from specific nectar sources e.g., oil seed rape, or optimising colony development to provide pollination services for particular crops, colony number increase, replacement queens or the adequate provision of winter food reserves. Indeed, beekeeping involves combinations of several, or all of these possibilities, and can include just letting the colony do its own thing and express its natural instincts.

Optimal colony size becomes particularly important in times of dearth and especially over the winter period because of the need to maintain cluster integrity, not to rapidly deplete the food reserves and not to have high losses of individual bees.

All of these courses can rely on the experience and judgement of the beekeeper and their understanding of the forage and other resources which the colonies need to fulfil these alternatives, including an appreciation of the characteristics of the bees they keep.

The British Isles have a complex geological structure and the soils derived from the basic geology are varied. Such variation is in part due to the topography and elevation of the land which in turn determines the land use, be it agriculture, uplands, forestry or, increasingly, subjugation to concrete, tarmac and bricks and the dead hand of the urban human with all the associated infrastructure. This determines the types and quantities of forage available. Many plant and animal species have been able to exploit these man-made habitats and to thrive.

Later in the book we will see how much forage resource is required by a honey bee colony to sustain itself, to reproduce, produce enough honey to enable the colony to survive periods of forage dearth such as over winter or periods when there is no forage available and for there to be a surplus which the beekeeper can consider appropriate to take from the colony as a crop.

It is worth reflecting on the different stages and types of honey bee colonies during the course of a beekeeping year as their needs vary considerably and there is opportunity and sometimes necessity for the beekeeper to influence this.

This however is not just the responsibility of the beekeeper. Current government and stakeholder programmes responding to the plight of insect pollinators recognise the fact that pollinators need homes and food and there is much effort taking place in promoting an understanding of the needs and attempting to influence policy makers, land managers and the public to be much more aware of the impact of what they are doing which is badly impacting on insect pollinators and to encourage them to actively do things which will radically improve the future prospects for insect pollinators. After all the human species owes its continuing existence to the activities of insect including many species of bees, wasps, hoverflies and butterflies which through pollination facilitate the production of many of the foods which we eat and determine the environment we live in.

12.2   Lessons from bees in the wild

Observational work by Seeley (2019) and others has shown that in honey bee colonies which can be considered as wild / natural, that is to say have a history of a significant number of generations with no interference from beekeepers and their bees and living in natural cavities (e.g. tree trunks with rotten centres or cavities, do not exhibit the same growth and colony development patterns in terms of numbers of bees produced. They do not swarm so readily. The amount of nectar collected, processed and set aside as honey stores are much less than those shown by colonies which are being actively managed by beekeepers, no matter how laissez-faire the beekeeper is.

Such wild colonies are usually found in a lower colony density than that of managed colonies and are therefore able to survive and persist on a lower total amount of forage resource. The amount of nectar and pollen resources required from the area around the wild colony are less per unit area than the same area having to carry the numerous colonies which a beekeeper may keep in an apiary or where there are numerous beekeepers who live nearby each other and whose bees may forage in areas also accessed by the bees from the other beekeepers.

The need for nutritional resources in sufficient quality and quantity throughout the year have an all-important effect on the health status of the colony.

It is also conceivable that colonies which are not kept in apiaries but are widely separated in the wild are less susceptible to the import / infection of pests and diseases through bees drifting between adjacent colonies or robbing activities which facilitate the transfer of diseases such as the foul broods or through the beekeeper acting as the disease transfer mechanism.

The excessive removal of honey (a complete and balanced nourishment) from colonies by the beekeeper as a harvest and its replacement with sucrose sugar solution, a mere source of carbohydrate which requires further processing by the worker bees before it can be made into a state which is fit for purpose to store and be utilised by the colony, will put colonies under stress. A further issue with the use of sugar syrup being that fermenting sugar syrup is harmful to bees.

As we shall see stored, processed sugar syrup is deficient in the many other substances which are contained in honey and which greatly contribute to the health of the individual bee; and the colony as a whole.

The rest of this section will consider managed honey bees and wild bees will only be referred to if necessary.

12.3 Colonies to be allowed to follow their natural cycle, including successful overwintering and reproduction (swarming)

We have already considered the typically annual cycle of the colony in terms of individual bee numbers in the colony through a year and the scenarios when the colony has or has not swarmed.

Numbers of bees in the colony will depend on the 'strain' of the bee, the local climate and abundance and availability of forage, the type of hive they are kept in and the beekeepers' management, whether active and interventional or letting the colony follow its natural instincts.

Beekeepers are often taught that the bigger the colony (i.e., number of bees) going into winter the better the colony will survive the winter and not fail. However, this is not always the case. It seems that there is an optimal cluster size which is a function of the density of the bees in the cluster and the ability of the bees on the outside of the cluster to maintain a high enough body temperature for them to be able to move and maintain physical contact with the cluster and not fall from the cluster onto the hive floor and then die.

The cluster also needs to move across the combs and gain access to fresh honey supplies as those under the cluster are progressively consumed. When the cluster is too large it may not be able to move sufficiently far enough in times when the ambient temperature has risen a little to enable the cluster to relocate on fresh supplies. If bees do not access fresh energy supplies, they will starve and fall from the cluster and die.

Tracking the movement of the cluster is as important as is knowing the disposition of the stores at the start of the winter. Crown boards with glass (referred to as glass quilts) or Perspex (harder to keep clean) are a way of achieving this if you are concerned about the status particular colonies. Leaving them on the hive overwinter is not usually recommended because of their poor insulation properties but there may be times when their use could help save a colony. Squares of the types of heat reflecting insulation sheets used behind radiators placed into the inside of the roof is a useful technique to improve the insulation. It may be that the stores are mainly concentrated on one side of the hive. Each hive should have an eke (an empty super or brood box being the most versatile to use) in place ready to utilise if it is necessary to place fondant or other feed, e.g., a rapid contact feeder directly over the cluster on top of the frames rather than just relying on the bees accessing the feed through the Porter Bee escape holes in the crown board.

Typically, a colony of 10,000 to 12,000 bees is considered an adequate size for successful overwintering in much of the British Isles and being a good platform for expansion in the new season. This of course depends on the type of hive you use, the characteristics of the bees you keep, the weather conditions likely to be experienced, potential forage availability and the quantity and quality of the food reserves held by the colony.

There are local strains of bees that will overwinter in small clusters and then an exhibit explosive growth in numbers in the spring.

Strains (typically Italian) will often rapidly convert food into brood and go into the winter in large numbers which mean they are not only vulnerable to losses from the cluster in extended periods of cold weather, but also quickly consuming their winter stores with the risk of colony starvation.

Enquiring about the winter thriftiness and typical colony sizes are questions worth asking the supplier of bees and queens.

12.4 Colonies to be managed to maximise honey production

Manipulating colony development to enable colonies to maximise honey production will vary according to the forage which will be anticipated to yield a good honey crop. Such crops may be agricultural or horticultural in origin, be from the wider environment or from the opportunities presented from urban and suburban gardens and habitats.

Colonies which are going to forage well (assuming other factors such as weather are favourable) require a strong foraging force and enough house and nurse bees to cover brood development and feeding activities as well as processing the nectar when it is brought back to the colonies. Colonies of the size of 50,000-60,000 bees would be suitable but colonies of this size can require a steady nerve and utter concentration when manipulating and managing them and a well thought out plan of action *before* opening the colony for inspection. Time spent ensuring easy and ready access to any spare equipment which may be required is time well spent. *Always* have a lit smoker to hand in case you need it. A hand - held spray filled with clean water is a useful way of stilling the activity of bees on a frame for closer inspection.

Changing agricultural practices and the impact of changing weather patterns are affecting the timing when crops are flowering. Such is the case with oil seed rape which in recent years has flowered progressively earlier, meaning that the beekeeper has to act early in the season to maximise the development of the colony if it is to take advantage of the crop. There is always that risk that the bees 'build up on the rape,' then have very little to do and with congested colonies and often very variable spring weather swarming preparations may begin.

Alternatively, it can be a good strategy to allow the bees to build up on the rape and carry out effective swarm tendency management and be ready if there is a following crop such as field beans.

Later in the year it may be that there is the risk that a colony being managed for a later crop requires careful swarm control management in order to maintain a strong foraging force.

If such colonies do swarm and the swarms are lost the new queens and their colonies may not be able to build up quickly enough to be able to achieve the correct colony demography and accumulate sufficient overwinter stores to survive the winter period. Such late queens may not be adequately mated and consequently drone – breeding queens become evident.

12.5   Colonies to be managed for pollination services

Some beekeepers, especially commercial bee farming operations, derive part or whole of their income from the renting of colonies or the provision of pollination services on a contractual and fee-paying basis and there are well established forms of contract to meet these requirements. These often describe the composition of the colonies (for example the number of frames of bees) being provided to carry out the pollination services and the timing when the colonies are to be brought onto and removed from the site.

12.6   Colonies to be managed for educational and instruction purposes

12.6.1   Observation Hives

Small colonies set up in Observation hives are an excellent way to introduce the subject of honey bees to any age of audience and have the advantage that within certain limitations they are portable and movable from one venue to another. The establishment a colony in an observation hive which may have two or three combs located vertically in the case is described in Showler (1978). A more recent book on the observation hive has been written by Linton (2017).

Observation hives can provide both educational and enjoyable experiences for the beekeeper who has the opportunity to watch bee behaviour at very close range, something not always easy to achieve when working with populous colonies. The behaviour shown by the foraging bees returning to the colony and disposing of their loads, the dance language and many other aspects of colony nutrition can be much more readily observed in an observation hive than in a full-sized colony.

The nutritional state of colonies kept in observation hives must be carefully monitored to ensure that the colony does not expand too much or fill the hive with comb thus making observations difficult and the hive interior and the glass panels hard to keep clean and to manage.

12.6.2   Nucleus colonies

For those interested in honey bees, maybe aspiring beekeepers and want to have the opportunity to look at honey bees close up can do so by being shown a 5 or 6 frames nucleus colony which is well-established with a laying queen, frames of brood at all ages well covered with worker bees and if possible drones and having good food reserves of honey and pollen / bee bread.

Nucleus hives made from high density polystyrene are excellent for the location of a nucleus colony and its thermal properties are particularly beneficial.

The observer can then be introduced to the fundamentals of the honey bee colony.

As expertise and experience is gained then larger colonies can be made available or the nuclei progressively expanded into a full-sized brood box and examined. But these colonies should not become too populous. The colonies should be given space as this will help in the observing process. It will also help the aspiring beekeeper to be able to manage their colonies and cope with a colony which is becoming quite a different proposition to handle and manage compared to that of the smaller nuclei. The aspiring beekeepers should be carrying out disease inspections and to look in depth at what is going on in the colony and its stage in the life cycle of the colony. Has the colony adequate food reserves should there be a period of no forage either through poor weather or the fact that there is insufficient forage (both nectar and pollen) available for the honey bees to utilise?

The temptation to remove honey from an expanding colony now located in a full-sized hive should be resisted as the colony may encounter a period of dearth during a time when you have not been able to visit the apiary and when you do the lack of food may be beginning to impact the colony through starvation. Some strains (e.g., Italian and their crosses) of honey bees do not naturally hold much honey in store and in the brood rearing season in the event of encountering a dearth period will consume food reserves and rear brood rather than reduce brood rearing, this often resulting in starvation and death of the colony.

Much beekeeping training is done in beekeeping association apiaries under conditions which are well-managed and controlled; the colonies docile, well- fed and all in all a delight to inspect and handle. Association apiarists carry out risk assessments to reduce or mitigate risk of injury to the students and observers. But they often do not give the aspiring beekeeper hands-on experience dealing with difficult colonies. Practicing the techniques for real under supervision is something which we believe should be provided during beekeeper training and assessed in the beekeeping certification assessments. Recognition of the reasons for a colony's behaviour is important as this will help in decision-making process in what can be done with the colony at the time and in its future. The sudden cessation of nectar flows, ambient weather conditions or demographic problems (e.g., queen failure) within the colonies themselves affecting colony behaviour are often related to nutritional aspects of the colony.

12.7   Colonies to be used for the breeding, raising, rearing and mating of queens

There is much literature about these subjects.

Woodward (2009) and Laidlaw and Page (1997) make good reading.

For the purposes of this book, we will briefly introduce the kinds of 'colonies' which are involved in the breeding and rearing / raising process and in queen- mating. They all have a specific purpose and getting the colonies to the state where they can fulfil these purposes involve providing the necessary nutrition.

The status of the colonies to be used at each stage depends on the demographic characteristics of the colony, for example the need for colonies with a large proportion of young nurse bees who will produce the royal jelly and continue with attending to the maturation process of the newly emerged virgin queen and her subsequent mating.

There are several Queen rearing techniques, and these can be grouped as follows:

12.8   Non-grafting techniques

- Artificial swarming.
- Miller – use of saw tooth cut comb for breeder queen to lay in.
- Alley – placing of a frame of newly drawn comb into the breeder colony between two combs of sealed brood.
- Queen confinement in box kits e.g., Jenter Cage to encourage the queen to lay eggs into small cups which are transferred when very young larvae to cell builder colonies.

12.9   Grafting-based techniques

- Breeder queen colony – strong colony good laying queen and colony in good health.
- Cell builder colonies- plenty of nurse bees.

12.10   Mating hives

- Full frame nucleus with from 2 to 6 frames of bees of all ages and well stocked with food or can be fed with thin syrup. The mating of virgin queen bees from a 2-3, or more full frame colonies with sealed brood, drones and plenty of stores and worker bees of all ages contain all the ingredients to support the new queen in her preparations for her mating flights and post mating development. Such frames will have come from a colony prepared and managed as described above.
- Micro-mating hives such as Apidea and Warnholz into which are poured cups of worker bees. Small frames with foundation and food are provided. Virgin queens can then be introduced and from which they can fly and then return after their mating flights. However, the preparation techniques used may mean that there are few nurse bees contained in a cup of bees taken from a colony to stock the micro-nucleus.

12.11   Breeder queen colonies

The choice of colony which will provide the queen material which can be the eggs / larvae at the correct age, unsealed queen cells or placing the queen directly in a confinement cage to lay eggs. The eggs, larvae for grafting or the queen cells should all come from a strong, disease-free (as far as can be ascertained from observations for disease and pest symptoms) colony with your chosen desirable characteristics. The chances of having a colony with such characteristics are much improved with careful attention paid to the nutritional management of the colony to achieve the optimal balance in the demographics of the colony.

12.12   Drone-rearing colonies

In recent years beekeepers are reporting early queen failure with the queens becoming drone layers often quite soon after mating, in the early spring or being unable to sustain the laying of fertile eggs for more than one season. This is due to the inadequate mating of the virgin queens and may have arisen because of poor weather at the time of the virgin queen emergence and her being unable to make her mating flights. Alternatively, or in addition to, the drones were too few or their state of fitness to be able to fly at speeds to catch up with the queen may be too poor. As we shall see inadequate nutrition affects the mating abilities of the drone, especially drones which have been reared from unfertilised eggs laid in worker cells and are usually smaller in size, as well as other effects.

Beekeepers often remove substantial amounts of drone comb and the destroy the larvae contained within them, for example when inspecting a comb with capped drone brood for the presence of *Varroa*. Drone foundation inserted in brood frames or the use super frames fitted with worker foundation (the bees draw drone comb from the bottom of the frame) may be inserted into the brood nest to encourage the building of drone comb and the raising of drones. The resulting drones may be used for varroa infestation inspections and removal of the varroa.

Colonies can be actively managed to produce large quantities of drones using drone foundation to boost the drone population numbers increasing the chances of better mating success with the mating queens. Such colonies must be strong, with a good laying queen and plenty of worker bees of all ages. There must be a good flow of nectar and pollen available, and the colonies should have plenty of stores. Done rearing is an intensive resource consuming activity for a colony and if not managed carefully could result in a colony whose population structure becomes affected with consequential impacts on its future and prosperity.

As we have noted queens can be reared in many ways and it is interesting to note that the size of the eggs which the queen lays into a natural queen cell is slightly larger than that laid into a worker cell which is subsequently adapted and nurtured to become a queen cell and mature queen.

Whilst the subsequent queens produced from modified worker cells are still queens phenotypically, presumably because of the differential feeding of royal jelly to these larvae, the body and the ovary size of queens produced in these cells are smaller than those produced in natural swarm cells.

The paper published by Hao Wei et al. (2019) also refers to other research which has demonstrated a relationship between queen weight, queen ovariole number and fecundity an in addition an influence of queen weight on colony honey production. Other work has shown that rearing queens from older worker larvae results in a significantly lower production of worker comb, drone comb and stored food compared to those queens reared from eggs.

12.13 Colonies to be managed for increasing colony numbers by 'splits'

There are several effective ways to increase in colony numbers and they usually depend on having a good -sized colony and ideally one which is in the process for preparing for swarming. It should have at least 8 frames of brood, a laying queen, the presence of mature drones and a strong worker bee population of all ages. Typical (but not exclusively) methods can include:

1. The removal of three frames of brood with stores, together with the attached bees and adding a newly acquired queen. Two such sets of brood combs can be removed from the original colony (which still retains the queen) with the colony being left on its original site. The two small nuclei can then be moved elsewhere additional combs (if available) added or frames of foundation. They should be fed a weak syrup. The entrance to the hive should

be made into one bee way width to discourage robbing and help the colonies to protect themselves.

2. If the colony contains queen cells with eggs or small larvae (note check they are not supersedure cells, in which case take no further action), then an artificial swarm technique can be employed whereby the queen is placed on a frame of sealed brood in a brood box below a queen excluder and the box is then made up with other combs or foundation. The frames containing the unsealed queen cells and the bees on them can them be put into two or three comb nuclei depending on the number of queen cells available. Two cells per nucleus, the rest destroyed or consider cutting out the piece of comb with the cells and making up a small nucleus shaking in some bees (a good- sized mugs worth) obtained from another colony. If some of the cells are sealed then have available a couple of queen cages and carefully remove the operculum (lid) of the cell and see if there is a young queen waiting to emerge, in which case catch her in the queen cage and consider utilizing her as in method 1 above. The colonies should be fed as in Method 1. Remember when handling combs with queen cells on them do not shake the combs to dislodge the bees as the shaking will move the developing queen larvae from contact with her royal jelly and she will die.

3. Do not kill queen cells until you are absolutely sure you don't want them or are not in a position to make use of them.

Having introduced the basic kinds of colonies required in the queen rearing process the reader is invited to consider how nutrition will impact on the ability of the beekeeper or the commercial queen supplier to be able to produce healthy, disease-free, well-mated, good egg-laying capacity queens.

13. The composition of a honey bee colony

13.1　The Basics

Having introduced the honey bee colony as an entity is now time to look in more detail at its constituent parts. There are many texts which describe honey bee colonies, the three adult forms (castes) and the developmental stages found in it in much more detail.

Such descriptions can be found in books such as Winston's *The Biology of the Honey Bee* (1987) and Seeley's *The Wisdom of the Hive. The Social Physiology of Honey Bee Colonies* (1995).

However, we need to establish and remind ourselves about the composition of the honey bee colony as we start to build up our knowledge and understanding of honey bee nutrition.

In this book we will focus on those aspects which are directly relevant to the nutrition of the honey bee, both as an individual and as a colony.

Beekeepers refer to the 'brood' and this a collective term for the three development stages of an individual honey bee, irrespective of whether it is a queen, worker or drone. The numerical majority of which are destined to become workers, a smaller number to be drones and even smaller number to become virgin queens and then the sole queen in a colony.

The three forms of honey bee have different development times from the time of the laying of the egg in the cell by the queen to the time when the adult (imago) honey bee emerges from her cell after the removal of its wax capping. The fate of the egg is determined by a combination of genetic and nutritional factors.

The following table summarises the development times of the queen, worker and drone honey bee to emergence from the capped cell as an adult.

Stage	Queen	Worker	Drone
Egg	3 days	3 days	3 days
Status of Laid Egg	Fertilized	Fertilized	Unfertilized
Larva	5.5	6	6.5
Pupa	7	12	14.5
Total	16	21	24

13.2 The worker honey bee and her roles in the colony

The worker bee is the most numerous form of the honey bee in a colony. They are females, are diploid and possess reproductive organs, but these are incompletely developed. They have a smaller body size compared to drones and the queen. They cannot produce the same pheromones internally to those of the other female form of honey bee present in the colony, namely the queen. They also have a specialised adaptation the referred to as the 'pollen basket', or corbiculum on each hind leg. Workers have well-developed hypopharyngeal, mandibular, wax and scent glands.

Workers also forage for food and have adaptations for conducting these tasks including relatively long tongues (proboscis), powerful mandibles and distensible honey stomachs. They forage and process food and communicate the location and profitability of the food sources available at any one time to nestmates in the colony so that more bees can be recruited to exploit this forage in accordance with the needs of the colony.

They will also forage for plant resins and water.

Honey bees collect nectar and water honey bees using their 'tongue', more correctly described as the proboscis and which consist of several parts. These include a lapping tip / scoop (flabellum) which works through capillary action when immersed in the nectar, a sucking tube and a suction pump which consists of the cibarium and its associated muscular which is a chamber in the front of the head. There are other structures in the composition of honey bee mouthparts and these are well illustrated in Goodman (2003).

The roles of adult worker bees in the colony change as the worker bees age and in response to the various needs of the colony. This is called temporal polyethism. Plasticity in behaviour is an important aspect of the division of labour and colonies respond to changes in the internal and external environments by adjusting the ratios of individuals workers engaged in the different tasks required to be carried out in the colony.

The following table illustrates the changes in roles and behaviour and their timing shown by honey bees reared in the spring and through the year until the mid-late summer. These are referred to as 'summer bees'.

Honey bee workers reared late summer to late autumn usually live longer and form the colony population of honey bees, typically workers plus one queen and no drones. The late season reared workers are specially adapted to survive for long periods of time over the winter period- hence they are referred to as 'overwintering bees'.

The worker bee has the capability of producing bee bread from the pollen, royal jelly and worker jelly, honey, beeswax moulded into honeycomb as well as carrying out other roles. These are listed below in the chronological order in which they are carried out during the life an adult worker starting from the time when she first emerges as an adult bee from the cell in which she has been raised.

13.3   Ageing, adaptation and changing roles in the colony

The summer adult worker honey bee life cycle based on days after emergence from the cell.

0 – 6 days	Cell cleaning, general hive cleaning.
3 - 10 days	Hypopharyngeal glands produce brood food. Feeding the brood.
3 - 15 days	Attending the queen, cleaning and feeding her.
6 - 18 days	Receiving and processing nectar.
12 - 20 days	Wax glands fully developed. Wax production and comb building.
15- 25 days	Flight muscles developing. Hive ventilation.
18 - 35 days	Venom gland and sting apparatus developed. Colony guard duty.
20 days - death	Flight muscles fully developed. Bee attracted to light; foraging after short orientation flights; nectar and pollen collection.
25 days - death	Water and plant resin (propolis) collection.

We have already noted that the timings when the individual worker bees start and complete these various functions vary according to the demands of the colony and they respond to the opportunities and the stressors it experiences. Beekeeping involves a degree of manipulation of these changes to meet the requirements of the beekeeper and the colony. It is the understanding by the beekeeper of the nutritional basis of this flexibility of the worker bees being able to adapt their functioning and behaviour to circumstances that the colony experiences that is key to understanding the importance of good nutrition.

Another approach is to use the term 'behavioural castes'.

Adult workers undergo major changes in their physiology as they progress through their lifetime. This is associated with a move in nutritional requirements away from protein and towards carbohydrates.

Johnson (2010) describes the division of labour in honey bees, in particular the workers and proposes a push-pull model which provides the mechanism thought to be causing the temporal polyethism and division of labour. Temporal polyethism is in part based on internal and physiological differences. In the honey bee temporally based 'castes' can be envisaged and these are now described.

13.3.1   Cell cleaners

This temporal caste is described as newly emerged worker honey bees that cannot fly or sting. They are therefore developmentally immature. The first days of a bee's life are spent continuing development and developing these abilities as well as cell cleaning with the remainder of time spent either inactive or grooming.

It appears there is no priming mechanism, only a course of development for the cell cleaners to transition to become nursing bees. If the colony has stores of pollen, then the transition to nursing bees occurs. If not, this development is compromised. There are likely to be insufficient protein stores for brood development and the worker bees begin to forage earlier than is usual in their life span.

13.3.2   Nurses

The nurse caste of worker honey bees feeds all the other colony members. This is achieved by either feeding honey and bee bread directly or by feeding them with glandular secretions (e.g., royal jelly) which the nurse bees have produced themselves after they have consumed honey and bee bread.

Nurse bees are the main consumers of pollen in the colony and are the nutritional centre of the colony.

The nursing caste phase lasts for about one week (between 4-12 days) during which the nurse bees feed a proteinaceous secretion (not pollen) to the larvae (brood) which are critical for the development and maintenance of the brood.

Nurse bees also tend the queen forming a retinue which can regulate queen behaviour through the route by which they feed her. They also act as messenger bees by spreading the queen's pheromones and guard the nest entrance.

There are two mutually reinforcing processes which enable a worker bee to have the physiological capacity to nurse and to stay as a nurse. When a worker bee feeds the brood she is exposed to both queen mandibular gland pheromone (QMP) and brood pheromone (BP). BP stimulates the hypopharyngeal glands, which trigger pollen feeding, which results in high levels of vitellogenin (and other nutritional stores) in the bee's body tissues, the production of brood food, and possibly the partial suppression of the normal age-based rise in juvenile hormone (JH).

The breaking of the cycles and priming processes controlling the levels of these pheromones to which the bees are exposed triggers the transition of the nursing bees to middle-aged bees.

Nurse bees undergo a major change in their physiology as they move through the behavioural castes from nurse to forager. Associated with the changes in behaviour is a reduction in the size of the nurse bee hypopharyngeal glands, changes in brain morphology, neurochemistry, gene expression and flight muscle biochemistry.

Nurse bee lipid reserves decline just prior to the initiation of foraging. The onset of foraging is delayed by the presence of foraging bees regardless of the nutritional state of the colony. It is thought that the foragers exert a social (possibly pheromonal) effect and not directly affect the nutritional state of nurse bees by trophallaxis. It is possible that the role of diet in the regulation of the division of labour in honey bee colonies may be less influential relative to those of social signals and juvenile hormone. (Wright et al. 2018).

13.3.3   Middle-aged bees

The duration of this caste phase lasts about one week, from ages 12-21 days. Bees of this phase have no direct interaction with the brood. Their activities comprise some 15 tasks ranging from nest building and maintenance, to nectar receiving and processing. Young middle-aged bees seem to spend more time on comb building and general colony maintenance. Older middle-aged bees then transition to nectar processing and other tasks which put them nearer the entrance of the nest cavity / hive entrance.

Some of the middle-aged bees receive nectar from the foraging bees near the entrance of the nest for processing it into honey and storage. The actual number of bees receiving the nectar must be adjusted so that it matches the current foraging rate and flow of nectar brought back to the colony. The tremble dance carried out by the forager bees is performed when they determine that there are insufficient receiver bees and more middle-aged bees need to be recruited to receive the nectar.

Middle-aged bees also build new comb at a rate sufficient to ensure that there is enough space available for the incoming nectar to be processed.

The factors influencing the transition of middle-aged bees to foragers involves a complex interaction between the middle-aged bees and the foragers. Johnson (2010) proposes that the oldest middle-aged bees are fully capable of foraging but may not do so because they may be required as nectar receivers, or there is insufficient forage available in the field to make it cost effective for them to switch.

Earlier, Johnson (2008) determined that young bees specialise in brood- related tasks whilst middle-aged bees specialised in nectar processing and nest maintenance and were observed in less than 1% of the observations to be caring for brood. Nurse bees stay within the brood nest area, while middle-aged bees move around the colony taking on work needed to be carried out.

13.3.4   Foragers

Normally, once the workers transition to foraging activities they no longer engage in within-nest tasks but focus on the sourcing of nectar, pollen, water and propolis. Forager bees have lower levels of lipids in their body at the onset of foraging than do nurse bees, (Toth and Robinson 2005). The results reported in this paper suggested that the worker nutritional state may be involved in the regulation of the division of labour in colonies. If the forager bees need to revert to nursing activities, they do not regain their body lipid stores, which indicated that high lipid stores were not required to carry out sustained brood care behaviour.

Most foraging bees specialise on either pollen or nectar on a given foray and appear to be generalists.

For forager-age bees that do not forage it has been proposed that these bees have a defensive function, but it is also likely that they are waiting to be recruited to foraging.

The maintenance of forager-aged bees in the foraging state may be a physiological mechanism based on the Juvenile Hormone (JH) which inhibits the expression of genes associated with key factors relating to the preceding stages in the life of the individual worker honey bee.

Juvenile hormone (JH) mediates the behavioural plasticity and worker to worker interactions influence both the levels of JH and the age of transitioning to become a forager.

Brillet et al. (2002) measured the age of the onset of foraging in colonies derived from three races of European honey bees, *Apis mellifera mellifera, Apis mellifera caucasia, Apis mellifera ligustica*. Their results showed that there were differences in the timing of the onset of foraging in the European races studied.

13.4   The drone & his roles in the colony

The drone is the male sex of the honey bee and is a haploid adult developing from an unfertilized egg. His reproductive organs are fully developed. His testes produce sperm and there are associated ducts and glands. The anatomy and composition of the body of the drone is specialised for producing sperm, holding and then transferring large quantities of sperm as well as mucus an essential factor in successful mating. These adaptations support the connection of the queen and the drone and probably stimulates oogenesis (development of the ova or eggs) and oviposition (the depositing of eggs) by the queen.

Drones lack hypopharyngeal (brood food) glands and wax glands. Its proboscis is shorter than that of the worker and does not have the structures associated with forage collection including the corbicula on the hind legs to hold pollen and propolis. The mandibular glands are small, and the drone's stomach is thinner compared with that of the worker honey bee.

The main function of the drone is to try to successfully mate with a virgin queen when she is flying on a mating flight. The virgin queen will mate several times on a single flight; and she may make several flights consequently mating with drones from a variety of genetic backgrounds. There may be drones from many colonies which may gather in the air in locations referred to as drone congregations.

The drone contributes other activities including using its large muscle masses in the thorax together with the actions of the wing is thought to have other important effects on the colony through heat generation and ventilation of the hive. The conditions within the hive are important in terms of nest homeostasis (temperature and carbon dioxide concentration) and assisting in the evaporation of water from nectar during the processes involved in the manipulation of nectar and its transformation into honey.

The number of drones reared by a colony of the size managed by beekeepers throughout the season has been estimated as many as 45,000 (Winston 1987). The numbers of drones are at their highest during the period when colonies are preparing to undergo reproduction through swarming. The numbers of drones at any one time in the season are dependent on the total colony population and the type of colony (e.g., prime swarms, newly established swarms and after swarms).

Sexually mature drones spend much of their time on pollen-filled parts of the comb around the edge of the brood area and leave the colony when the weather is favourable and there are likely to be virgin queens flying on their mating flights. A drone which successfully mates subsequently dies during the copulation process having transferred his sperm to the 'virgin' queen.

Drones become sexually mature around 34 – 38 days after the egg was laid and 10-14 days after emergence and the drone subsequently lives, unless it successfully mates, for about one month. As we shall see the colony makes a large investment of resources in producing drones and beekeepers should make a particular note of this.

Drones are normally driven outside the hive towards the end of the queen rearing season and the presence of drones in a colony outside this time period often indicates that all is not well with the colony. For example, drones seen in a colony in late winter / early spring might indicate a drone laying queen; or drones seen late in the season may mean a failing queen that the colony is attempting to supersede. It is not unusual for colonies with well- functioning queens to carry a few drones over winter.

13.5 The queen & her roles in the colony

13.5.1 Eggs for colony development and reproduction

The queen is also a female honey bee, but in contrast to the adult worker bee she has fully developed and functional reproductive organs; her ovaries, which produce eggs, the spermatheca which stores the sperm passed over on copulation with the drones together with the associated ducts and glands. In colonies which are not being managed by beekeepers, queens usually live for 2-3 years, sometimes more. Where colonies are managed by beekeepers this very much depends on the practices of the beekeeper and the behaviour of the queen in a colony and she may be actively replaced every two years, assuming the colony has not swarmed or become a drone laying queen in that period.

Usually there is only one mature, fertile queen present in a colony but there can be times when this changes. For example, at the time of queen supersedure when a mother and daughter queen can be seen co-existing in the colony until the time when the young queen takes over and the old queen is disposed of by the colony. On other occasions at the time of swarming a colony may have several immature virgin queens at large in a colony until one succeeds in becoming the sole young queen, the original queen having already left the colony with a prime swarm.

The queen's principal functions in the honey bee colony are firstly to lay eggs whose fate is determined by whether or not they are fertilized as they pass the valve fold to become female bees (workers and queens) as in the case of fertilisation taking place; or whether the eggs become males (drones) in which case they remain unfertilised. Differentiation of the females into workers or queen larvae and adults are largely determined by nutritional differences in the nourishment fed to the larvae of the worker honey bees and the larvae selected to become raised as queens. The queen may lay of the order of 1500 eggs per day, sometimes more, in the height of the brood-rearing season which equates to around one million eggs in her lifetime.

13.5.2   Colony cohesion

The second principal function is to produce a number of chemical substances, including pheromones, which exert influences on the population of the honey bee colony to maintain colony cohesion and also regulate the activities of the worker bees. The queen substance pheromone (trans-9-keto-2-decenoic acid) is secreted by the queen's mandibular salivary glands and stimulates worker foraging behaviour. It also contains ((2 E-)-9-oxodecenoic acid (9-ODA) and 9- hydroxy-(E)-2-decenoic acid (9-HDA). 9-oxodecenoic acid stimulates the olfactory receptors of drones and inhibits the development of the worker bees (sterile females) but it is entirely effective only when it acts in conjunction with another inhibitory pheromone, 9-hydroxydecenoic acid.

This specialization comes at a price and the queen honey bee has to be fed and groomed by worker bees. Once mated and established in a colony she does not go outside the hive until the time when she leaves with a prime swarm to set up a new colony in a new location.

13.5.3   End of Life

Usually, some two to four years after mating the queen will run out of the sperm which has been stored in her spermatheca and increasing amounts of drone brood (drone larvae develop from unfertilised eggs) are produced. Depending on the time of year unless the queen is superseded and killed by workers the colony in effect becomes a drone breeding one. Queens in large colonies including colonies which have been artificially stimulated by supplemental feeding by the beekeeper often rapidly use up their sperm bank contained in the spermatheca and this shortage of sperm can then be exacerbated if the queen was inadequately mating post emergence as a virgin queen. In contrast queens in wild or unmanaged colonies may survive longer because she has not been required to lay to her maximum capacity and additionally if she had been well-mated then she may be able to survive longer meeting the needs of the colony.

14. Some important aspects of insect and honey bee structures and physiology – with respect to nutrition

14.1   Hypopharyngeal glands

These are a pair of glands located inside the head of the adult worker honey bee and have a long central duct attached to which are acini which give an appearance of small bunches of immature grapes.

The glands develop after the emergence of the adult bee in the first 7-10 days and require the availability of a lot of pollen for consumption and processing of the pollen's components. These glands secrete the protein-rich component of royal jelly and worker jelly in the young bees and then as the worker bees age and become foragers the glands secrete invertase which is an enzyme that converts sucrose to the simple sugars of glucose and fructose.

14.2   Mandibular glands

These are simple sac-like structures which are located at the base of each adult worker bee mandible. In young bees they secrete the lipid-rich components of the royal jelly and then as the bees age and become foragers they secrete the alarm pheromone 2-heptanone.

Excellent images of these glands and their location in the honey bee can be found in Goodman (2003).

The mandibular glands of the queen are larger than those of the worker caste. They produce the queen mandibular gland pheromone containing the queen substance 9-oxo-2(E)-decenoic acid (9-ODA).

In the drone the mandibular glands are small and the production of glandular secretions (a pheromone which attracts other drones flying in close range to gather and form drone congregation areas) finish at the time (9 days after emergence) when drones begin leaving the hive for nuptial flights.

14.3   Worker jelly

Worker 'jelly' is a mixture of mandibular and hypopharyngeal gland secretions. Mandibular gland secretions are white and made up of secretions of lipids (fats). Hypopharyngeal gland secretions are clear and made up of proteins. It is similar in composition to Royal Jelly, but Worker jelly has 12% sugars, mainly sucrose and has only half the sugar content compared to Royal Jelly.

14.4   Royal jelly

Royal jelly is 67% water, and 32% dry matter. The dry matter is composed of 12.1% carbohydrates, 4.0% lipids, 12.9 % proteins and 1.1% ash. These proportions vary slightly depending on the season and the time in the season. There are also many trace minerals, antibacterial and antibiotic components and trace amounts of vitamin C. No fat- soluble vitamins (A, D, E and K) are present in royal jelly.

Wright et al. (2018) reported a comparison of several sources of royal jelly and concluded that on average it is about 63% water, 14% protein, 18% carbohydrates (mainly glucose and fructose) and 6% fats (expressed as the wet weight).

Royal jelly also contains sterols (approximately 0.5% wet weight of royal jelly) with 24-methylene cholesterol being the most common found followed by β-sitosterol, and Δ5-avensterol, and also found are cholesterol, stigmasterol and isofucosterol. All of these sterols have to be acquired from plant sources as the honey bee is unable to synthesise sterols

The majority of the proteins contained in royal jelly are called major royal jelly proteins (MRJPs) and are termed 1 to 6, each of which have many variations. These are specifically produced by the nurse bees in their hypopharyngeal and mandibular glands and are the developing larva's only source of essential amino acids. Some argue that these play a central role in caste determination, but this is still to be determined.

The fats of royal jelly include the two important compounds 10-HDA ([E]-10-hydroxy-2 decenoic acid]) and 10-HDAA (10-hydroxydecenoic acid)

The enzymes glucose oxidase, peroxiredoxin and glutathione- S- transferase are also present.

Royal jelly is highly nutritious for the growth and development of the honey bee larva and the precise roles of each of the components, including the major royal jelly proteins, play in honey bee larval nutrition are still the subject of research.

It is still unclear whether royal jelly is qualitatively different to the jelly fed to worker larvae during their first three days of life (Wright et al. 2018).

14.5   Digestion and excretion

The digestive system of the worker honey bee consists of the mouth after which is a muscular pharynx drawing the food (nectar or honey) further through the honey bees' thorax through the oesophagus and into organs in her abdomen and into the so-called 'honey stomach' in which no digestive processes take place…hence its alternative name of the honey sac. The distal end of the oesophagus is closed by a muscular valve called the proventriculus and the nectar does nor progress further down the digestive tract.

Instead, this nectar is passed to a receiver bee when the forager returns to the hive for subsequent processing into honey.

In contrast when the bee is feeding (being fed) the food, be it honey or pollen, is passed through the opened proventriculus to the stomach (or ventriculus) for digestion and assimilation.

The digestive tract ends in the anus at the end of the rectum. The rectum is capable of reabsorbing water and other reusable solids such as valuable ions and minerals.

Connected to the digestive tract between the ventriculus and the intestine are numerous Malphigian tubules which remove waste products from the haemolymph and these waste products are moved into the rectum for excretion and elimination from the honey bee's body.

A book of this kind cannot cover all aspects of the honey bees' natural history and its physiology and these are described more fully in Lipinski (2018) and in Winston's earlier book (1987).

It can however introduce the reader to the subjects and focus on those aspects which beekeepers should understand and appreciate how their beekeeping practices work with the biology of the honey bee or thwart it or at worst might lead to harming of the colony.

14.6   Storage and utilisation of carbohydrates

Honey bees store the large quantities of the carbohydrates (sugars) in the hive as honey and in their bodies as glycogen, glucose and trehalose circulating in the haemolymph.

In order for there to be sugars constantly available to meet its needs the individual retains nectar or diluted honey (honey mixed with water) in its honey crop for as long as is needed.

Adult honey bees can live on a pure carbohydrate diet for a long time, however without a protein intake they are unable to raise brood nor enable the progressive glandular development which enables the adult worker honey bee to fulfil her life cycle and the various stages of behaviour and colony functions which she goes through during her life.

When flying, foraging honey bees have a big demand for energy which they obtain through the biochemical pathways involved in the Krebs cycle which yields the high energy containing ATP (adenosine triphosphate) which on hydrolysis releases the energy required, e.g. muscle contraction. The energy demands of a foraging honey bee carrying pollen loads is 50 times that of the honey bee at rest.

Carbohydrates such as glucose is also converted in the fat body into fats and lipids and as these have a higher energy content, they are the main honey bee body energy store.

14.7   Storage and utilisation of proteins and amino acids

The degradation of proteins into amino acids and then rebuilding (recycling) into new proteins into tissues which are undergoing structural changes e.g., during metamorphosis, or skeletal muscles during starvation is part of the flow of protein origin materials through the individual bees' body and Lipinski (2018) summarised them as follows:

1. A build-up of the mandibular and hypopharyngeal glands to produce both types of bee milk (Royal and Worker jelly) in workers aged 1 to 12 days. The more rapidly young workers assimilate free amino acids the larger the food glands become, the higher the quality and the quantity of the bee milk produced.

2. The processing of proteins in the food glands into amino acids for the move into the wax glands to produce wax in workers aged 12 to 18 days. The quicker they assimilate the free amino acids the larger the wax glands and better the quantity and quality of wax.

3. The proteins sequestered in the wax glands move to the flight muscles in bees >18 days old, for flight (foraging) activity due to faster muscle development.

14.8   Storage and utilisation of lipids

Honey bees obtain natural lipids from pollen and are important in their role in the development of the honey bee and its reproduction.

Lipids (fats) are stored in the tissues of the fat bodies in the form of droplets (adipocytes) in the cytoplasm of the cells which comprise the lobes of the fat bodies.

In the brood-rearing season the worker bees (so-called summer bees) utilize the majority of their lipids in brood-related activities such as nursing bees, production of brood food and feeding the developing larvae or producing wax.

As the worker bees age and by the time they move to the point in their life cycle when they start foraging, they will have used up over 50% of the lipid content stored in their fat bodies and abdomen.

In comparison the bees which will form the overwintering colony are physiologically different and enrich the dorsal and ventral layers of their fat bodies and these also act as thermal insulators aiding survival of the bees over the winter period.

If there is a shortage of early spring pollen, or bad weather, substantial lipid depletion is observed in nurse bees before they transition to full scale foraging.

14.9   The Fat Body

Lipinski (2018) describes the Fat Body as a dynamic, peripheral, nutrient-sensing, multifunctional organ analogous to the vertebrate liver and adipose tissues. It plays a central role in the lipid, protein and carbohydrate metabolism of the honey bee.

During the larval stage the fat body is found in most parts of the larva and at the end of metamorphosis when the adult bee emerges the Fat Body is only found in the abdomen.

14.9.1   Functions

It is involved in many metabolic activities which concern:

*   Growth and metamorphosis.
*   Water balance and osmoregulation.

- Temperature regulation.
- The synthesis of lipids, carbohydrates and proteins e.g., vitellogenin- a storage protein, including antimicrobial acting peptides.
- Nutrient storage, which is often correlated with behaviour.
- Nutrient storage and mobilisation related to hormonal signals e.g., insulin.
- Energy mobilisation especially relating to nutritional and other manipulations that affect ageing and health.
- It prolongs the longevity of queen bees.
- The involvement of the adipocytes in the fat body synergistically working together with the wax glands to produce a complex mixture of hydrocarbons, fatty acids and proteins (lipophorins).
- Detoxification of pesticides.

'Fat bodies' consist of tissues which contain lipids (fats), glycogen (a form of sugar), triglycerides which are energy rich and some albuminoidal (protein) granules.

Glycogen is a polysaccharide comprising around 100,000 linked D-glucose molecules. It accumulates in the fat bodies and to a lesser extent in muscle tissue.

Fat bodies store and release energy according to the demands of the honey bee and are also involved in the detoxification of xenobiotics. A xenobiotic being a chemical found within an organism that is not naturally produced or expected to be present within the organism. Pesticides found in body tissues are considered to be xenobiotics.

The energy stored in the fat bodies is especially important during larval growth and when the cell is capped at 9 days after the egg is laid it amounts to 65% of the larval body weight.

The cream colour of the healthy larva is due to the whiteness of the fat body tissue pressing against the transparent larval skin.

In the pupal stage the fat tissue disintegrates and the proteins, carbohydrates and lipids which were contained in the fat tissue are dispersed into the haemolymph. New fat body cells are formed in the adult where they function as an important reservoir of nutrients and energy.

In the adult honey bee most of the fat bodies are found just beneath the dorsal and ventral sinuses in the abdomen just below the cuticle.

14.9.2   Structure

The size and content of fat body tissue changes over the course of the bee's life. Both newly emerged bees and adult worker forager bees have depleted fat body tissue (from the demands of metamorphosis in the former and changes associated with changing roles after feeding larvae in the latter) likely to contributing to both life stages functioning as nutrient -poor host resources. Nurse bees have substantially larger and therefore more nutritionally dense fat bodies than other stages of the worker bee caste.

The thin tissues of the fat body have lobes which consist of rounded or polygonal cells called adipocytes (also referred to as trophocytes, lipocytes, or just fat cells) and form the greater part of the fat body. They are present in the fat bodies in all stages of the bee's life. They store lipids, glycogen and protein.

Attached to the fat body tissue are oenocytes – (also referred to as 'oil cells'). These are also present in all stages of the bee's life larval, pupal and adult but their form changes. The cells are destroyed in the pupal stage and replaced by new cells in the adult phase where large numbers of oenocytes are formed over the wax glands and are at a maximum at the time of wax production. They contain lipids. They are associated with the trophocytes. Oenocytes are larger in the queen and are also found in the egg yolk.

Also attached to the fat body tissues are a third type of cells, urocytes (urate excretory cells) and these are only found in larvae and pupae. Their function is to store the nitrogenous waste in the form of uric acid and urates and once the Malpighian tubules are formed the urocytes disappear from the fat bodies.

Finally attached to the adipocytes are haemocytes which are involved in the metabolism of fats.

The condition of the fat bodies in workers which are produced later in the season and will form the colony population which will overwinter (winter bees) is of vital importance. The fat bodies of the winter bees should be plump and enlarged filling much of the space in the worker bee's abdomen. These fat bodies will contain the albuminoids and little fat which will enable the storage of high amounts of protein which can be used for early brood rearing in the late winter / early spring when there is little or no fresh pollen available and or the weather is poor and the bees are unable to forage for pollen.

In drones the abdominal fat body present as aggregations of cells is only found in adults up to 2 days old. Some fat cells have been described in the head.

Lipinski (2018) provides a more detailed description of the structure of the fat body.

14.9.3   Fat Bodies and Varroa

Recent research has shown that varroa feed on fat body tissues and not haemolymph as originally believed. Beekeepers need to acquire good knowledge and skills to manage and control the parasitic mite *Varroa destructor*. Varroa parasitism can have a dramatic effect on the nutritional status of an individual honey bee and of the colony as a whole depending on the degree of infestation and the period of time the parasitic mite has been present in the colony. This is an important subject which is further discussed in the section on 'Important Threats to Honey Bee Colonies'.

15. Honey bee pheromones and their role in nutrition

Pheromones play a key role in the organisation and functioning of the honey bee colony and enable the colony to deal with unforeseen events and changing environmental conditions as well as being integral to the development processes and the communications which take place between members of the colony. This section of our book will only deal with selected aspects of pheromones and will introduce the reader to the role of pheromones in the nutritional aspects of the development of the individual and the colony.

It is a huge subject, however there is a very readable introduction to the subject pheromones of the honey bee colony can be found in Bortolotti and Costa (2014).

15.1   Some definitions

Honey bee pheromones are chemical substances secreted by the exocrine glands (glands which secrete substances onto an epithelial surface by way of a duct) that provoke a behavioural or physiological response in another individual of the same species. They enable communication and interaction between all the castes of the honey bee in the colony.

There are two types of pheromones:

Primer pheromones which act at a physiological level, triggering complex and long-term responses in the receiving bee and can produce both behavioural and developmental changes. In the honey bee the main primer pheromones are the queen signal and brood pheromones responsible for social cohesion in the colony and maintaining colony homeostasis.

Releaser pheromones that produce simple and short-lived responses that only influence the receiving bee at the behavioural level. Most worker pheromones are releaser pheromones.

15.2   Queen pheromones and the queen signal

The queen is the principal regulator of the colony and she achieves this in the main through a variety of pheromones which are produced by several glands and released into the colony as a complex blend of chemicals known as 'the queen signal'. This queen signal elicits both primer and releaser effects on the colony nestmates.

These can be summarised in the following table:

Queen signal		
Type of effect	**Stimulates**	**Inhibits**
Releaser effects	Retinue behaviour	
	Swarm clustering	
	Drone attraction	
	Mating	
Primer effects	Comb building	Queen rearing
	Brood feeding	Worker reproduction
	Guarding	
	Foraging	

The Queen Mandibular Pheromone (QMP) secretion consists of at least five chemicals and are the main components of the queen signal. The chemicals consist of (E)- 9- oxodec-2-enoic-acid (9-ODA), two enantiomers (a pair of molecules that are mirror images of each other but not identical) of 9-hydroxydec-2-enoic acid (9-HDA), methyl p -hydroxybenzoate (HOB) and 4-hydroxy-3 methoxy-phenylethanol (homovanillyl-alcohol or HVA).

The QMP partly regulates the temporal polyethism (time related changes in behaviour of the worker bees). These changes in behaviour are also affected by the environmental factors that modify the requirements of the colony. For example, the loss of older foragers that die in the field can result in the quicker development of young bees into foragers whilst a lack of young bees (as a result of the interruption in the queen's egg-laying) can mean a slower rate of development or the reversal of foraging behaviour back to nest / nurse bee activities.

QMP also functions as a sex pheromone for drones and by the 6th day after emergence the young queens are producing enough QMP to attract drones for mating. The laying queen produces twice this amount of QMP. The lack of QMP seems to attract robber bees to a colony. Foraging worker bees do not seem to be attracted to QMP.

15.3   Worker honey bee pheromones

Worker bee age polyethism, whilst effected to some extent by the queen signal, is mainly under the control of the worker bees themselves. The main primer pheromone is ethyl oleate (EO) and exudes from the epithelium in the honey crop where it is produced from fermented nectar to the exoskeleton from where it is transmitted amongst the workers by evaporation, in close proximity to and physical contact. When the number of foragers in the colony secrete sufficient EO the development of the young bees is inhibited so these bees remain longer in their nursing and nest related activities. As and when foragers are lost through age and / or death in the field the removal of the inhibition allows the young bees to develop and become foragers.

The Nasonov Gland is only found in the worker honey bee and it produces the Nasonov Gland Pheromone. This is a secretion of seven volatile compounds, namely geraniol, nerolic acid, geranic acid, (E)-citral, (Z)-citral, (E-E)-farnesol and nerol. In the nutritional context it is used to mark forage sources thereby helping the recruitment of foragers, especially to sources of water when the gland can be seen to be exposed. Release of the Nasonov Gland Pheromone also recruits foragers to nectar sources that have particularly high sugar levels.

15.4 Tarsal glands

The worker footprint pheromone is an attractant, oily, colourless secretion produced in the worker tarsal glands and is deposited on forage sources and works in conjunction with the Nasonov Gland Pheromone, the latter being more volatile and detectable over a wider area and the former being more precise in its location.

15.5 Pheromones, the Waggle Dance and foraging

Foraging bees which return to the hive and then conducting waggle dances to communicate forage location produce and release from their cuticles four hydrocarbons (tricosane, pentacosane, Z-(9)- pentacosene and Z-(9)- tricosene) and these are believed to play a role in the recruitment process of bees to exploit the forage resources which have been located.

15.6 Worker mandibular glands and 2- heptanone (2-HPT)

This is deposited on forage sources and has a repellent effect. Its function may be to act as a repellent forage-marking pheromone that may help to discourage subsequent forager bees not to waste time and effort working a recently visited flower.

15.7 Brood pheromone (BP)

The Brood Pheromone (BP) of the worker larvae of the honey bee *(Apis mellifera)* acts both as a primer and a releaser pheromone and regulates worker development and colony growth. BP is a complex blend of 10 fatty- acid esters (methyl palmitate, methyl oleate, methyl stearate, methyl linoleate, methyl linolenate, ethyl palmitate, ethyl oleate, ethyl stearate, ethyl linoleate, ethyl linolenate. Each component on its own, or in combination, induces one or more releaser effects on adult bees.

In the presence of brood, worker honey bees begin foraging earlier compared to brood less colonies which should mean the gain of adequate food for the colony's growth. Colonies with high levels of brood pheromone have workers with low levels of vitellogenin. This shortens the workers' life expectancy and may be one of the mechanisms which control the longevity of workers in the winter months when little or no brood is reared. On the other hand, too many nurse bees could cause a decrease in food collection, shortage of reserves and a decline in brood nourishment.

Brood pheromone (BP) and another highly volatile pheromone, E-β-ocimene (EBO), a terpene, regulate this nurse / forager balance in the colony. Young larvae emit principally E-β-ocimene (EBO) and low levels of BP and in this way because young larvae have lower nursing needs this promotes workers to forage and collect pollen. BP production increases as the larvae age (and produce less E-β-ocimene) reaching its highest concentration at the cell capping stage. Older larvae have higher nursing needs and the higher levels of BP delay the transition to foraging and promote increased brood care (cleaning, feeding, cell capping, development of the hypopharyngeal glands).

Even from this short description of pheromones it can be seen that the honey bee age related behaviour and function is complex and flexible. The combined effects of queen signals, worker and brood pheromones enable the worker honey bees to respond to the needs of the colony, which in turn vary depending on the stage in the colony's development and on changing environmental factors.

16. Honey bee hormones and their role in nutrition

16.1   Introduction

Hormones are organic substances secreted in glands and transported in the circulatory system to the target organs and function in the regulation of physiological activities and in maintaining homeostasis. These functions evoke responses from specific organs or tissues in the body of the individual bee that are adapted to reacting to minute quantities of them. They may be stimulatory or inhibitory. A single hormone may have multiple targets in the bee's body with different effects in each target.

The honey bee has three important endocrine glands (glands which are ductless and secrete directly into the haemolymph) that secrete hormones. One of them is the corpora allata present in the head of the bee behind the brain and in front of the connection to the neck (oesophagus). This endocrine gland produces a hormone called Juvenile Hormone (JH).

16.2   Juvenile Hormone (JH)

This hormone controls development in the larva and pupa.

During metamorphosis, a larva is kept as a larva (i.e., in a juvenile state) whilst the level of JH is high. It will become a pupa and then an adult when the JH concentration is low and ecdysone (a moulting hormone) is present. The JH achieves this by inhibiting the expression of genes associated with adult development. It regulates caste differentiation in the honey bee.

In reproduction in adult honey bees juvenile hormone (JH) acts as a gonadotrophic hormone with higher levels of JH inducing reproduction. It may exert functions in spermatogenesis and mucus production by the accessory glands in the drone.

JH plays a role in the adult worker polyethism (age-related division of labour) as it sets the physiological conditions for the age specific tasks performed by adult bees. In young workers the haemolymph JH levels are low, and these bees carry out tasks within the brood nest. In the main this is feeding the developing brood with secretions from their well-developed hypopharyngeal glands. As the levels of JH increase the worker bees change their activities away from the brood nest, progressing to the entrance area to the hive or cavity entrance thence to outside the hive and forage for nectar, pollen and water. JH plays a role in the flight behaviour of drone and queen bees and promotes flight activity in worker bees.

JH levels are also related to aggressive (behaviour which is defensive of the nest), whichever perspective you take. Aggressive bees have higher JH levels.

Higher JH levels may also be involved in regulating other behaviours and levels of JH in workers show daily changes with time with peak levels at around 23:00 h.

The other two endocrine glands are the Prothoracic Gland, a small leaf-like structure located between the first two segments of the thorax, which under the influence of prothoracicotropic hormone (PTTH), which is secreted by two pairs of cells in the larval brain, produces the steroid hormone ecdysone. Ecdysone controls moulting in the larva and pupal stages of development but is not found in the adult bee. The PTTH and ecdysone trigger every moult (larva to larva and pupa to adult).

Secondly, the corpora cardiaca (a pair of glands) whose function is not completely understood; but is known to produce neurohormones (as does the corpora allata) and some secretions from endocrine cells. The neurohormones are known to mobilize lipids and carbohydrates for energy-consuming activities including reproduction.

17. Key nutrients required for growth / reproduction & existence

17.1 Some basic considerations

Before beginning to look at each of the key types of nutrients that the bees in a honey bee colony require it is worth setting the scene to appreciate some basic considerations that have to be borne in mind.

Worker bees need to ingest and process enough food, in the form of nectar, honey, pollen, bee bread, royal and worker jelly to provide themselves with enough nutrition to fulfil their functions in the colony. Unless any shortfall is spotted by the beekeeper and attempts made to remedy the situation there is a real risk that the colony could succumb to disease and even perish.

In the case of pollen, the digestion of much of this food source goes towards the production of proteinaceous secretions produced in their hypopharyngeal glands (only present in worker honey bees) and is fed to larvae in their cells, to other worker bees and to the queen and drones. Drones and queens do not produce these secretions.

Pollen consumption by workers increases as more brood is required to be reared.

Workers carry out the collection and consumption of pollen, its digestion through enzyme activity and the transfer of proteinaceous jelly by workers and are thus the main providers of digested food for the colony. Queens and drones rely on this pre-processed food.

17.2 Carbohydrates

Carbohydrates in the form of sugars are a complex subject and Lipinski (2018) provides a comprehensive description of the types of carbohydrates that are utilised in the plant and animal kingdoms.

In general terms sugars are classified into monosaccharides, disaccharides, tri-saccharides and polysaccharides.

Simple sugars (monosaccharides) (e.g., glucose and fructose) are also referred to as hexoses, whilst dihexoses include sucrose.

Carbohydrates are necessary for the energy requirements of the individual bee and the colony. Adult bees need carbohydrate to move muscles to carry out the hive tasks, energy to fly for foraging, for moving muscles to generate heat both for brood-rearing and survival in cold weather.

They exert a stimulatory effect on bee behaviour. Beekeepers will be taught that sugar feeding can kick start bees into action. This is because the 'taste cues' which enable the bee to differentiate edible from non-edible foods and the corresponding patterns of behaviour are considered to exist in the genetic memory of the honey bee. The main source of carbohydrates is secured by the colony in the form of nectar.

17.2.1   Nectar

17.2.1.1   Functions

Whilst this section is concerned with nectar as a source of carbohydrates as a nutrient it is essential that we refer to other functions of nectar which have been described and have an impact on the health of the honey bee and this is dealt with in the section on the relationship of nutrition, vitality and bee health.

Nectar principally contains water, sugars and amino acids and is the flowering plant's way to attract pollinators and recruit defenders (e.g., in the case of ants protecting plants (myrmecophytes) from herbivores) and is protected from nectar robbers and microorganisms by secondary compounds and antimicrobial proteins.

In the case of the honey bee a nectar flow can cause the honey bee colony to respond in one or more of the following ways:

- Causing the queen to lay eggs.
- Cause young workers to rear brood.
- Stimulate young workers to synthesise wax and construct the comb.
- Encourage older workers to forage for pollen.
- Encourage hygienic behaviour in young workers with the removal of weak and dead brood.
- Process and store the nectar as honey stores for periods when there is no forage, e.g., to overwinter.

17.2.1.2   Nectar composition

The main natural source of carbohydrates for honey bees is nectar which may originate from phloem sap, but the carbohydrates may also be obtained from direct synthesis in parenchyma cells of the nectary tissue itself.

Freshly collected nectar principally consists of sucrose in varying concentrations in water, including some enzymes and mineral content.

Plants secrete nectar in a range of colours including, yellow, orange, red, blue, green, brown and black as well, but it is the paler and clearer colours our honey bees' experience and collect in the British Isles.

Nectar is water containing a variety of dissolved substances that range between 3 and 87% of the total dry weight, and 90-95% of the total solid matter. It is usually acidic with a pH range of 2.7 - 6.4. The amount and sugar content of nectar is dependent on the plant species and the environment, especially weather conditions.

A typical nectar composition is as follows:

Water	30-90 % by weight
Sucrose	5-70%
Fructose	5-30%
Glucose	5-30%
Other constituents	up to 2%

More detailed analyses have identified the following substances in samples of nectar, although the composition varies depending on the plant source.

- Sugars: such as sucrose, glucose, fructose, xylose, raffinose, melezitose, trehalose, melibiose, maltose, dextrin, and rhamnose.
- Vitamins: principally vitamin C and some of the vitamin B complex.
- Amino acids: including aspartic acid, glutamic acid, serine, glycine, alanine (the most important to bees of 13 isolated from nectar samples).
- Minerals: e.g., potassium, calcium.
- Organic acids: e.g., gluconic and citric acids.
- Pigments.
- Aromatic compounds: e.g. alcohols and aldehydes.
- Enzymes: including invertase, diastase (α amylase), transglucosidase, transfructosidase, phosphatases, and oxidases e.g., tyrosinase.
- Mucus, gums, ethereal oils and dextrin.
- Particulate constituents including pollen, fungi, yeasts and bacteria.
- Antioxidants mainly ascorbic acid (Vitamin C).
- Occasionally lipids and alkaloids and proteins (see below).
- Secondary metabolites such as traces of flavonoids, carotenoids, alkaloids (caffeine and nicotine), glycosides (e.g., iridoid glycosides).
- Derivatives of phenol, eugenol, indole, morphine, various carboxylic acids.
- Anti-microbial, anti-fungal, antioxidant, soporific, sedative, narcotic, attractant (volatile aromas), hormonal, neurotoxic (e.g., aesculin).

Nectars which have a higher phenolic content are bitter and deter honey bees.

Nectars containing high levels of sugars tend to have higher levels of amino acids, detectable lipids and/or antioxidants. If alkaloids are present, these constituents plus protein are usually found.

The influence of the lesser ingredients becomes more obvious as water is evaporated from the nectar and it becomes more concentrated during conversion into honey. It is the concentrated constituents that give the aroma and flavour to honey.

Honey bees prefer to collect warmer and less viscous nectar, regardless of the sugar concentration.

Honey bees avoid nectar which has been colonised by some species of lactic acid bacteria because the bacteria have made chemical changes to the nectar. Such contamination of the nectar occurs after the initial period when the flower is new and freshly opened by airborne microbes and other visiting insects.

17.2.1.3   Types of nectar

The sugar composition of nectar in terms of the relative proportions of sucrose, glucose and fructose may not be physiologically important because the honey bees rapidly digest sucrose and the nectar sugars are efficiently assimilated, (Gmeinbauer and Crailsheim 1993).

Nectars can be categorised into the following types:

*   Sucrose-dominated nectars which are associated with tube flowers and with protected nectaries.
*   Fructose and / or glucose (with little sucrose) dominated nectars are associated with open flowers and unprotected nectaries, and usually containing more fructose than glucose.
*   Nectars containing roughly equal proportions (referred to as 'balanced') of fructose, glucose, and sucrose. These are less common.

The type of nectar affects the granulation characteristics of the honey that is derived from it.

It is sometimes a feature of the plant family with closely related families having similar kinds of nectar.

Pure sucrose dominated nectar is found in members of the Ranunculaceae (e.g., hellebores), Berberidaceae and some in the Fabaceae.

Glucose and fructose dominated nectars are found in representatives of the Brassicaceae, Apiaceae and some Asteraceae. Most Brassicaceae nectar contains no sucrose, and oil-seed rape nectars yield more glucose than fructose; dandelion nectar is also high in glucose.

Most of the nectars found in the Fabaceae contain equal quantities of all three sugars, although red clover and false-acacia nectars contain more fructose than glucose.

Within individual flowers of several of the Brassicaceae (rape, turnip, sweet alison, garden radish, charlock, thale cress), nectar is produced from both the lateral nectaries (associated with the short stamens) and the median nectaries (outside the long stamens). On average 95%

of the total nectar carbohydrate is produced from the lateral nectaries and this nectar has a higher glucose: fructose ratio (usually 1.0 -1.2) compared with that from the median nectaries (0.2 - 0.9). Anatomical studies of these brassica flowers show that the lateral nectary glands are supplied with relatively rich amounts of phloem tissue (food-conducting), whilst the supply to the median nectary glands is poor.

Sugar concentration is highly dependent on the anatomy of the vascular system that supplies the nectary. Highly concentrated nectar is characteristic of several plant species where the nectar originates from phloem tissue as demonstrated in the anatomical studies of the brassica flowers described above. Plants that produce high volumes of dilute nectar have limited connection to phloem tissue, but a greater degree of connection with xylem (water conducting) tissue.

Controlled experiments with honey bees have shown that nectars containing sucrose, glucose, and fructose in the same ratios are the most attractive to them. However, in nature, the usual ratio in flower nectar actually used by the honey bees is 2:1:1. An equal mix is less common.

Examples of Nectar sugar concentrations abstracted from Table 3 in Percival (1965).

Common name	Botanical name	% Sugar concentration
Wild Cherry	*Prunus avium*	12
Rosebay Willowherb	*Chamerion angustifolium*	13
Raspberry	*Rubus idaeus*	12-24
Wild Plum	*Prunus domestica*	16
Cultivated Plum	*Prunus domestica (cultivated vars)*	10-28
Blackthorn	*Prunus spinosa*	24
Great Willowherb	*Epilobium hirsutum*	25
Bird's-Foot- Trefoil	*Lotus corniculatus*	25-50 plus
Gooseberry	*Ribes uva-crispa*	30
White Clover	*Trifolium repens*	41
Crab Apple	*Malus sylvestris, various vars*	50
Rape	*Brassica rapus*	51
Dandelion	*Taraxacum officinale*	51
Willows	*Salix* spp.	60
False-acacia	*Robinia pseudacacia*	63
Horse-Chestnut	*Aesculus hippocastanum*	33-74

Some of the more important carbohydrates found in nectar and honeydew, and their value or otherwise to bees are described in the following table.

17.2.1.4   Types of sugars and their importance to honey bees

Type of Carbohydrate	Type of sugar	Important properties
Monosaccharides (70% of honey by weight)	Glucose (dextrose)	Energy
	Fructose (laevulose)	Energy
	Galactose	Inedible to the honey bee
Disaccharides	Sucrose	Broken down by invertase (sucrase) to glucose and fructose
	Maltose	Secondary sugar resulting from breakdown of sucrose. Bee can metabolize maltose
	Trehalose	Not important as a blood sugar in honey bees, c.f. other insects. Bee can metabolize trehalose
Trisaccharides (in the case of melezitose 0.3% of nectar honey and up to 8% honeydew honey)	Melezitose	Found in honeydew, bee can metabolize melezitose.
	Raffinose	Bee cannot metabolize raffinose

17.2.1.5   Sugars which are harmful to the honey bee

Worker honey bees prefer sucrose > glucose > maltose / fructose.

When fed solutions of sugars and their derivatives raffinose, glucose, sucrose, maltose trehalose and melezitose were found to be highly attractive.

Glucose is 62% as sweet as sucrose.

Some sugars are harmful to honey bees and some nectars and honeydew often contain one or more of these sugars and they might include mannose, galactose, arabinose, xylose, melibiose, raffinose, stachyose and lactose.

Barker (1977) lists some carbohydrates which can be found in pollen and pollen substitutes and were found to be toxic to caged honey bees. He found that they could be diluted to safe levels by sucrose and concluded that collected nectar dilutes the toxic sugars in the pollen thus permitting the assimilation of essential nutrients from pollen.

17.2.1.6 Examples of types of nectar and sugar values

The following table is a compilation of information on the types of nectar that are found in a range of plant species. Not all of the plants listed are visited for nectar by honey bees and these are shown in italics in the Plant Name column, but are included for interest.

Plant name	Species name	Sucrose predominates	Sucrose, glucose and fructose about equal	Glucose and fructose predominate
Wallflower	*Chieranthus cheiri*			+
Sweet Violet	*Viola odorata*			+
Heartsease	*Viola tricolor*	+		
Yellow Corydalis	*Corydalis lutea*	+		
Columbine*	*Aquilegia vulgaris*	+		
Wild Cherry	*Prunus avium*		+	
Blackcurrant	*Ribes nigrum*	+		
Gooseberry	*Ribes uva-crispa*	+		
Purple Loosestrife	*Lythrum salicaria*	+		
Red Campion*	*Silene dioica*			+
Common vetch	*Vicia sativa*	+		
Primrose	*Primula vulgaris*	+	+	
Periwinkle	*Vinca minor*	+		
Eyebright	*Euphrasia officinalis*			+
Spotted Deadnettle	*Lamium maculatum*	+		
Harebell	*Campanula rotundifolia*		+	
Dandelion	*Taraxacum officinale*			+
Coltsfoot	*Tussilago farfara*			+
Ragwort	*Senecio jacobaea*			+

Plant name	Species name	Sucrose predominates	Sucrose, glucose and fructose about equal	Glucose and fructose predominate
Hemp Agrimony	*Eupatorium cannabinum*			+
Dwarf Thistle	*Cirsium acaule*	+		
Hardhead	*Centaurea nigra*	+		
Bluebell	*Hyacinthoides nonscripta*			+
Daffodils	*Narcissus* sp. *Trumpet Daffodil*			+
Marsh Helleborine	*Epipactis palustris*	+		
Twayblade	*Listeria ovata*			+
Pasque Flower	*Anemone pulsatilla*	+		
Meadow Buttercup	*Ranunculus acris*	+		
Stinking Hellebore	*Helleborus foetidus*	+		
Greater Stitchwort	*Stellaria holostea*			+
Mountain Cranesbill	*Geranium pyrenaicum*			+
Bramble	*Rubus fruticosus*		+	
Sea Spurge	*Euphorbia paralias*			+
Raspberry	*Rubus idaeus*			+
Land cress	*Barbarea verna*			+
Rowan	*Sorbus aucuparia*			+
Hogweed	*Heracleum sphondylium*			+
Cicely / Sweet chervil	*Myrrhis odorata*		+	

Plants listed with an asterisk * are secondarily robbed for nectar by the honey bees accessing the nectar through holes made at the base of the flowers by short-tongued bumble bees.

Crane (1979) contains an interesting table (2.123/1) which tabulates the sugar values of nectars from different plant species. The sugar value being defined as the milligrams of sugar produced per flower each 24- hour period. This is an interesting basis of comparing different species and assessing their importance as sources of nectar and honey.

Botanical Name	Common Name	Sugar value
Borago officinalis	Borage	Variable 0.2 to 4.9, average approx. 1.2
Cynoglossum officinale	Common hound's-tongue	0.38 -1 .3
Echium vulgare	Viper's bugloss	0.09 - 1.64, average approx. 1.3
Helianthus annuus	Sunflower	0.12 - 0.30
Brassica napus	Oilseed rape	0.03 - 2.10
Sinapsis alba	Charlock	0.4
Raphanus raphanistrum	Radish	0.04 - 0.56
Calluna vulgaris	Ling	0.12
Aesculus hippocastanum	Horse-Chestnut	2.08
Phacelia tanacetifolia	Phacelia	0.069 - 0.98
Lavandula spica	Lavendar	0.26
Lamium album	White Deadnettle	0.12-0.72
Lamium purpurea	Red Deadnettle	0.07
Salvia splendens	Salvia	0.70
Thymus vulgaris	Thyme	0.01-0.09
Lotus corniculatus	Bird's Foot Trefoil	0.08
Medicago sativa	Meddick	0.04 - 0.12
Melilotus albus	Ribbed Melilot	0.06
Melilotus officinalis	Melilot	0.0005 - 0.004
Onobrychis vicifolia	Sainfoin	0.24

Botanical Name	Common Name	Sugar value
Robinia pseudoacacia	Rowan	0.22 - 2.3
Sophora japonica	Sophora	0.3 – 1.0
Trifolium hybridum	Alsike Clover	0.01
Trifolium incarnatum	Crimson or Italian Clover	0.003 – 0.07
Trifolium pratense	Red Clover	0.03 – 0.19
Trifolium repens	White Clover	0.012 – 0.14
Allium schoenoprasum	Chives	0.37 – 0.48
Chamerion angustifolium	Rosebay Willowherb	0.02 – 4.26
Fagopyrum esculentum	Buckwheat	0.05 – 2.68
Malus sylvestris	Apple	0.28 – 1.37
Prunus armeniaca	Peach	0.31 – 0.84
Prunus avium	Wild / Sweet cherry	0.08 – 1.50
Prunus cerasus	Sour cherry	0.15 – 1.31
Prunus domestica	Plum	0.13 – 1.47
Prunus persica	Peach	0.54 – 1.38
Pyrus communis	Pear	0.05 – 0.30
Ribes sylvestre / rubrum	Red Currant	0.04 – 0.70
Ribes uva-crispa	Gooseberry	0.13 – 5.41
Rubus idaeus	Raspberry	0.18- 8.1
Salvia officinalis	Sage	0.3 - -3.3
Salvia pratensis	Meadow Sage	0.60
Salvia splendens	Scarlet Sage	0.70
Tilia cordata	Small-leaved lime	0.01 – 3.06

Botanical Name	Common Name	Sugar value
Tilia x euchlora	Lime	0.12 – 2.9
Tilia platyphyllos	Large-leaved lime	0.31 – 7.7
Tilia tomentosa	Silver lime	0.07 – 5.00
Tilia vulgaris	Common Lime	1.04

17.2.2 Honey

17.2.2.1 The transformation of nectar into honey

The honey stomach (crop) can hold approximately 100 mg of nectar, but most foragers return to the colony after a forging trip with approximately 40 mg nectar, which is about half the body weight of a forager bee (90 mg).

Other data suggests 25.5+/-15 mg of nectar compared to average worker weight of 120 mg.

The proventriculus (a tiny valve with four flaps) is located at the posterior end of the honey stomach and opens and allows a small amount of the nectar for the foraging bees to pass into her midgut for her own consumption. The remainder begins the processes of converting the nectar into honey and is subsequently passed to the receiver bees in the colony for completing the process into finished honey which is then capped and stored; or goes directly into the flow of honey through trophallaxis between all the bees of the colony to ensure their needs are met.

There are several phases during the transformation from nectar into honey:

1. The floral nectar – as collected.

2. In the honey crop of the foraging worker bee.

3. In the honey crop of the worker house bees.

4. In the unsealed honey containing cells of the comb.

5. In the sealed honey cells.

There are two principal processes involved in the conversion of nectar to honey:

- The conversion of the disaccharide sugar sucrose to the monosaccharide sugars glucose and fructose (a chemical process).
- The increased concentration of sugars in the nectar due to the evaporation of water (a physical process).

However, the transformation of nectar into honey also involves physical, chemical, biochemical, physiological, behavioural and ecological factors.

The potential for deterioration of the nectar resource during its processing into honey is controlled by:

- Low water activity.
- Acidic pH – acid or neutral reaction pH 2.7-6.4, more rarely alkaline up to pH 9.1.
- High oxidation – reduction potential (hydrogen peroxide).
- The presence of competitive microorganisms e.g., bacteria e.g., *Lactobacillus* spp.

17.2.2.2   Physical transformation

Floral nectar contains a range of sugar concentrations. There is a reduction in the water content of nectar increasing the nectar sugar concentration to around 40% and reducing nectar volume (foraging bees regurgitate nectar on the hive return flight and during foraging).

Receiver bees in the hive further reduce moisture content by drying the nectar on their mouthparts and / or repeatedly regurgitating a large drop held between the proboscis and the mandibles as well as fanning which moves air over the nectar.

When the moisture content has been reduced to the desired range between 16 and 20% the bees seal (cap) the honey-containing cells.

The conversion process continues as the receiver bee regurgitates a small drop of the nectar into the fold of her partly extended proboscis (the sucking apparatus of the mouthparts). By repeatedly exposing it to the air the moisture content of the nectar droplet is reduced; she will do this 80-90 times over a period of 20 minutes with the same droplet of nectar.

This evaporation of water gives a supersaturated sugar solution which has a high osmotic potential and in which bacteria and yeasts find it difficult to thrive. At hive temperatures the solubility of glucose increases with the fructose concentration resulting in a very concentrated sugar solution that inhibits microbial growth (Donar 1977).

Nectar is passively evaporated within the ripening cells and its relocation between cells before arriving in its final storage location is essential to the ripening process.

After placing into a cell in the comb the processed nectar, now in the form of honey, may remain liquid or granulate in the comb cells depending on the composition of the carbohydrates in the floral nectar.

17.2.2.3   Chemical and biochemical transformation

The workers utilise enzymes to change the various sugars in the nectar into simpler sugars.

The enzymes are produced by lactic acid bacteria resident in the worker honey bee crop and enzymes such as sucrase (called invertase in plants) are secreted by the hypopharyngeal, post-cerebral, thoracic, labial and mandibular glands. There are however other enzymes which are also produced by the worker bee in the enzymatic breakdown of other sugars and carbohydrates as part of the transformation of nectar into honey.

Plant invertase and worker bee sucrase convert the sucrose (the primary sugar in nectar) into glucose and fructose.

The sucrose in the nectar is inverted by the action of the enzyme sucrase in the honey stomach of the foraging bees in transit from the flower to the colony and in the colony itself. The sucrose is inverted to fructose (laevulose) and glucose (dextrose). As the ripening process progresses, as well as the inversion of the sucrose the moisture content of the nectar is progressively reduced usually somewhere in the range 12-21%. Once the honey is ripe in the comb it is capped (sealed) with beeswax.

Glucose is a very essential sugar because the brain cannot function without it, and it also supplies energy to the muscles and is important in the function of body cells. Fructose also supplies energy but is not used by the bee brain.

17.2.2.4   Nectar consumption by the colony

Each year a colony requires around 120 kg of nectar. Approximately 70 kg is consumed during the summer months to provide food for the brood and the adults, for their own nutrition, for extra energy to keep the brood nest warm (35-36⁰C) and to provide energy for foraging. The balance of nectar is converted to around 20 kg of honey that is stored in the colony and consumed during the cold flowerless months or when the weather conditions are unsuitable for foraging. This amount of honey may be more depending on the resource availability and the strain of the honey bee.

A colony's heat production in mid-winter is of the order of 40 watts, which is enough to keep the surface bees from falling from the cluster and dying even when the temperatures outside are –30°C or less. Such intense heat production consumes a lot of energy and a colony will use nearly a kg of honey per week throughout the winter months (Seeley 1995).

17.2.2.5   Honey Granulation

Honey granulates because it is a supersaturated solution containing more sugars than can normally remain in solution. The supersaturated solution is unstable, and in time can only return to the stable state by some of the sugars being thrown out of solution i.e. they granulate. Glucose is less soluble in water than fructose and this property has an important bearing on the granulation of honey. A high glucose: fructose ratio in the nectar will produce a rapid granulation, however it is the glucose: water ratio, for example in oil seed rape (Brassica napus) (canola) that is more closely related to the granulation tendency of the honey. Honey with a ratio of 1 water:1.7 glucose or lower will not granulate; whereas a ratio of 1 of water:2.1 glucose or higher usually means there will be rapid granulation. The optimum granulation temperature for honey is 13-15⁰C.

The sugar crystals in the granulated honey melt between 36 - 39⁰C.

Granulation can be promoted by the presence of minute crystals, air bubbles, particles of wax, pollen grains and dust, both in honey stores in the hive and in honey after extraction from the comb.

The granulation characteristics of the honey are of key importance to the colony and determine how much of the carbohydrate in the honey can be solubilised, ingested and utilised by the colony. If the crystals remain too large, they cannot be ingested and so the honey cannot be utilised by the colony especially in the overwinter period when there is little or no fresh nectar available. If the colony is merely hefted (lifted) to check the weight and hence the amount of food reserve in the colony estimated the beekeeper runs the very real risk of misunderstanding the true status of the food reserves if much of the stores consist of honey containing large crystals and granulated, which cannot be utilised by the colony.

17.2.2.6   Honey Fermentation

If the nectar is very dilute it may have been gathered from an open form of flower and the nectar has been diluted with rain, or from water absorbed hygroscopically from a very humid atmosphere. If it is available in significant quantities and the bees have a shortage of capacity to fully process the nectar the bees may not reduce the moisture below 21%, the value below which honey is at risk of fermentation.

The moisture content of the honey is important because if it is between 17% and 20%, in the presence of osmophilic yeasts (active between 10 and 20^{0}C), there is a risk the honey will ferment. The glucose and fructose will be metabolised to produce alcohol and water.

Above 22% moisture content there is an increased risk of fermentation, even if osmophilic yeasts are absent.

When honey granulates there is increased water content between the sugar crystals allowing the osmophilic yeasts to become active. Some of the wild yeasts found in flowers and in soil are osmophilic.

Osmophilic yeasts are adapted to be able to survive in substances with high sugar concentrations and a high osmotic pressure. They are sugar tolerant and unlike many other microorganisms able to thrive in them. Examples include *Saccharomyces rouxii*, *S. bailii*, *S. cerevisiae* and *Debaromyces* spp.

Non-granulated honey will also ferment if the moisture content and yeast count are the same as that found in fermenting granulated honey.

Storing honey at less than 10^{0}C discourages granulation and fermentation but if stored at more than 27^{0}C it will be damaged. This can be detected by measuring the increased content of hydroxymethylfurfuraldehyde (HMF) and the reduced diastase level. Fructose, if heated, will decompose and become the source of HMF. HMF is toxic to honey bees.

17.2.3   Honeydew

17.2.3.1   What is Honeydew?

Honeydew should not be confused with nectar. Nectar is actively secreted by the tissues which constitute the nectaries, whereas honeydew requires the intermediary of a sap-sucking insect. Honeydew also differs in its composition from nectar.

17.2.3.2   Origins of Honeydew

Honeydew is produced through the activities of plant - sucking insects (Hemiptera) which include aphids and scale insects. These insects have mouthparts that are modified to pierce into the tissues of the host plant and feed on the sap inside the plant's phloem tissue. The sap comes out from the injected hole due to the internal pressure in the plant's vascular system and the action of the insect sucking, speeds up the process. The ingested sap passes into the gut of the insect and, after digestion, it is excreted as droplets of honeydew. These droplets are collected by a number of insects including honey bees. Honeydew may be produced in significant quantities in certain seasons when the weather is hot and dry favouring the build-up of populations of sap-sucking insects.

17.2.3.3   Plant sources

In the British Isles, trees are the most important sources of the honeydew that are exploited by honey bees. Species such as beech, oak, poplar, fir, scots pine, and larch only produce honeydew, whilst trees such as sycamore, small and large-leaved limes, and sweet chestnut will yield both nectar and honeydew. Currants, gooseberries and sunflowers are also a source of honeydew.

17.2.3.4   Composition of honeydew honey

Honeydew honey contains catalases, amylase, and acid phosphatase which are enzymes derived from the plant. Added to this are peptidase and proteinase enzymes from the sap-sucking insect. It usually contains significant amounts of the trisaccharide sugars melezitose, erlose, raffinose, and dextran, a polyglucose.

Maltose (also called maltobiose and malt sugar) is a phloem sugar produced from the feeding activities of coccids (Scale insects). Maltose can also be formed when glucose is caramelised. The disaccharide sugar trehalulose is produced by the feeding activities of sap-feeders, including whiteflies. It can constitute around 30% of honeydew sugars.

Honeydew typically consists of:

- Fructose approx. 32%.
- Glucose approx. 26%.
- Sucrose approx. 0.8%.
- Maltose approx. 8.8%.
- Higher sugars e.g. dextrins 4.7%.
- Melezitose 2.3%. May exceed 20% and granulate in the comb.
- Erlose.
- Raffinose.
- Protein / amino acids and amides up to 1%, usually around 0.6%.

At least three types of honeydew are recognised:

- Erlose containing– honey from this type of honeydew does not granulate.
- Melezitose containing.
- Containing both erlose and melezitose.

The amounts of glucose and fructose in honeydew honey are usually less than in honey produced from plant nectar. Honeydew honey is characterised by the presence of algae and sooty moulds as well as dust and soot particles. The presence of these materials can affect the colour, the flavour, and the keeping qualities of the honey. Their presence is used to characterise the honey as from honeydew when examined under the microscope.

17.2.3.5   Use by honey bees

When honeydew is being collected in large amounts by worker bees many of the bees are seen to have a shiny black abdomen and waist. This can easily be confused with a paralysis virus such as Chronic Paralysis Bee Virus. If the bees are working honeydew formed on pine and spruce trees, more so than say beech, oak or maple, the body hairs are abraded by the contact of the body hair with the foliar needles of the tree.

Honeydew honey is unsuitable for feeding over-wintering bees because it may have a high enough protein or ash content that can cause a dysenteric condition if bad weather prevents them from flying frequently to void faecal matter. Honeydew honey, if it contains more than 8-12% melezitose, is considered unsuitable for overwintering stores. However, most honeydew honeys are considered safe for the colony to use to build-up in the spring.

17.2.4   Plant toxins in honey bee nutrition relevant to humans and honey bees

It is estimated that from 9-55% of nectars from floral sources contain plant-synthesised glucosides, flavonoids and polyhydric alcohols as well as some indigestible sugars which can be toxic to honey bees if present in sufficient levels. All these toxins can be present in nectar, honeydew, pollen and plant sap. Some which are toxic to honey bees have been used as 'botanical' insecticides for plant protection products. They include citronella oil, eucalyptus oil, garlic extract, neem oil, rotenone and andiroba oil (Johnson 2015).

The nectar and pollen of *Rhododendron ponticum* contains alkaloids and some species of Rhododendron contain neurotoxins called grayanotoxins.

A toxic saponin may be found some years in the nectar and pollen from the horse chestnut *(Aesculus hippocastanum).*

Viper's bugloss *(Echium vulgare),* Ragwort *(Senecio jacobaea)* and Butterbur *(Petasites hybridus)* produce nectars contain pyrrolizidine alkaloids and essential oils which are toxic to man.

Other plants with nectars containing alkaloids include:

Azure *(Aconitum carmichaeli);* Black Hellebore (*Veratrum nigrum*); Common Thorn Apple *(Datura stramonium)* and Plume Poppy *(Macleaya cordata)*

17.3   Proteins and amino acids

We have already identified proteins and carbohydrates as being essential nutrients for the honey bee colony. We will now consider the role of proteins and amino acids in more detail.

17.3.1   Proteins

Proteins are composed of usually more than 20-30 amino acid residues linked together by peptide bonds (-CONH).

Proteins and amino acids are vital in the composition of the cytoskeleton (complex mesh of fibres made from proteins) which provides the support of cellular shape and the physical integrity of the bee body. They are also essential for the nucleus, for muscles, for the multiplicity of enzymes and for structure and functioning of chemoreceptor cells which respond to internal hormone changes and to external environmental stimuli. The individual bee can thus develop properly and contribute to acquisition of nutrients, the colony functions of reproduction, the rearing of brood, helping in disease resistance, alerting and participate in the co-ordination of colony defence.

Honey bees obtain the majority, if not all, of their protein requirements from pollen.

The nutritional value of any given pollen for the honey bee is determined not only by the amount of protein content, but also by its amino acid composition.

17.3.2   Amino acids

Lipinski (2018) states that there are more than 300 known amino acids, but only 22 of these are proteinogenic (protein-building blocks) amino acids. These are linked into peptide chains which in turn form proteins. The thousands of different types of protein found in the honey bee are largely composed of 20 standard L-amino acid monomers (a molecule that can be bonded to other identical molecules to form a polymer). Honey bees (as well as all other animals) cannot synthesise 10 of these 20 amino acids in sufficient amounts and they must be taken unchanged directly from the food they eat.

17.3.2.1   Essential Amino acids

The basic building blocks of proteins are amino acids and in 1953 de Groot published a list of amino acids in honey bee collected pollens which were considered to be essential for honey bee nutrition.

Pollen also contains non-essential amino acids which the bee can transform metabolically from one to another or which it can synthesise from certain essential amino acids.

The following table lists these essential amino acids and the level of the amino acid in the pollen considered satisfactory for honey bee nutrition.

Essential amino acids	Bee requirements (g/16 g N)*	Ideal composition of essential pollen amino acids
Threonine	3.0	11%
Valine	4.0	14
Methionine	1.5	5
Isoleucine	4.0	14
Leucine	4.5	16

Essential amino acids	Bee requirements (g/16 g N)*	Ideal composition of essential pollen amino acids
Phenylalanine	1.5	9
Histidine	1.5	5
Lysine	3.0	11
Arginine	3.0	11
Tryptophan	1.0	4

*This measurement is a ratio of the amino acid to the total nitrogen content in the pollen and not a quantitative measurement based on the dry weight of the pollen. When a pollen source is said to be low in one or more amino acids particular attention should be paid to the crude protein content or total nitrogen content. Pollen with a low crude protein content and limiting in one or more essential amino acids is a much greater concerns in terms of honey bee nutrition than pollens with a high crude protein content and limited in one or more essential amino acids.

The amino acids which are especially needed by honey bees are leucine, iso-leucine and valine.

Honey bee longevity is reduced by feeding diets high in essential amino acids and proteins with free amino acids shortening the lifespan more than whole proteins. The protein: lipid, (P:L) ratio of pollen collected by honey bee colonies in the field and on several sites was found to be 10:1, (Reported in Wright et al. 2018).

17.3.3   Pollen – more than just proteins and amino acids

17.3.3.1   What does pollen contain?

Honey bees obtain the majority, if not all, of their protein and amino acid requirements from pollen.

Brodschneider and Crailsheim (2010) estimated that a single ten frame colony requires between 13 and 18 kg of pollen per year.

A colony will collect more pollen when the level of brood pheromone in the colony is high and there are high levels of queen pheromone. Some colonies have a genetic disposition to collect more pollen.

Pollen is consumed primarily by young adult workers (nurse bees) who can convert pollen to worker jelly and royal jelly in their hypopharyngeal glands. These 'jellies' are a nutrient-rich food source that is fed to larvae and the queen and is required for brood rearing.

Pollens from insect-pollinated plants are heavier and larger compared with plants which are wind-pollinated.

Pollen also provides a number of other chemicals / nutrients of key importance to the development and survival of the honey bee.

The composition of pollen varies according to the plant source, and the following types of substances have been identified in pollen samples and these are listed below.

- Lipids (essential for brood food production).
- Carbohydrates and related compounds such as cellulose and lignin.
- Major minerals including potassium, sodium, calcium, and magnesium.
- Organic acids including phenolic acids.
- Free amino acids (main source of nitrogen for bees).
- Nucleic acids (DNA and RNA).
- Terpenes.
- Enzymes.
- Vitamins, including B2 (riboflavin), B6 (niacin), pantothenic acid, biotin, C (ascorbic acid) and E.
- Nucleosides.
- Pigments, carotenoids, and flavonoids.
- Plant growth regulators.

Over 250 biologically active substances have been found in pollen grains from different plant species.

Some pollens, as well as nectars, also contain sugars which are toxic to honey bees and these include mannose, galactose, arabinose, xylose, melibiose, raffinose, stachyose and lactose. The toxicity of these sugars is reduced when the foraging bees have access to plenty of nectar from a variety of sources.

Most pollen proteins are enzymes which are concerned with the pollination process and the growth of the pollen tube.

17.3.3.2   Colours of Pollen Loads

The colour of pollen loads is determined by the presence of pigments such as flavonoids (e.g., red, yellow, purple) and / or carotenoids (e.g., yellow and orange). There is a strong correlation between the colour of pollen and its preference by foraging bees.

Hodges' classic book *Pollen Loads of the Honey Bee* was first published in 1952 and reprinted in 1987. It includes a series of colour charts presenting the different colours of the pollen between plant species and the seasons of the year in which you might expect pollen loads of the various colours to be being brought into the hives by the foraging bees.

Watching foraging bees return to the hive with pollen loads or different colours or examining the pollen loads placed into the comb cells for their different colours is an enjoyable pastime and gives an indication of the sources of forage being exploited by the colonies in the apiary. It is often the case that the pollens being collected by colonies in the same apiary are quite different indicating that the colonies are foraging over different areas.

Common Name	Botanical name	Pollen Colour
Maple	*Acer* spp.	Light yellow
Grey Alder	*Alnus incana*	Brownish - yellow
Hawthorn	*Crataegus* spp.	Yellow - brown
Apple	*Malus domestica*	Yellow - white
Plum	*Prunus* spp.	Light grey, grey
Wild cherry	*Prunus avium*	Yellow or light brown
Sour Cherry	*Prunus cerasus*	Dark yellow
Peach	*Prunus persica*	Reddish Yellow
Pear	*Pyrus communis*	Red - yellow
Blackberry	*Rubus* spp.	Light grey
Raspberry	*Rubus idaeus*	White grey
Willow	*Salix* spp.	Lemon
Oil seed rape	*Brassica napus*	Lemon
Snowdrop	*Galanthus nivalis*	Orange, red
Sainfoin	*Onobrychis viciifolia*	Yellow - brown
White Mustard	*Sinapsis alba*	Lemon
Dandelion	*Taraxacum officinale*	Red - yellow, orange
Blueberry	*Vaccinium myrtillus*	Red - yellow, orange
Fireweed	*Epilobium angustifolium*	Blue
Buckwheat	*Fagopyrum esculentum*	Light yellow to light green
Sunflower	*Helianthus annus*	Golden
Alfalfa	*Medicago sativa*	Khaki
Poppy	*Papaver somniferum*	Grey
Alsike Clover	*Trifolium hybridum*	Yellow - brown
White Clover	*Trifolium repens*	Caledonian brown
Field Poppy	*Papaver rhoeas*	Near black, purplish
Squill	*Scilla siberica*	Deep blue
Horse Chestnut	*Aesculus hippocastanum*	Brick-red
Bluebell	*Hyacinthoides non-scripta*	White
Balsam	*Impatiens balsamifera*	Off-white

Food consumption by worker bees is not related to pollen quality (e.g., protein concentration) but is influenced by their physical or chemical cues, (Schmidt and Johnson 1984).

17.3.3.3 How honey bees use pollen

Some 50% of a pollen grain is of no use to the bee, however honey bees are very proficient in their removal of the content of the pollen grains as they pass through the bee's gut. Gut transit time for pollen varies between 2 and 24 hours.

Honey bees are very efficient at removing the protein content of the pollen grains which have been reported in the range 77-83% (Schmidt and Buchmann 1985).

Protein is essential for adequate fat cell development in fat bodies. Fat, glycogen, protein, and albumin are stored in the fat bodies that are located along the inside of the dorsal part of the abdomen. Pollen also contains up to 5% lipids, these are important as bees cannot synthesise cholesterol without the essential sterol precursors obtained from pollen. Bees that must over winter in the colony are physiologically different from summer bees. They must develop good fat bodies to enable them to survive the winter, and able to sustain the provision of brood food during the early part of the season, when pollen sources are scarce or bad weather prevents foraging. In August / early September, the adult worker bees in the colony, and the newly developing bees feed very heavily upon pollen; this brings their hypopharyngeal glands back into the plump condition of the young nursing bee to enable them to feed the spring brood. The amount of pollen required to rear a single worker has been estimated at between 125-145 mg, and on average this contains about 30 mg of protein. The adult worker nurse bee consumes on average 4 mg pollen / day with a maximum intake of 9.5 mg / day. The importance of a continual supply of pollen throughout the beekeeping season is the most important factor in whether a particular apiary location will support healthy and productive colonies. In the summer, a colony rears approximately 150,000 bees, and therefore the quantity of pollen required to be collected is around 20kg. However, colony requirements vary and figures ranging from 15 –55 kg have been recorded.

Pollen is essential for young bees in the first 5-6 days after emergence to ensure their hypopharyngeal glands develop. Hypopharyngeal gland development is necessary for young adult bees, both to produce the special brood food to feed the brood, and later to secrete wax during the comb-building phase of their lives (12-18 days after emergence as adults).

17.3.3.4 Nurse bees and their key role in protein availability

Nurse bees are of crucial importance in the processing of pollen and the distribution of the proteins and other nutrients around the honey bee colony. They process the proteins in the pollen into a protein- rich jelly which is fed as follows to other members of the colony:

- The early stages of worker, drone and queen larvae.
- Workers just after emergence.
- The queen who also shares this material with her attendant worker bees.
- Young drones.
- Worker bees returning from foraging activities.

Nurse bees are referred to as the 'social stomach' of the colony.

See Corby-Harris et al. (2018) and Somerville (2000).

17.3.3.5   Consequences for the colony when nurse bees do not have access to pollen

If nurse bees cannot access pollen or bee bread, they will consume their own body proteins to make jelly and continue brood rearing until the residual protein level reaches 20% of their body weight. Once there is no proteinaceous material the nurse bees can continue to nurse the developing brood for about one week after which the larvae fail to mature and are removed by worker bee from the colony. The queen also stops laying and the colony population begins to decline.

17.3.3.6   Regulation of pollen collection by the colony

Most of the pollen consumption takes place during the spring and summer months when brood rearing is most intense. The intake of pollen by a colony undergoes far greater fluctuations than does its demand. To manage this balance a colony builds a stockpile of approximately 1 kg and this will last at least a week, usually sufficient to buffer the colony and keep it nourished in the event of a failure in the external pollen supply, or during a period of bad weather. A shortage of pollen significantly reduces the amount of brood reared and thus fewer worker bees emerging 2-3 weeks after the pollen shortage.

Bee bread and / or freshly collected pollen is required for feeding the larvae and the nurse bees up to the age of 15 - 18 days.

A colony adjusts its collecting rate with respect to the pollen reserve in the brood nest partly through changes in the total number of active pollen foragers, and partly through changes in the per capita collecting rate. The mechanism whereby the pollen foragers obtain information on their colony's pollen reserves is probably through excitatory feedback in the case of too little pollen in the hive which causes the nurse bees to prepare more cells for pollen storage. As a result, the returning forager bees have more space to unload their pollen. The size of the pollen reserves seems to serve as a stimulus in the regulation of foraging behaviour.

The adult worker bee which is at the foraging phase in its life cycle and collecting pollen contains little pollen in its gastrointestinal tract. Most of pollen is consume by the nurse bees (see section of nurse bees). Foraging adult worker bees cannot digest pollen directly and their protein needs of met by receiving proteinaceous feeds (jelly) from nurse bees who are 7-14 days old.

17.3.4   Pollen quality and proteins

17.3.4.1   Protein Content and its determination

The quantitative determination of the protein content of the pollen of a forage plant species or the nutritional status of the bees in a colony can be made several ways (Lipinski 2019).

* 'True' protein (TP).
* Crude (also referred to as overall, raw or total) protein (CP).
* Total content of essential amino acids.
* General calorific value of pollen.

The crude protein expressed in % age dry weight ranges from 66-74% of the dry weight of a healthy adult honey bee. During the first few days of the bee's life the protein content increases as proteins are synthesised and then decrease as the honey bee workers age.

17.3.4.2   Nutritive Value expressed as Crude Protein Content

The nutritive value of honey bee collected pollen can be expressed in terms of its crude protein content and a range of 20-25% is considered as the minimum level. This range is good for enabling the colonies to meet their protein requirements. Examples of such pollen sources include oil seed rape, faba (broad) bean, vetch, common knapweed and spear thistle.

17.3.4.3   Crude Protein Content of some Plant Species

Pollen sources with higher crude protein values (late 20's to early 30's percentage crude protein content) are even better value to a colony and can help mitigate any in-balance in the amino acids of the pollen or the shortage of pollen.

The following table is based on table 4 from Lipinski (2018) with additional information from the Somerville (2001) report on bee collected pollens. Further data can be found in Liolios et al. (2016).

Crude Pollen protein content values for a selection of Plant Species		
Common Name	**Taxonomic Name**	**% Crude Protein content**
Kiwifruit	*Actinidia deliciosa*	42.2
Wormwood	*Artemesia vulgaris*	13.0
Cornflower	*Centaurea cyanus*	26.2
Yellow Star Thistle	*Centaurea solstitialis*	20.6-22.4
Ox-eye daisy	*Leucanthemum vulgare*	28.0
Dandelion	*Taraxacum officinale*	13.2-22.69
Spear thistle	*Cirsium vulgare*	17.6-31.8
Cotton thistle	*Onopordum acanthium*	36.4
Cats Ear	*Hypochaeris radicata*	9.2-20.4
Sunflower	*Helianthus annuus*	13.8-30.6
Ragworts	*Senecio* spp.	16.8
Tansy	*Tanacetum vulgare*	11.7
Elm	*Ulmus* spp.	12.37
Buckthorn plantain	*Plantago lanceolate*	17.3-23.9

Crude Pollen protein content values for a selection of Plant Species		
Common Name	**Taxonomic Name**	**% Crude Protein content**
Catalpa	*Catalpa speciosa*	37.8
Sainfoin	*Onobrychis vicifolia*	32.53
Broom	*Genista* spp.	20.6
Locust 'Robinia'	*Robinia pseudoacacia*	22.19
Alsike clover	*Trifolium hybridum*	13.7
Red clover	*Trifolium repens*	22.5-35.4
White clover	*Trifolium repens*	22.5-35.4
Lucerne / Alfalfa	*Medicago sativa*	14.1-22.5
White lupin	*Lupinus albus*	28.0-34.7
Narrow-leaved lupin / Blue lupin	*Lupinus angustifolius*	28.0-34.7
Big flower vetch	*Vicia grandiflora*	42.8
Broad bean / Faba bean	*Vicia faba*	22.3-24.4
Yellow melilot	*Melilotus officinalis*	24.0
Dogwood	*Cornus* spp.	19.0-21.0
St John's wort	*Hypericum perforatum*	30.3
Wintergreen	*Pyrola minor*	29.4
Spreading bellflower	*Campanula patula*	23.6
Pinks	*Dianthus* spp.	21..5-21.92
Buttercup	*Ranunculus repens*	17.4-17.83
Lavendar	*Lavandula* spp.	19.4
White mustard	*Sinapsis alba*	24.5
Charlock mustard / Wild mustard	*Sinapsis arvensis*	23.4-33.8

Crude Pollen protein content values for a selection of Plant Species

Common Name	Taxonomic Name	% Crude Protein content
Rape / Canola	*Brassica napus*	22.8-31.9
Black mustard	*Brassica nigra*	33.6
Turnip	*Brassica rapa*	44.1
Field Mustard	*Brassica rapa* subsp. *campestris*	25.2
Wild radish	*Raphanus raphanistrum*	25.2
Annual Bastard cabbage	*Rapistrum rugosum*	21.6-25.3
Greater celandine	*Chelidonium majus*	22.2
Poppy	*Papaver somniferum*	21.4
California poppy	*Eschscholzia californica*	43.1
Horse-chestnut	*Aesculus hippocastanum*	26.7
Maple / Red maple	*Acer* spp. / *Acer rubrum*	21.44 / 39.4
Lacy Phacelia, blue tansy	*Phacelia tanacetifolia*	28.1
Desert bells	*Phacelia campanularis*	59.0
Purple viper's - bugloss	*Echium plantagineum*	Mean value 25.9 (some values up to 37.4%
Bittersweet (Nightshade)	*Solanum dulcamara*	51.4
Sweet cucumber	*Solanum muricatum*	39.9
Buckwheat	*Fagopyrum esculentum*	9.0-11.4
Rhubarb	*Rheum* spp.	19.52
Bistort	*Polygonum bistorta*	15.3
Knotgrass / Knotweed	*Polygonum aviculare*	13.91
Peach	*Prunus persica*	28.5
Bird cherry	*Prunus padus*	38.0

Crude Pollen protein content values for a selection of Plant Species		
Common Name	**Taxonomic Name**	**% Crude Protein content**
Almond	*Prunus dulcis*	23.3-30.7
Black cherry	*Prunus serotina*	>25.0
Stonefruits	*Prunus* spp.	25.1
Common pear	*Pyrus communis*	26.2
Wild pear	*Pyrus pyraster*	23.3
Raspberry	*Rubus ideaus*	19.0-21.3
Apple	*Malus* spp. *Malus domestica*	24.0 24.7
Blackberry	*Rubus fruticosus*	14.8-20.0
Common agrimony	*Agrimonia eupatoria*	21.74
Strawberry	*Fragaria* spp.	21.6
Cow Parsley	*Anthriscus sylvestris*	29.0
Grand fir	*Abies grandis*	24.1
Mountain pine	*Pinus mugo*	13.7
Lodgepole pine	*Pinus contorta*	13.0
Ponderosa pine	*Pinus ponderosa*	11.7
Aleppo pine	*Pinus halepensis*	13.2
Asparagus	*Asparagus officinalis*	37.0
Star of Bethlehem	*Ornithogalum umbellatum*	8.6
Sedges	*Carex* spp.	9.3
Lime tree	*Tilia intermedia*	18.22
Lesser Canary Grass	*Phalaris minor*	10.4
Maize	*Zea mays*	14.0-23.9

Crude Pollen protein content values for a selection of Plant Species		
Common Name	**Taxonomic Name**	**% Crude Protein content**
White willow	*Salix alba*	17.0
Goat willow	*Salix caprea*	36.8
Crack willow	*Salix fragilis*	14.8-15.1
Golden willow	*Salix alba var. Vitellina pendula*	26.0
Creeping willow	*Salix repens*	38.7
Willow herb family	*Onagraceae (average for the family)*	21.51-25.0
Cranberry, Blueberry, Bilberry (Whortleberry), Lingonberry (Cowberry)	*Vaccinium* spp.	13.9-19.0
Pink asphodel	*Asphodelus fistulosus*	14.0-22.5
Hedge Mustard	*Sisymbrium officinale*	22.0
Gorse	*Ulex europaeus*	28.0
Brooms	*Cystus* spp.	11.9
Poplar	*Populus* spp.	23.9
Verbascum /Mullein	*Verbascum* spp.	29.8
Mignonette	*Reseda* spp.	28.4
Deadnettle	*Lamium* spp.	25.0
Wild Clary	*Salvia verbenaca*	22.9
London Rocket	*Sisymbrium irio*	22.3
Thistle	*Carduus* spp.	22.2
Rockrose	*Cistus* spp	17.1
Chinese Privet	*Ligustrum lucidum*	16.6
Sow thistle	*Sonchus* spp.	16.6

Crude Pollen protein content values for a selection of Plant Species		
Common Name	**Taxonomic Name**	**% Crude Protein content**
Sea-lavender	*Limonium comarianum*	16.0
Purslane	*Portulaca olercea*	15.8
Chicory	*Chichorium intybus*	15.3
Kiwi	*Actinidia chinensis*	13.8
Walnut	*Juglans nigra*	13.0
Fungal spores	*In general*	7.31

The following table contains UK data abstracted from a paper by Hanley et al. (2008) detailing the relative protein content (% by dry weight) of pollen collected from 23 plant species native to southern England. It also contains data on their essential amino acids expressed as the proportion of the protein content which was comprised of essential amino acids. The essential amino acids being arginine, isoleucine, leucine, methionine, phenylalanine, threonine and valine. Histidine was not included because this can be metabolically replaced by the α-keto acid carnosine.

Botanical Name	Common Name	%mean protein content (+/- SE)	Essential amino acids (%)
Centaurea nigra	Common Knapweed	27.5(1.5)	31.1
Centaurea scabiosa	Greater Knapweed	24.2(1.3)	32
Cirsium arvense	Creeping Thistle	21.9(0.3)	31.9
Cirsium vulgare	Spear Thistle	22.1(1.4)	30.2
Senecio jacobea	Ragwort	17.2(1.5)	34.9
Echium vulgare	Viper's-bugloss	44.1(2.2)	39.9
Symphytum officinale	Common Comfrey	17.5(1.3)	35.1
Scabiosa columbaria	Small Scabious	24.8(1.7)	34.2
Calluna vulgaris	Heather / Ling	13.9(0.2)	35.3
Genista tinctoria	Dyer's Greenweed	22.8(0.5)	34.5

Botanical Name	Common Name	%mean protein content (+/-SE)	Essential amino acids (%)
Lotus corniculatus	Birds Foot Trefoil	35.8(1.1)	40.6
Melilotus altissima	Tall Melilot	39.2(1.1)	40.5
Onobrychis vicifolia	Sainfoin	37.5(1.8)	40.2
Trifolium pratense	Red Clover	40.8(3.6)	40.6
Trifolium repens	White clover	35.2(0.8)	41.6
Chamerion angustifolium	Rosebay Willowherb	16.2(1.8)	36.1
Papaver rhoeas	Field Poppy	19.1(0.5)	36.6
Potentilla erecta	Tormentil	16.3(0.3)	35
Rosa canina	Dog Rose	14.5(0.4)	32.8
Rubus fruticosus	Bramble	15.1(1.4)	34.5
Digitalis purpurea	Foxglove	20.9(0.4)	36.2
Rhinanthus minor	Yellow-rattle	20.8(0.9)	32.1
Odonites vernus	Red Bartsia	36.6(0.60	40.8

17.3.4.4   Pollen quality and the season

The literature usually categorises pollen collected by honey bees into 'excellent' and 'poor', sometimes other terms are used such as 'good'.

Honey bees forage for nectar and pollen throughout the active season searching for nutrients which will sustain their colony demands whether for brood-rearing in the spring or in the preparations for winter. The question arises as to whether the nutrient composition of the pollen sourced during the foraging seasons changes in accordance with the demands of the colony. Pollens gathered in the spring may have nutrients which support brood rearing whilst those collected in the autumn may be rich in nutrients which can be stored and readily mobilised or that they contain key components of immune systems which enable the colony to withstand pathogen pressures during their time clustered overwinter, (DeGrandi-Hoffman et al. 2018).

Liolios et al. (2016) examined a range of bee collected pollens in Greece ranking them according to their contribution to bee nutrition and taking into consideration seasonal variation in the pollen. They concluded that the most significant factor was the quantity of the pollen collected for each plant species studied and that the richness of the protein content in the pollen was insufficient to define whether the pollen collected was a good one. Of the 46 pollens from different plant species studies some 88.8% of the nutrition was provided by only 14 plant species. Seasonal variation in the pollen, the flowering period and the season were also important. They also found in the places they studied that whilst the bees collected pollen from a wide variety of plant species only a few of the species contributed significantly to the nutritional requirements of the colony. The protein content of plants flowering in the spring were higher (20-24%), those from summer flowering (15.1-19.9%) and autumn flowering (19.3-23.1%).

There is evidence that foragers show preferences for certain micronutrients and that these choices vary according to the season, (Bonoan, O'Connor and Starks 2017).

17.3.5   Bee bread

17.3.5.1   The importance of bee bread

Most bees prefer to eat freshly stored pollen and consume it within 3-5 days of storing it, preferably in cells close to brood cells. Because of this there are only relatively small amounts of stored pollen mostly in a narrow ring around the edge of the brood nest.

Pollen which is processed and packed into cells for storage is called bee bread.

Honey bees obtain the majority of their nutrients from pollen but pollen is resistant to enzymatic digestion and so honey bees deal with this by having symbiotic microbes in their intestines which convert a mixture of pollen and nectar into the source of protein for the honey bee, namely bee bread. These symbiotic microbes also protect the pollen from spoilage organisms, some of which could be pathogenic to the various stages in the life of the honey bee.

The ring of freshly collected pollen, or that of pollen that has been processed into bee bread, is found around the outside of the brood rearing area on the brood combs, usually on the upper parts. During most of the year approximately 1kg of bee bread is stored read for use by the colony.

The critical importance of the role of bee bread in the honey bee colony is often poorly understood by beekeepers and this needs to be remedied as we attempt to adapt our beekeeping practices to meet the requirements of our honey bees in the coming years as we all adapt to the changes which climate change will bring including the types and availability of forage.

17.3.5.2   Preparation of bee bread

After deposition of the pollen pellets by the foraging bees into cells around the edge of the brood nest the pollen this is further manipulated by hammering it into a paste-like consistency by the foraging bees by including mixing the pollen with regurgitated honey and saliva, both of which contain enzymes which will make the sugars in the pollen more accessible to the bees and to the microbes which live in the honey bee gut and digestive system.

This mixture is then inspected by nurse bees and if acceptable it is then processed further.

Honey bees prefer to eat fresh pollen which has only been stored in the comb for less than 3 days, but this seems subject to debate as others say the pollen can continue to be eaten until the nurse bees start the process of fermentation (a form of lactic fermentation). Once the fermentation process is complete the packed pollen, now bee bread, is usually covered by a thin layer of honey which will preserve the bee bread for many months and isolate it from any pathogens. If the pollen is not consumed fresh nor the fermentation process begun in the first few days after collection then other worker bees called food handlers will convert the accumulated pollen stores by a form of lactic fermentation and packed in the comb under a thin layer of honey into a form of bee bread which is preferred less compared to that produced by the nurse bees from fresh pollen, even though it is nutritionally similar.

In the colony the bees mix pollen with regurgitated nectar, honey and glandular secretions to produce bee bread. This differs from freshly collected pollen in having a lower pH and less starch and is more nutritive. This is due to lactic acid bacteria belonging to the genera *Lactobacillus* and B*ifidobacterium* being involved in the fermentation process of bee bread and responsible for improving the nutritive value by producing vitamins. There are many (29 on average) kinds of bacteria present in the bee bread and bee gut. These are described in more detail in Lipinski (2018).

Young workers acting as nurse bees convert the bee bread into brood food which is used to rear the larvae (queen, drone and worker) of the colony and to feed other adult worker bees through trophallactic (transfer of regurgitated liquid) interactions.

The fermentation process through the actions of bee-associated fungi and bacteria such as *Lactobacillus* spp., in particular *Lactobacillus kunkeei, Pseudomonas* and other microbial organisms (such as yeasts *(Saccharomyces)* and moulds) makes the bee bread nutritional content more easily digestible and can be readily assimilated. It reduces the level of starch from 2% to 0%. Nearly all the proteins and vitamins needed by the honey bees in a colony are derived from pollen stored as bee bread inside the hive when there are periods of time when pollen is not otherwise available in the outside environment.

Colonies without pollen supply can only maintain brood rearing for a short time as they successively use up their stored bee bread and later by depleting their body reserves.

The colonies respond to changes in the ratio of pollen supply and the protein demand of the brood. They progressively cannibalize the younger, then the older larvae, and if the pollen dearth continues no more brood can be produced.

17.3.5.3   Composition of bee bread

The following table (adapted from Lipinski 2018) compares the general compositions of pollen and bee bread.

Component	Pollen	Bee bread
Proteins	24.06%	20.3-21.7%
Fats	3.33%	0.67-1.58%
Carbohydrates	18.5%	24.4-34.8%
Lactic acid	0.56%	3.06-3.20%
pH	6.3	4.3

Bee bread also contains vitamins, minerals and polyphenols as well as enzymes which are not present in fresh pollen.

Yeasts found in bee bread can synthesise B vitamins and the presence of lactic acid could enhance the availability of vitamins B7, B11 and B12. Consequently, bee bread may have a higher nutritional value to the colony than just pollen.

Whether the microbes found in bee bread increase its nutritive value compared to collected pollen is still being debated. See the review in Brodschneider and Crailsheim (2010).

The pollen and the additional materials and microbes mixed with it go through a fermentation process which is considered to have several component processes on the way to becoming bee bread. The pollen and additional materials placed in the cells and the cell sealed with an impervious cap under which anaerobic and then acidic conditions develop which help preserve the pollen and additions under the capping. Lactic acid is generated by microbial activities and this creates the low pH conditions (around pH 4). At this point *Lactobacillus* species of bacteria predominate and kill or inactivate any remaining bacteria from previous phases in the conversion process. This makes the bee bread sterile and resistant to further microbial decomposition and can be stored until required by the colony for brood rearing.

Molecular finger printing methods that target plant DNA can be used to detect and identify the plant species where the pollen of single plant species occurrence is found inside the hive. The situation becomes more complex when considering a typical hive where there are a number of plant species whose pollen has been collected by the honey bees during their foraging activities. Understanding this is achieved through using next-generation sequencing (NGS), commonly used to evaluate DNA, RNA and DNA- protein interactions.

The combined benefits of the presence of proteins, carbohydrates, lipids and amino acids gives a bee bread which has varying nutritional value depending on the properties and diversity of the pollens collected and processed into the bee bread, (Carroll et al. 2017).

17.3.5.4   Nutritional value of bee bread in response to local land use and forage

Donkersley et al. (2017*)* published a paper reporting the analysis of stored pollen (the bee bread) samples taken from colonies owned by hobbyist beekeepers or were colonies used in training courses at local beekeeping associations across the NW of England. The study assessed the bee bread nutritional content and the plant species whose pollen had been used to produce these stores. Protein was the most abundant nutrient by mass (67%), followed by carbohydrates (26%). Protein and lipid content of the bee bread (but not carbohydrate) contributed significantly to the ordination the floral diversity, linking dietary quality with forage composition. The analytical technique involved DNA sequencing of the IT2 region of the nuclear ribosomal DNA gene. The number of distinct plant genera identified was 89, with each bee bread sample containing between 6 and 35 pollen types. Ordination is a widely used family of methods used in ecological work to relate the composition of different stands of vegetation (ecological communities) to each other using one or more environmental axes giving a representation of the relationship (associations) of the vegetation stands / communities reflecting their similarities and differences.

Dominant genera included *Taraxacum* (dandelion) which correlated positively with the protein content of the bee bread and *Prunus* (cherry) which was negatively correlated with the amount of protein. Proportions of amino acids (e.g., histidine and valine) varied as a function of the floral species composition.

The result of the study also quantified the effects of individual plant genera on the nutrition of honey bees.

The conclusions drawn from the study were that pollens of different plants act synergistically to influence host nutrition; the pollen diversity of the bee bread being linked to its nutrient content.

Reference has already been made to the contribution made by Liolios et al. (2016) in their study on the ranking of pollen from bee plants according to their protein contribution to honey bees.

Diverse environments compensate for the loss of individual forage plants and diversity loss may, therefore, destabilise consumer communities due to restricted access to alternative sources.

Donkersley et al. (2017) contains data on the composition of the bee bread being produced by honey bees in the NW of England.

The major nutritional constituent of bee bread was protein, followed by reducing sugars and then non reducing sugars. Reducing sugars include glucose, fructose, galactose, lactose, maltose and glycogen. Non-reducing sugars include sucrose and trehalose.

Substance	Mean mg g-1	+/- SD mg g-1
Protein	629	290
Reducing sugars	130	63
Non reducing sugars	119	85
Lipids / fats	38	2
Starch	13	8
Water	290 (29%)	180

SD refers to Standard Deviation

In Bee bread, linolenic (Omega 3) acid and arachidonic acid accounts for a high proportion (48%) of the total fat content and linoleic (Omega 6) acid is lower at 34%.

Some 17 amino acids were present in 49 bee bread samples.

Amino acids present		
Aspartate	Threonine	Methionine
Glutamate	Arginine	Tryptophan
Asparagine	Alanine	Phenylalanine
Serine	GABA (gamma-aminobutyric acid)	Isoleucine
Glutamine	Tyrosine	Leucine
Histidine	Cysteine	Lysine
Glycine	Valine	Proline

We have already noted that phenylalanine is a phagostimulant and it is also interesting to note that proline stimulates and initiates feeding activities.

Sixteen plant genera accounted for more than 95% of the total number of sequence- reads from the bee bread samples obtained from the hives in the NW of England.

In order of importance

Common name	Botanical name
Clovers	*Trifolium* spp.
Impatiens	*Impatiens* spp.
Bramble	*Rubus fruticosus*
Sycamore	*Acer pseudoplatanus*
Thistle	*Cirsium* spp.
Euscaphis	*Euscaphis japonica*
Honewort	*Cryptotaenia* spp.
Soybean	*Glycine max*
Coriander	*Coriandrum*
Rose	*Rosa* spp.

Common name	Botanical name
Wild Cherry (Sweet Cherry) and Gean	*Prunus avium*
Dandelion	*Taraxacum officinale*
Camelina	*Camelina sativa*
Buttercups etc	*Ranunculus* spp.
Willows	*Salix* spp.
Andira	*Andira* spp.

Higher protein content bee breads were associated with plant communities dominated by *Acer, Trifolium, Impatiens and Coriandrum.*

It is always questionable whether you can immediately go from the particular to the general, but the findings of this report do indicate that the nutrition that the honey bees derive from their foraging is clearly linked to the environment, habitats and species assemblages both spatially and temporally.

Land use composition surrounding hives impacts the 'nutritional value' of bee breads.

Donkersley et al. (2104) found that there were temporal variations in the nutritional content of the of the bee bread produced in their study colonies. The protein content varied significantly and non-linearly through the season peaking in late July. In terms of reducing sugars (e.g., glucose, fructose and galactose) these varied non-linearly through the season declining from spring to mid-summer before increasing to a peak in August -September. Non-reducing sugars (e.g., sucrose and trehalose) increased through the season, doubling from a low in early April to a peak in late August. The lipid content varied non-linearly peaking in September. Starch levels increased through the season. The moisture content of the bee bread did vary but not temporally.

Studies of the variation within and between 'boxes' containing frames in which bee bread was being produced suggested that the mixed composition of the pollens contained in cells on the same frames suggested that this variation was due to groups of bees from different cohorts of foraging worker bees working in one hive box at any one time. Such mixture of pollens may contribute to the ability of the nurse bees to access and utilise bee bread from different pollen sources of varying nutritional content and be able to provide the nutritional resource that will be important to the overall fitness of the honey bee colony

Donkersley et al. (2014) reported that the bee bread in hives close to agriculturally improved grasslands and those near coniferous woodland contained lower bacterial diversity. The diversity was the greater in hives near habitats such as broadleaf woodland, rough grasslands and coastal landscapes. Hives located near urban landscapes also demonstrated lower diversity in the microbiome of the bee bread which may in part be linked with the planting of non-native species. Nutrition derived from bee bread and the microbiome therein directly affects the health of bees it is possible that this demonstrates an indirect link between landscape composition and bee fitness.

17.3.6   Vitellogenin

Vitellogenin is a very important substance in the life of the honey bee and the honey bee colony.

Vitellogenin is a phospholipoglycoprotein.

It is produced by females and is normally referred to as being an egg-yolk protein precursor and a storage protein. It is now recognised as having a wider set of functions in the honey bee, including drones.

It is involved in the nutrition, the innate and direct- action immune system and longevity of honey bees. Please see the section on the Honey Bee Immune System. It may also play a role in neurophysiology as it is found in the region of the neuroglia tissue in the honey bee brain.

Its levels in the honey bee body are associated with behavioural maturation.

A reduction in its levels in the bee body leads to precocious foraging.

Such reduced levels indicate physiological stress, accelerated behavioural maturation and can destabilise the demography of the colony potentially leading to its decline and death.

Vitellogenin is a specific glycoprotein and is the precursor molecule from which lipoproteins and phosphoproteins that are the main constituents of the protein of the yolk of the egg and the fat body of nurse bees and the queen.

It is found in the ovaries, hypopharyngeal glands and head fat bodies of adult worker bees as a high-density glycol-lipoprotein consisting of fat, sugar and protein.

Vitellogenin and hexamerins (another form of storage protein to support gonad development, egg production and to support foraging activity; supply amino acids during the non-feeding pupal development and different functions in the adult stage) accumulate in the haemolymph of adult workers.

This is the main storage protein in the haemolymph and the precursor for many other proteins in honey bee metabolism.

In the egg-laying queen it comprises 50-70% of the total haemolymph protein.

The control of vitellogenin is affected by other endocrine hormones such as insulin and juvenile hormones. Conversely vitellogenin acts as a negative regulator of Juvenile Hormone.

The amount of vitellogenin increases in the worker bees which are destined to overwinter as the autumn progresses.

If winter bees have suffered from varroa attack they are unable to store enough vitellogenin in their fat bodies resulting in a reduced life span and early spring losses.

Queen longevity is achieved by her synthesising vitellogenin (a yolk protein) in her abdominal fat bodies during her laying lifetime which is released into the haemolymph and taken up by the developing oenocytes. This synthesis of vitellogenin is much greater in the queen than in the adult worker honey bee.

Vitellogenin also acts as an antioxidant, scavenging free radicals from the bee's body and this prolongs the longevity of the queen and the overwintering bees as they suppress oxidative stress damage.

A deficit of vitellogenin accelerates forager worker bee ageing and reduces longevity as the worker benefits less from its immunoregulatory properties.

See the following publications for more information, (Eisenhart, D. Guirfa, M., (eds) 2011; Corona M et al. 2007; Ramsey, S.D et al. 2017).

17.4 Determining the nutritional status of a honey bee colony

It has been suggested that the determination of the total protein or vitellogenin concentration in the haemolymph from 13 day old bees and the protein concentration of fat bodies from 9-day-old bees could be a good indicator of the nutritional status of honey bees, (Basualdo, et al. 2013).

17.5 Essential fatty acids

17.5.1 Source and Function

Some 73 different fatty acids have been identified in pollen, some of which are essential fatty acids and are ingested in their food regime because they cannot synthesise them. The fatty acids in pollen are unsaturated fatty acids and their content in pollen ranges between 1 and 20%.

The three most common fatty acids in pollen include palmitic, linoleic (omega-6) and alpha-linoleic (omega-3) acids and make up 60-80% of all the fatty acids.

Two fatty acids, α- linoleic acid (highest in bee body) and linoleic acid (highest in bee brain) are essential in the bees' development and in foraging performance.

Deficiencies in the diet of omega-3 have been shown to impair honey bee learning. (Arien et al. 2015).

Honey bees require essential polyunsaturated fatty acids in their diet and polyfloral pollen diets can provide a balance of essential amino acids and fatty acid, (Doke, Frazier, and Grozinger 2015).

Linoleic and alpha linoleic acid together with decanoic, dodecanoic and myristic acids have antimicrobial properties and pollens rich in these fatty acids may also be important in hive hygiene as well as for their nutritional value, (Manning 2001).

Zarchin et al. (reported in Wright et al. 2018) found that forager bees danced more vigorously to recruit other bees to work pollen sources that complemented the essential fatty acid deficiency of the colony.

17.6 Lipids

17.6.1 Composition

Lipids are a group of substances defined more by their physical properties than by their chemical ones. They are macrobiomolecules soluble in nonpolar solvents. They include fats (solid at room temperature), oils (liquid at room temperature), waxes, terpenoids, cannabinoids and some steroids.

These are often referred to as 'fats', but lipid is composed of fatty acids, sterols (steroids) and phospholipids. Phospholipids are important in the structural integrity and function of the cellular membranes of insects.

Other functions of lipids include energy storage, insulation, cellular communication and protection.

Lipids also act as phagostimulants. Pollens with high lipid levels but low protein levels seem more attractive to foraging honey bees.

The primary source of natural lipids is from pollen and under normal conditions pollen will meet honey bee requirements.

The composition and quantitative content vary from one plant species to another, (Roulston and Cane 2000).

The total fat / lipid content taken from the corbicula of foraging honey bees has been reported to range between which have levels in the range 6-20% for Brassica spp. pollen.

17.6.2 Function

Lipid substances are oxidised in the honey bee body for use in a number of ways:

- As a source of energy especially during the brood stage and as precursors for further biosynthesis.
- In the synthesis of cell membrane phospholipids.
- As a storage mechanism for fats when not feeding e.g., during pupation.
- Overwintering.
- In maturing queens for egg-laying.
- Cuticular lipids are critical in enabling terrestrial insects such as honey bees to resist desiccation.

Lipids are mainly metabolised during the brood stage of the honeybee development and are an important energy source. For example, the calorific values of triglycerides are more than twice those of glycogen and protein.

The dietary needs of honey bees for fats or lipids are unknown, other than the requirements for cholesterol.

Foraging bees have been observed to prefer pollens with higher lipid levels.

Pollen rich in lipids can be found in dandelion (18.9%) and mustard whilst pollen from birch or common buckthorn are much less.

17.7   Sterols

Phytosterols, also known as plant sterols, and stanols are naturally occurring compounds in plants and are similar in structure to cholesterol found in humans.

Honey bees cannot synthesise sterols and have to obtain them from their diet.

Sterols found in pollen are essential to honey bees. Workers convert the ingested phytosterols to 24-methylene-cholesterol (the honey bee's major sterol), into β-sitosterol and isofucosterol. One of the main sterols is cholesterol and developing larvae obtain this and the other sterols by the worker honey bees selectively transferring the sterols to the larvae through brood food secreted from the hypopharyngeal and mandibular glands and / or the crop of the worker bees.

24-methylene-cholesterol (also known as campestrol) is found in the fats on leaf surfaces.

The worker bees covert the campestrol into makisterone A, whilst cholesterol is converted into 20-hydroxyecdysone both of which are ecdysteroids. These hormones trigger moulting, regulate growth, metabolism and fertility.

Nurse bees have high abdominal lipid stores which are depleted prior to the start of foraging.

17.8   Vitamins

Vitamins are necessary for the growth and development of honey bees and in particular in enzyme activities.

17.8.1   Sources

Honey bees do not have the ability to synthesise most vitamins (Vitamin C, a derivative of glucose, is an exception) and most are obtained from plants (mainly from pollen but also nectar) and from vitamins produced by intestinal bacteria and yeasts in the bee's gut microbiome. The role of such microbial gut flora is now the subject of increasing research interest.

Vitamins are essential co-factors involved in all physiological processes including the synthesis of many different chemical compounds including sugars, proteins, amino acids, fats, fatty acids, phospholipids, neurotransmitters, and nucleic acid.

Water soluble vitamins (cf. fat soluble) such as the B Vitamins are common in pollen. They can vary, for example vitamin C varies throughout the season depending on the different floral sources. The symbiotic microorganisms found in the honey bee gut are assumed able to synthesize vitamin C to supplement that being collected in the pollen.

A mixture of fat-soluble vitamins A, D, E and K in an experimental diet substantially improved the amount of brood produced, however these vitamins are not regarded as essential.

The importance of vitamins is linked to brood development rather than the longevity of the adult honey bee and are especially needed when bees begin producing jelly for the young larvae and the queen.

As with all other insects the Vitamin B complex are essential, and pollen is an excellent source of these vitamins.

17.8.2 Functions

The following table illustrates the important functions associated with each vitamin.

Vitamin name	Chemical name	Mg in 100 g pollen	Important sources	Important functions
Water-soluble vitamins				
B complex	See below			Vitamin B complex crucial for hypopharyngeal gland development
B1	Thiamine	0.6 – 1.3		Co-factor for the enzymes which release energy from sugars and amino acids
B2	Riboflavin	0.6 – 2.6		Brood rearing
B3	Nicotinic acid / nicotinamide / niacin	4 – 14.4		Co-factor for the enzymes which release energy from fats
B5	Pantothenic acid	0.5 – 2.0		
B6	Pyridoxine	0.2 – 0.7		Essential for the larval development
B7 (H)	Biotin	0.05-0.07		
B8	Inositol			Not a vitamin but a type of sugar
B9	Folic acid	0.3 – 1.0		
B12	Cyanocobalamin		Honeydew	
C	Ascorbic acid	7 - 56	Willow, pear, apple and dandelion	

Vitamin name	Chemical name	Mg in 100 g pollen	Important sources	Important functions
Fat-soluble vitamins				
Provitamin A	B-carotene	1 - 20		Brood rearing, hypopharyngeal gland development
Carotenoids				
D	Calciferol	0.1 -3 µg / 10g pollen		
E	Tocopherols and tocotrienols	4 - 32		Brood rearing
K (K1 and K2)	Naphthoquinones		Plants and bacteria in bee bread after lactic acid fermentation	Aids protein biosynthesis, brood rearing, hypopharyngeal gland development

The vitamin content of pollen varies and is influenced by:

- Source.
- Season.
- Changing weather.
- Pollen stores in the hive.
- Intestinal bacteria and microflora.
- Any supplemental feeding.

17.9   Inorganic elements

These are mainly obtained from pollen collected by honey bees, but other sources of minerals such as nectar, water or even body reserves in adults are also important. They are discussed further in the section headed Dietary Elements (Minerals) and Various Other Substances.

17.9.1   Other sources of inorganic elements

17.9.1.1   Micro-organisms

Honey bees are also able to form loads from spores of fungi and fungus-like eukaryotic micro-organisms that contain protein and more fat than starch. Bees can collect spores from several genera of rusts including *Uromyces, Puccinia, Candida, Caeoma*, and *Melampsora*, which are plant pathogens. Some rust fungi produce sugar droplets amongst the maturing spores, and these are collected by honey bees.

17.9.1.2   Fungal spores

Honey bee foragers have also been observed collecting spores of the orange rust (*Gymnoconia nitens*), whose host plant is blackberry, as well as other fungal spores and to make pellets from them placing them in the corbicula and transporting them back to the hive, (Lipinski 2018).

17.10 Water

17.10.1 Requirements of a colony

Water is crucial for a honey bee colony, yet barely gets a mention when new beekeepers are being started on their beekeeping.

Water is required for survival and may be collected at high cost to the colony in adverse weather and exposed location, (Nicholson 2009).

Seeley (1995) has estimated that an average single wild colony in a temperate region annually requires 25 litres of water, in addition to the 120 kg of nectar and 20 kg of pollen.

17.10.2 Functions of water in a colony

Lipinski (2018) suggests honey bees forage for water for the following reasons:

- To satisfy thirst, especially young nursing bees which produce large quantities of royal jelly, which in turn is fed to all larvae, the queen and nest mates.
- To dilute honey on which the older larvae are fed. Utilisation of stored honey requires it to be mixed to a 50:50 w/w basis with water so that it can be metabolised.
- To dilute and convert crystalline honey back into liquid honey rich in glucose in periods when foraging is difficult, mostly during overwintering.
- To 'irrigate' and cool the nest to maintain the temperature and humidity of the hive through evaporative cooling when the ambient hive temperature exceeds 35^{0}C.

17.10.3 Quality and Availability

Many species of bees do not collect water separately but *Apis* spp. are an exception.

When the ambient temperatures are not excessively high and nectar is available, the colony will obtain most of its water requirements from the collected fresh nectar. If however, there is no forage (nectar) and the bees have to rely on their stored honey or the weather is dry and hot then the ratio of field bees in a colony significantly changes to increase water collection.

If there are no natural water sources near the apiary, or in the immediate likely foraging area then provision of water should be made. This could be important either in hot dry weather or in the winter to enable the colonies to utilise their stored honey. They can spend less time foraging for a water if a source is available nearby the hives. In adverse weather when the temperature is cold and windy, foraging bees spending a long time away from the colony are at an increased risk and may perish in their attempts to bring water back to the colony. Water is needed especially in the early spring and summer when the colony is rearing a lot of brood.

It is not unusual to see bees collecting water from the surface of moist soil, sand, stones and bricks and they seem to prefer to collect water covered by green slime, or by sitting on the surface of leaves, sticks or rotting leaves. Rainwater and pond water are preferred to pure water.

Honey bees seem to use mineralised water to supplement the micronutrients obtained from the floral sources they forage on and may be limited by the availability of calcium and potassium.

Salty water <0.1% per cent salt is favoured, but it is toxic if > 0.5% salt.

When first supplying a new source of water add a pinch of table salt and this will encourage the bees to use it.

Studies also suggest that honey bees may seasonally seek specific micronutrients associated with preparations for overwintering, (Bonoan, O'Connor, Starks, 2018).

In the colony there is a large turnover of water through:

- The collection of nectar.
- The production of metabolic water in bees' bodies which evaporates and then condenses in the hive or vents to outside the hive.
- Exhaled air.
- Water excreted in faeces.

17.10.4   Provision of water

Hot dry weather or long periods with little or no rainfall reduces the variety of sources of water which the bees can access. This may mean bees seeking water from places which are close to people, children and animals. These can include ornamental ponds and pools, bird baths, swimming pools, compost heaps, wet washing on the line, dripping and leaking taps and animal water troughs. The need for the provision of a water source and where it is best located should be borne in mind.

Bees will readily drown and so preferentially take up water from moist surfaces rather than directly from the body of wate, so providing a surface for them to walk and stand on to access the water helps them.

Once established, a watering place is accepted by the bees they will come to rely on it. The water should be clean and replaced if it becomes contaminated e.g., with bird or other animal faecal material, or leaf detritus.

17.11 Dietary elements (minerals) and various other substances

17.11.1   Dietary Elements

These are also referred to as dietary minerals or mineral nutrients.

These are mainly collected from pollen by honey bees, but other sources of minerals such as nectar, water or even body reserves in adults are also important.

There is little published information on the dietary element requirements of honey bees, however in looking across the requirements for other insects on which there is data indicates that substantial amounts of potassium, phosphate and magnesium are required by all insects.

Bee bread, honey, propolis and water are considered the main storage sources for these substances for the colony to access on demand, however it may be possible that there is also a reserve pool being carried in the adult worker bees.

Dietary elements required in quantity can be defined as macro, micro, and ultrabiogenic elements and are usually present in biological systems as ions.

Macrobiogenic elements include C (carbon), O (oxygen), H (hydrogen), N (nitrogen), S (sulphur), P (phosphorus), Na (sodium), K (potassium), Ca (calcium), Mg (magnesium), Cl (chlorine), and Fe (ion). In general, these are the building blocks of life.

Dietary element	Function
Ca (calcium)	Essential for cell signalling, nerve and muscle activity
Na (sodium), K (potassium), Cl (chlorine)	Components of cells which determine their electrolytic and osmotic status
P (phosphorus)	Key constituent of the energetic system of cells and in nucleic acids
Zn (zinc)	Deficiency can affect bee immunity systems
Cu (copper), I (iodine), Mo (molybdenum), Mn (manganese), Zn (zinc), Co (cobalt)	Essential for many life processes e.g., muscle contractions, maintenance of pH and osmotic pressure of body fluids, including ion equilibrium

Microbiogenic elements include Cu (copper), I (iodine), Mo (molybdenum), Mn (manganese), Zn(zinc), Co (cobalt). In total they constitute less than 0.1% by weight.

These elements in general have a catalytic function, e.g., as a component of enzymes.

Ultrabiogenic, or trace elements, include Al (aluminium), As (arsenic), B (boron), Br (bromine), F (fluorine), Li (lithium), Ni (nickel), Se (selenium), Si (silicon), Ti (titanium), V (vanadium) and are less than 0.001% by weight. They are usually part of enzyme systems.

Excessive levels of sodium, sodium chloride and calcium are known to be toxic to honey bees, as are heavy metals such as Cd (cadmium), Pb (lead), Ni (nickel), Hg (mercury), V (vanadium) and metalloids such as As (arsenic). Their toxicity depends on the amount taken into the bee's body.

All these dietary elements are vital for the overall structure and function of their membranes, organelles and the cytoplasm in general.

17.11.2 Dietary elements in honey and honeydew

Honeydew honey usually contains higher amounts (up to 1%, typically 0.32-0.52%) of microbiogenic elements than does a clear honey (0.05-0.35%).

Dark honeys contain more magnesium (Mg) and iron (Fe) than clear, lighter honeys and these high levels together with sugars which can be toxic and reduce the life expectancy of honey bees.

The most commonly occurring elements in honey are K (potassium), P (phosphorus), Mg (magnesium), Na (sodium), and Ca (calcium). Also found are Al (aluminium), Cu (copper), Si(silicon), Ni (nickel), Fe (iron), and Ba (barium). The trace elements include Sr (strontium), Mn (manganese), Zn (zinc), Cr (chromium), Ti (titanium), Ag (silver), Co (cobalt), Mo (molybdenum), S (sulphur), Pb (lead), Cd (cadmium), As (arsenic), Se (selenium), V (vanadium) and Li (lithium).

17.11.3 Dietary elements in pollen

These are mainly collected from pollen by honey bees but other sources of minerals such as nectar, water or even body reserves in adults are also important.

Pollen is considered to contain 2-4% ash on a dry weight basis, or 1-7% when expressed in mineral form.

The following table is from Bogdanov (2016).

Minerals	Mg in 100g
Potassium (K)	400 - 2000
Phosphorus (P)	8 - 600
Calcium (Ca)	20 - 300
Magnesium (Mg)	20 - 300
Zinc (Zn)	3- 25
Manganese (Mn)	2 - 11
Iron (Fe)	1.1 – 1.7
Copper (Cu)	0.2 – 1.6
Selenium (Se)	0.05 – 0.005
Sodium (Na)	28 – 93

These values vary depending on the plant species and the mineral levels found in the pollen loads vary throughout the year, in particular for potassium, magnesium, calcium, manganese and iron, whilst the levels of zinc and copper appear to remain more constant.

Honey bees raised on a synthetic diet containing various concentrations of pollen ash reared the greatest amount of brood at 0-5-1.0% ash levels.

Conversely when fed on a synthetic diet exceeding 2% ash there was a decline in brood rearing which almost ceased when the ash content in the diet reached 8%.

One dietary recommendation to help brood rearing is as follows:

Potassium	1000ppm
Calcium	500ppm
Magnesium	300ppm
Sodium / zinc / manganese / iron and copper	50ppm

17.11.4   Ballast Substances

Food substances collected by honey bees also contain raw fibres from the walls of pollen grains. These are not digested by enzymes secreted from the bee's gastrointestinal tract. They are instead fermented by gut-inhabiting microbes (e.g., *Gilliamella apicola*). These fibres may constitute up to 20% by weight (expressed as crude fibres) of the pollen grain and their function is to fill up the bee's intestine and to facilitate the digestion of its total content.

17.11.5   Crude fibre in pollen

Values for the crude fibre content of pollen vary considerably which is explainable by the use of different analytical methods as well as the plant species being examined. Values from pollen analyses made in France gave a range 9.2 – 14.4 g / 100 g, (Percie du Sert, 2009).

Analysis by basic and acidic hydrolysis of pollens have given the following ranges in composition:

50-80%	cellulose
10-50%	lignin
20%	hemicellulose

17.12 Beeswax

17.12.1 Composition

Beeswax (*cera alba*) is a complex mixture of about 300 long-chain molecules which include wax esters, fatty acids and hydrocarbons. It consists of about 80% melissyl palmitate, about 15% free cerotic acid and small quantities of the aromatic substance, cerolein. The complex wax glands of the worker honey bee metabolize honey but will not do so if there are poor honey stores and / or there is a lack of nectar and so comb construction and foraging flights for pollen become very much reduced. Estimates of the conversion ratio of honey to wax vary considerably (more than an order of magnitude). A conversion ratio of between 5 and 10 units of honey to 1 of wax is considered appropriate. As the energetic content of beeswax is more than 40 kJ/g it is clearly a resource that the bees wish to utilise effectively. Beekeepers should also respect this in terms of our expectations of colonies to be able to draw foundation in times of forage dearth or when the colony demographics mean there are relatively few workers in the right age group, and physiological condition to synthesise the beeswax and fashion it into comb.

Beeswax is brittle at low temperatures, but at higher temperatures it becomes plastic and malleable with a melting point of 62-64^{0}C. The specific gravity of beeswax varies from 0.958-0.970. It is insoluble in water, sparingly soluble in alcohol, soluble in warm ether chloroform, fixed and volatile oils.

Wax- producing bees are about 2-3 weeks old (after emergence) and after the time the worker bees become foragers at about three weeks the wax glands degenerate.

17.12.2 Functions of beeswax in the colony

We tend to equate bees wax with the constructions of combs. However, there is another key function. It is the main component of the protection mechanism that honey bees have to prevent the loss of water through their exoskeleton is from the secretion of cuticular lipids which contain 58% of wax. Comb wax in comparison only contains 13-17% of these lipids.

Pollen is essential for wax secretion.

The requirement for large quantities of sugars in nectar or honey is not only for the source of fatty acids but also the energy to enable wax synthesis to occur.

Cowan (1908), Coggshall and Morse (1984) and Crane (1999) are interesting references for the subject of beeswax.

17.13 Plant resins and propolis

17.13.1 Composition

Resins are naturally occurring, usually dark brown coloured, sticky substances obtained by specialist collecting honey bees which have ways of working and then offloading the resin where it is needed in the hive.

The composition of resin varies according to the plant species from which it is collected, but it consists of many chemicals, the precise number is still not defined.

Resin is collected the scales which enclose and protect the leaf buds of a large variety of trees, shrubs and plants, especially *Populus balsamifera*. Other examples include sunflowers, birch, alder, beech, willows, horse chestnut, plum, and cherry. It is usually collected during very warm weather when the resins are soft enough for the bees to manipulate. Resin collection is a specialist task, and a worker will not collect resins at the same time as pollen or nectar. After collection, the resin is carried back to the colony in globules on the pollen collecting apparatus of the bee's hind leg and deposited in the places where it is required or stored for further use.

Propolis consists of naturally occurring resins that honey bees and other bees collect from living plants and use alone or mix with beeswax and saliva in the construction and adaptation of their nests. It is a resinous substance varying in colour from light tan to black. It is heavier than water and water repellent. It is sticky at hive temperatures (around 350C), becoming quite hard below 150C and brittle below 50C.

There are several descriptions of the composition of propolis and two are given below.

The composition of propolis depends upon its source and the time of year.

- 55%   balsams, resins, flavonoids, phenolic esters and acids.
- 30%   waxes.
- 10%   ethereal oils.
- 5%   pollen.
- <5%   fatty acids, essential oils, and other organic chemicals and minerals.

Propolis can be described as consisting of resins and gums. Mixed in with these substances are waxes, pollen and bee exudates (saliva).

- Resins and vegetable balsams 50-80%
- Essential oils 4-15%
- Waxes 12-40%
- Pollen 5-11%
- Tannins 4-10%
- Physical impurities 5-20%

Physical characteristic	Temperature ^{0}c
Brittle	5
Hard	15
Pliable	30
Melts	60

Propolis is an abundant source of micro-nutrients for bees and plays an important role in the health of the honey bee colony.

Terpenes are an important component of propolis and are polymers of isoprene. They are synthesised and emitted by many species of trees e.g., oaks and poplars. They are biologically active against a wide range of micro-organisms including bacteria, viruses and fungi.

Other important compounds found in propolis are polyphenols (e.g., flavonoids, phenolic acids and esters), terpenoids, steroids and amino acids as well as variable amounts of vitamins B1, B2, B6, C, E), bio-elements (Mg, Ca, I, K, Na, Cu, Zn, Mn, Fe) and numerous fatty acids and enzymes. The lining of comb cells with propolis results in the migration of some of the phenolics into nectar, honey, winter carbohydrate stores, bee jelly and royal jelly which are then ingested by larvae and adult bees acting in a nutritional and protective function.

17.13.2 Use of resins and propolis by honey bees

On returning to the colony the foraging worker bee with her corbicular loads of resin makes her way into the colony to the 'cementing' site where the resin will be used and then the unloading process begins.

The resin-foraging honey bee requires her nestmates to remove the resin from her corbiculae.

Trembling by resin foragers appears to be an unloading signal similar in the way it stimulates nest honey bees to receive and store nectar from nectar foragers.

Waiting at the cementing site are cementing worker bees (forager- aged workers with atrophied wax glands) which bite chunks of resin from the corbiculae of the resin forager and mix it with varying amounts of beeswax into a form of cement, which is termed propolis. The off-loading bee who then immediately attaches the resin to a site along the hive or cavity wall and then smooths the propolis with their mandibles into holes and gaps which the bee has located through the probing of her antennae into the space. There appears to be no chemical modification of the resins when they are applied as propolis.

The unloaded resins may also be placed in storage areas from which middle- aged workers take and use it in comb-building and comb-strengthening.

Cementing behaviour takes place mostly during the afternoon with the resin foraging bees also joining in the cementing activities when their corbicular loads of resin have been removed.

Resin / Propolis is not fed directly to bees but contains vitamins and micronutrients which the bee requires and can be acquired from close physical contact with hive parts and through trophallaxis. It is used as collected, or mixed with beeswax and saliva by the worker bees for

- Sealing cracks and restricting air movement helping to maintain the temperature of the colony.
- Repairing and strengthening the edges of the comb and cell walls.
- Reducing hive vibrations.
- Reduction of the size of the entrance as a defence measure.
- Smoothing rough surfaces.
- Polishing the inside of brood cells sealing in debris left by previous occupants.
- Covering the internal surfaces of the hives (with a resin layer of 0.3-0.5 mm), the combs and frames and biologically active constituents of resin help protect the colony from the spread of diseases, mould fungi, parasites and other infections.
- Embalming dead organisms, e.g., mice which have accessed the hive and died or were killed by the bees before they could escape and are too big for the bees to move and expel from the hive.

17.14 Venom production

Honey bee venom is produced in the venom (acid gland). Hussein *et al* (2019) and Carpena et al. (2020) are recent papers which describe the current understanding of its composition.

Its composition changes with the worker's age and its protein composition varies through the season, (Crane 1999). The venom sac is full at about 2 weeks after the bee's emergence, about the time some become 'guards' defending the hive or cavity entrance. Once they reach the age of 6 weeks, they can no longer produce venom.

Queen venom differs in composition to that of workers and in its age-related patterns of change.

Pollen is essential for venom production.

On a practical beekeeping note, rubbing the hands with leaves of Balm (Lemon Balm) (*Melissa officinalis*) is claimed to, and probably does help in preventing stings (Howes 1979).

18. Metamorphosis and aspects of the physiological & nutritional requirements of each caste thoughout the individual's life cycle

Honey bees have a rigid exoskeleton and like all insects can only grow by the periodic shedding of this exoskeleton and moulting allowing the insect to enlarge before the new exoskeleton dries hard and restricts expansion.

18.1   Metamorphosis - definition

Metamorphosis is defined as a profound change in form from one stage to the next in the life history of an organism.

Firstly, let us summarise the main things that happen from a nutritional perspective during the individual honey bee's life.

It is during the larval stages in the metamorphosis of the individual honey bee when the individual obtains most of its nutritional requirements and energy stores which will enable it to complete its growth, development and in the colony reproduction phases.

An instar is a developmental stage in insects between each moult (ecdysis) until sexual maturity is reached.

In the early larval instar, the nutrients come exclusively from the worker bees, in particular royal / worker jelly that is provided by the young nurse bees that support the larvae (worker, drone and queen) as well as the other adults in the colony.

Larvae inform their nurse bees of their nutritional needs through brood pheromones and these will promote foraging. These brood pheromones consist of ten fatty acid esters e.g., methyl linoleate that stimulates workers to increase the deposit of royal jelly, or methyl palmitate which informs nurse bees about the increase of larval weight.

Royal Jelly is formed from secretions produced in the nurse bee's hypopharyngeal and mandibular glands. In later larval instars worker jelly contains pollen and nectar as well as the secretions of the glands whereas royal jelly only contains secretions from the nurse bee glands. The royal jelly provides the larval stages with optimal nutrition for growth and development.

The larva does not defaecate during its first five days and this only happens when the midgut and the hindgut are connected at the time of the last moult into a mature larva.

During the last larval instar changes the brood cell containing the pre-pupa in its cocoon is sealed with a capping by the worker bees with no food being provided.

Complete metamorphosis takes place in the seemingly dormant stage of the pupa. The cells of most of the larval structures die and nutrients are released and utilised in the further developments of areas of cells called the imaginal discs. These grow and differentiate into the structures of the adult bees including the wings, legs and compound eyes.

During the pupal stage, no food is eaten, even very young adult worker bees (<12 hours old) do not eat. However, during this time any surplus nutrition not consumed during the larval and pupal metamorphosis is stored in the fat body of the developing bee.

Adult worker bees can survive on a diet consisting mainly of carbohydrates and little or no protein. For protein requirements, the adult bee relies on protein, lipid and other micronutrients stored in the fat bodies to enable survival and maintenance.

Earlier in the book we referred to the role of hormones and it is useful to mention them again.

If there is enough Juvenile hormone and ecdysone present in the larval tissue this promotes the changing larval stages.

If there are lower amounts of Juvenile hormone relative to ecdysone this promotes pupation.

If there is a complete absence of Juvenile hormone this results in the formation of the adult bee.

18.2   Caste determination

We have already noted that the drone is produced from an unfertilized egg. But how do queens and worker bees develop into their two different forms and functions when they both develop from fertilized eggs? This is referred to as caste differentiation. Egg size and the subsequent feeding programmes given to the larvae by the nurse bees affect larval growth rates through nutrient signalling pathways and metabolic state. But there is much more to it. Subsequent levels of effectors (e.g., juvenile hormone) then establish gene regulatory networks that fine-tune differential gene expression for caste-specific tissue and organ differentiation during the larval and pupal stages resulting in the two types of diploid honey bees, the queen and the worker. (Cervoni and Hartfelder 2019). The understanding of differentiation is further complicated by the actions of a commonly found flavonoid, *p*- coumaric acid. Mao et al. (2015) demonstrated that larvae which had *p*- coumaric acid in their diets (found in pollen and bee bread) had significantly reduced ovaries compared to those not receiving a diet containing *p*- coumaric acid. A diet lacking in *p*- coumaric acid results in the expressions of genes which cause differentiation into a queen whilst a diet with an abundance of *p*-coumaric acid plays a role in the differentiation of the larva into a worker honey bee.

Poor nutrition of the honey bee colony reduces the supply of replacement bees which leads to a reducing hive population having low vitality and increased sensitivity and vulnerability of the colony as a whole to diseases such as Nosema, the foulbroods and varroa mites together with their associated viruses, (Hrasnigg and Crailsheim 2005).

Having made some general observations about nutrition and metamorphosis we will now look in more detail at the nutritional requirements of each caste through the individual bee's life cycle.

18.3   The worker

18.3.1   Egg

The honey bee egg consists of a micropyle, chorion, delicate vitelline membrane and a cortical layer of cytoplasm. The major protein in the egg yolk is vitellogenin.

The egg weighs in the range 0.12 and 0.22 mg and its weight is, in part, determined genetically. The size of the egg varies significantly between subspecies and between queens of the same subspecies. Its weight decreases by about 30% during incubation. The eggs are laid on end on a small quantity of royal jelly.

18.3.1   Larva

Days	Developmental state	Weight mg	Length mm	Volume Cubic mm	Food source
1	egg	0.132	1.2 (Range 1.12-1.85)	0.10+/-0.12 (Range 0.06-0.15)	Yolk in the egg
2	egg	not listed			Yolk in the egg
3	egg	0.09			Yolk in the egg
4	larva	not listed			royal jelly
5	larva	3.4			royal jelly
6	larva	33.3			royal jelly/honey and pollen (bee bread)
7	larva	100.1			honey and pollen (bee bread)
8	larva	134.5			honey and pollen (bee bread)
9	larva	155.2			honey and pollen (bee bread)

The following is adapted from Lipinski (2018.)

Developmental Stage	Fresh Weight (mg/bee)	Protein (mg/bee)
Egg	0.05	
Larva	0.3-1.5	0.56-13.8
Pupa	150-117	13.8-11.2
Newly emerged	80	11.9
Nurse Bee	72	16.3
Forager bee	66	15.0

The larvae are tended and visited by the adult bees in particular the 'nurse' bees and the larvae are each visited about 10,000 occasions for feeding, inspection and finally the capping of the larva prior to pupation. Rearing one worker bee from egg laying to emergence requires 25 - 37.5 mg protein equivalent which is in the range of 125-187 mg pollen, (Hrassnigg and Crailsheim 2005).

At the pre-pupal stage the final weight of the worker larva is about 140-150 mg, some 900 times the weight of an egg.

Worker larvae require about 142mg of honey for development (Seeley 1985).

The average weight of adult worker bees emerging from their cells is around 120 mg, however nutritional and environmental factors can affect this value.

The complete food requirements of worker larvae have not been precisely determined, and no artificial diet has been found to completely replace the honey, pollen, and glandular secretions fed to the larvae.

For the period 24-36 hours of larval development worker and queen larvae are fed similar food.

During days 1 and 2 in the worker life of a larva destined to become a worker the nurse bees begin to feed worker jelly brood food and this light feeding continues for three days. Worker jelly is a mixture of mandibular and hypopharyngeal gland secretions. Mandibular gland secretions are white and made up of secretions of lipids (fats). Hypopharyngeal gland secretions are clear and made up of proteins. It is similar in composition to Royal Jelly, but Worker jelly has 12% sugars, mainly sucrose and is only half the sugar content compared to Royal Jelly. On or around days 4-5 the amount of brood food from the hypopharyngeal secretions is reduced and honey and pollen is added to the diet with the peak of pollen feeding on day 5. The sugars content of the food given increases to around 47% and the main sugar is fructose. The worker larvae receive ten times less food compared to queen larvae as well as receiving a lower calorie content diet during the first days of larval development. The direct feeding of pollen to worker larvae does not seem necessary for normal development.

The maximum total amount of protein in the worker larva is reached at the time the cell is sealed.

At the point when the larva is ready to be capped it produces a pheromone ('cap me' pheromone) which contains high levels of methyl linolenate, methyl linoleate, methyl oleate, and methyl palmitate which causes the nurse bees to respond by sealing or capping the cell in which the larvae is developing.

18.3.3   Adult ... various stages

Emerging adult worker honey bees do not have sufficient reserves of glycogen and lipids to survive without feeding, even though these substances are present in high levels in the various larval stages. Adult bees contain little glycogen compared to larvae (ranging from 0.05 to 0.47 mg per worker) and after emergence the bee tries to rebuild its glycogen reserves.

The hypopharyngeal glands of 3-9 days-old nurse bees are stimulated by digesting the protein -rich pollen and bee bread and synthesise and secrete a proteinaceous substance called jelly. This is rich in the glycoprotein, vitellogenin.

Nurse bees are well- adapted to digest pollen and because they are rich in protein and produce large quantities of jelly (royal and worker) and distribute the proteins being digested from the incoming pollen together with other nutrients during their feeding (trophallactic) activities (Crailsheim 1990).

During the first 3-10 days after emergence the young adult workers need pollen for complete post-emergence glandular development and the growth of internal structures. During this period, her hypopharyngeal and mandibular glands are used to produce brood food to be fed to developing larvae. After that, pollen is not essential unless the older workers have to restart producing brood food and feeding larvae because the colony circumstances require it.

Worker bees that are raised in the autumn have little or no brood to feed, and they live for six or more months through the winter before starting their brood food production early in spring, after which they are replaced by spring reared worker bees.

The level of body protein in honey bees varies considerably, ranging from 21-67%. Ideally beekeepers should aim to ensure their adult worker bees have a minimum body protein level of 40% because if the level is below 40% the adult worker bee lifespan is 2026 days, but if it is more than 40% the life expectancy is 45-50 days.

Worker bees which have a high body protein content can maintain a relatively good physical condition. They can rapidly rebuild their protein levels after foraging on a rich nectar source and maintain it on an average nectar source.

Adult worker bees can survive on carbohydrates alone however during brood rearing proteins, lipids, vitamins and minerals are required. Sugar is required in the production of royal jelly and beeswax.

Seven kilograms of honey are required by nurse bees to produce 1 kg of beeswax.

Worker bees raised in weaker colonies which experienced inadequate access to pollen during larval development (i.e., malnutrition) can lead to smaller and lighter workers with shorter lifespans, and impaired productivity and performance later in their lives.

Workers which were 'pollen stressed' during their development become poor foragers and poor waggle dance performers, as well as having:

- Fewer stressed workers were observed foraging.
- Those that did forage started foraging earlier in their lives.
- They foraged for fewer days.
- Were more likely to die after only a single day of foraging.
- The information they conveyed about the location of forage resource was less precise.
(Scofield and Mattila 2015).

When worker bees mature their diet changes from a high protein to a high carbohydrate one and a halving of their lipid storage in the form of vitellogenin.

Worker bees reared at the end of the season (for both brood rearing and forage collection) have higher vitellogenin levels. Without sufficient stores honey bee colonies cannot survive for any length of time without a continual availability of carbohydrates and proteins (this will be principally found in the body storage proteins).

Worker and drone bee flight muscles are not fully developed at the time of emergence of the adult and in the case of workers their flight capacity increases as they reach their foraging age. The body weight of the worker decreases by about 40% with the greatest weight losses being in the head, mid gut and rectum as by this time the main requirement of the worker is carbohydrates for meeting the energy requirements of flying.

18.4   The Drone

The honey bee colony invests and expends a lot of resources into the rearing and maintenance of a relatively small number of drones during the honey bee colony's reproductive season. The drones being primarily there to find and mate with newly emerged queen bees which themselves are seeking out collections of drones thereby increasing their chances of successful multiple mating. This prime function of the drone is sustained by the intensive feeding and care provided to the drones by the nurse bees as they themselves have no ability to collect and digest nectar nor to produce nutrient jelly, enzymes or wax.

Lipinski (2018) cites a reference to the fact that a colony consumes around 7kg honey in the production of 1000 drones.

The lack of fresh pollen being brought into the colony has a major negative effect on the rearing of drones (Somerville 2005).

18.4.1   Egg

The drone egg is the same size and weight as that of the worker (please refer to section on workers for information).

18.4.2   Larva

Drone larvae are fed greater quantities of worker jelly than worker larvae because of their larger size. Their diet also contains a wider range of proteins. The diet of the older drone larvae changes as the workers add more pollen and nectar to the drone larva's diet. The amount of pollen fed to the drone larva in days 1-3 is negligible. The composition of the jelly fed by the nurse bees changes after 108 hours of larval growth. The direct feeding of pollen to larvae does not seem necessary for normal development.

The total amount of carbohydrates consumed from oviposition to the emergence of the adult drone is approximately 98.2 mg whilst the total amount of protein consumed during the same period is in the range 65-97.5 mg.

It is estimated that the total amount of protein needed to rear one drone larva is in the range 325-487.5 mg and is of course very dependent on the nutritional value of the pollen which has been collected and used.

The pre-pupal weight of a drone larva is 346 mg, some 2,300 times the weight of an egg and in the range 1.8-2.6 times heavier than an equivalent worker larva.

The maximum total amount of protein in the larva is reached at the time the cell is sealed.

The weight of newly emerged adult drones is about 280 mg, but as in the weight of workers nutritional and environmental factors effect weight. Drones which have been infested with the Varroa mite *(Varroa destructor)* have significantly lower weights on emergence.

18.4.3   Adult

Young adult drones from one to eight days after emergence are fed with a mixture of brood-food secretions from nurse bees, and pollen and nectar. Then after about 8 days they gradually begin feeding themselves with honey taken from honey cells which will provide the energy for mating flights. Drone larvae are fed more than worker larvae with brood food that contains a greater range of proteins. The production of spermatozoa is not dependent on the amount of protein fed to young drones, but brood food and pollen feeding may have an influence on drone longevity and mating ability. Drones with a higher protein content reach sexual maturity sooner than those with a lower protein content and they also produce more sperm, (Nguyen 1999).

Drones reach sexual maturity about 12 days after emergence.

Older drones feed themselves exclusively from honey taken from the comb, and this provides the energy for mating flights.

The protein content of young drones (between 5 and 12 days old) increases with accumulations in the thorax where most of the flight muscles are located. Nurse-bees attend to the drones to ensure they receive good protein nutrition for the maturation of reproductive organs and associated mucus glands to take place in the drone.

As in workers the flight musculature of drones is not fully developed on emergence, but by the time the drones are ready to leave the hive their body weight has not changed. Drones need to get their thoracic temperature to a higher level compared to that of the worker before they are able to lift-off the ground and fly.

Drones which have not died through successful mating with a queen can survive into the autumn and in some cases overwinter. This increased longevity is thought to be due to decreased flight activity.

During autumn workers in colonies which have queens cease to feed adult drones and the drones are driven towards and out of the hive entrance. Drones are not usually present during the winter, but they may be present when the weather is mild, and the colony has ample food. Note that the presence of drones over the winter, especially the production of drone brood in the winter / early spring period can be a sign of the presence of a drone-laying queen. These drones often being raised in modified worker cells.

Hunger in the early spring can result in the number of drones being produced by a colony to be reduced. Drone larvae which survive cannibalism and emerge as adults are of low vitality and this results in the poor insemination of virgin queens. This in turn can lead to drone-laying queens, or new queens that start the season well but become drone layers because of shortage of sperm and which may subsequently be superseded by the colony.

18.5   The Queen

18.5.1   Egg

Amiri et al. (2020) have shown that queens from different genetic stocks of *Apis mellifera* produced significantly different egg sizes under similar environmental conditions, indicating some genetic variation for egg size that allows for adaptive evolutionary change. Further investigations revealed that eggs produced by queens in large colonies were consistently smaller than eggs produced in small colonies. They found that the queens dynamically adjusted egg size in relation to colony size. Similarly, queens increased egg size in response to food deprivation. They considered that these results could not be solely explained by different numbers of eggs produced in the different circumstances but instead seemed to reflect an active adjustment of resource allocation by the queen in response to colony conditions.

Larger eggs resulted in higher subsequent survival than smaller eggs. This suggested that honey bee queens might increase egg size under unfavourable conditions to enhance brood survival and to minimize costly brood care of eggs that fail to successfully develop. This would conserve energy at the colony level with implications for the nutritional state of the colony.

This extensive plasticity and genetic variation of egg size in honey bees has important implications for understanding the interaction of the individual bees and at the colony level. Such effects may be trans-generational.

18.5.2   Larva

Queen larvae are fed with plentiful amounts of royal jelly from the first larval instar until they metamorphose at the end of the fifth instar. For the first three days larvae only receive secretions from the mandibular glands of the nurse bee. The mandibular gland secretions in royal jelly contain 18 times higher levels of biopterin and 10 times more pantothenic than is found in worker jelly. Royal jelly contains 34% sugars with glucose being the main sugar.

During the fourth and fifth days the queen larvae are fed with copious amounts of royal jelly in which the ratio of mandibular: hypopharyngeal gland secretions is 1:1. The nurse bees feed the queen larvae for an average of 4.5 days before the queen cell is sealed and larval feeding ceases.

The pre-pupal weight of a queen larva is about 250 mg, some 1,700 times the weight of an egg.

Queen larvae are fed a different diet compared to worker and drone larvae. It is a more concentrated food with a different composition called royal jelly. One of the key differences is the higher amounts of sugars in the royal jelly fed to the queen larvae, which act as a phagostimulant to increase their appetites. Queen larvae are also mass-provisioned in their cells so have access to food after the cell is capped on day 9. A larva up to 3 days old can become either queen, or worker, and this different feeding regime results in a series of physiological processes that produce a queen.

18.5.3 Adult

Queen and worker honey bees are genetically identical but are fed a different diet as larvae. Specific histone protein patterns play an important role in determining which caste they develop into. Histones can switch genes on or off, a process known as epigenetics and diet can determine which switches are activated. At 96 hours of larval growth the queen-specific pattern of histones is fixed whereas it can change in workers. Time point is critical for promoting queen development from worker larvae.

19. Gross nutritional requirements of the honey bee colony during its annual cycle: Carbohydrates

19.1   Factors affecting the productivity of the colony

The productivity of a colony includes its ability to grow and reproduce and to maintain adequate stores of honey and pollen (bee bread) to overwinter to survive periods when there is little or no forage or the weather is too bad for the bees to forage successfully.

There are three main factors that influence the productivity of a colony, namely:

- The life expectancy of each bee.
- The average daily production of brood.
- The productivity of each worker.

See Brodschneider and Crailsheim (2010)

19.2   Nectar consumption by a colony

Estimates of the amount of nectar required by a colony vary and depend on the size of the colony the sugar concentrations of the nectar collected, the environmental conditions where the hive is kept, and the amount of foraging carried out by the adult worker honey bees.

Each year a colony requires around 120 kg of nectar. Approximately 70 kg is consumed during the summer months to provide food for the brood and the adults, for their own nutrition, for extra energy to keep the brood nest warm (35-36°C) and to provide energy for foraging. The balance of nectar is converted to around 20 kg of honey that is stored in the colony and consumed during the months of the winter when there is no forage or during periods of dearth in the brood-rearing season.

A colony's heat production in mid-winter is of the order of 40 watts, which is enough to keep the surface bees from falling from the cluster and dying even when the temperatures outside are −30ºC or less. Such intense heat production consumes a lot of energy and a colony will use nearly one kg of honey per week throughout the winter months, (Seeley 1995).

Other estimates have given a nectar with 50% sugar concentration annual consumption rate of a 50,000 sized colony of around 318 kg nectar per year.

The annual honey requirements for a good-sized colony have been estimated at between 60-80kg. It will probably consume more than 20 kg of honey to overwinter (Seeley 1985).

19.3   Larval nutrition

A larva is regularly inspected by nurse bees and fed as required so that the larva is always sufficiently provided with food.

The sugar content (fructose and sucrose) of brood food is 18% in the first three days of larval development, then 45% in the last two days.

It has been calculated that 59.4 mg of carbohydrates are fed to one worker larva during its development, however it is also necessary to include the carbohydrates needed by the adult bees to ensure successful brood rearing, namely comb building, foraging and thermoregulation of the brood nest.

A lack of carbohydrates limits the number of larvae reared in the spring when nectar sources are poor and overwinter stores are already depleted. This may also result from the beekeeper who has harvested honey without due consideration of the consequences and the adequate replacement of carbohydrate.

19.4   Adult workers

Adult bees in comparison to worker bee larvae have low glycogen stores (0.05-0.47 mg per worker) and do not survive long periods of time without feeding as they do not have large carbohydrate, protein or lipid reserves in their bodies.

Consequently, they are very dependent on the colony food stores when they need energy, e.g., in preparation for foraging flights, which they obtain either from the honey stores or through trophallactic contacts with other worker bees in the colony.

An adult worker bee needs about 4 mg of utilizable sugars per day for survival.

20. The gross nutritional requirements of the honey bee colony during its annual cycle: protein

20.1 Colony nutrition

Pollen is the only natural source of protein for honey bees. Estimates of how much pollen colonies collect vary from 10-26 kg / year and it has been estimated that 10 frame colonies require in the order of 13.4-17.8 kg / year, (Somerville 2001).

Other estimates have ranged from 25-55 kg per year, however estimating likely requirements is not easy because of the interaction of climatic conditions, nectar availability, brood area, pollen quality, availability and variations in genetic based propensity to collect and hoard pollen.

The amount of effort to acquire such amounts of pollen is large bearing in mind the weight of an individual pollen grain. Homitzky (2004) determined the average weight of an individual pollen grain of canola (*Brassica napus*) to be 0.0059μg ± 0.00155μg.

Demand for protein and therefore pollen is particularly during periods of high brood rearing and wax production.

20.2 Larval nutrition

One worker larva requires 25-37.5 mg protein (or 125-187.5 mg pollen). Pollen is only fed directly to larvae in small amounts. It is mainly supplied as processed brood food produced by the adult worker bees in the hypopharyngeal glands of the nurse bees which contain the enzymes able to synthesise the protein derived from the pollen into the highly nutritious larval / brood food. The rates of feeding (time and amount) are varied by the nurse bees according to the age of the developing larvae. Older larvae are visited and fed more often by the nurse bees than the younger larvae.

In cases of malnourishment colonies may terminate brood rearing rather than produce malnourished larvae as this maintains the quality of the larvae that are produced. This may not always be the case as under experimental conditions malnourished larvae have been reared into adults with impaired quality. Effects on life span, body dry weight, wing and body size, ovary development, lifetime flight performance or the timing of the onset of foraging have all been found.

When colonies collect small quantities of pollen whatever its crude protein content or when the crude protein content is below 20% the colony will respond by reducing the amount of brood being reared.

20.3   Adult nutrition

Proteins make up 66-74% of the dry matter of adult honey bee workers.

The body protein level has a seasonal variation and has been cited as being from 21 – 67% (Lipinski 2018).

The protein content of adult worker bees increases during the first few days due to protein anabolism (the metabolic pathways that construct molecules from smaller units for growth and repair) and slightly decreases in older bees.

Adult bees have age-specific protein requirements, e.g., the maturation of flight muscles in the young adult bee and if their pollen diet is inadequate this delays the time by which the adult bee reaches its maximum thorax size and weight. Worker bees require pollen and its proteins for hypopharyngeal gland and ovary development. Protein nutrition affects the development and mass of the fat body and this is an indirect indicator of the immunocompetence of the individual bee.

A worker bee consumes on average 3.4 - 4.3 mg pollen per day peaking at the age of when the bees are nurse bees.

The protein content of the haemolymph increases in the first few days after emergence and has been found to be in the range of 11.6-27.6 μg/μL whilst that of bees experimentally fed for 6 days on various pollens or protein diets was determined between 6.0 – 9.4 μg/μL.

Worker bees do have the capacity to rebuild their body protein stores and this process can take up to 12 weeks. There are several factors which influence this time- period, including:

- The state of their body reserves.
- How much brood the bees have fed.
- Protein content of the pollen, pollen sources containing < 20% crude protein do not meet the nutritional requirements of the honey bee.

This rebuilding process can be shortened if the bee body protein levels do not fall below 40% and then may be between 14-30 days, (Somerville 2005).

If colonies have been working on a crop of sunflowers and their body protein content is low it will deteriorate further and the colony is then subject to the risk of diminishing or at worst collapsing if the colony is then moved to an average or good nectar source but only has an average pollen protein content. Recovery of the colony may take 4 months before it is big enough to work another nectar flow.

20.4   Overwintering bees

In late summer the last generations of bees produced in the colony show considerable physiological differences to those short-lived bees produced in the summer. These late bees are often called overwintering bees and have a much longer life span surviving throughout the winter and into the early spring. They are heavier and their bodies have a higher protein content.

The main difference between the two types of adult workers is that the overwintering bees build up high levels of proteins in the haemolymph and these storage proteins enables the overwintering bees to survive the winter months feeding on carbohydrates alone…the principal content of overwintering stores in the colony laid down by the bees as honey.

Overwinter bees are reared during the late summer and autumn. Usually there is much less pollen available to forage for however if the summer has been good and the bees have had access to a wide range of forage and to nectar there will also be bee bread inside the hive which is available to feed these bees which are going to overwinter. Fewer larvae are required to be reared for the colony to reach the desired number of overwinter bees.

The strength of a honey bee colony is dependent upon the protein vitellogenin and other protein reserves of the nurse bees.

Overwintering bees are physiologically different. They contain more fat bodies, have higher levels of fat, sugar and protein levels in their haemolymph and lower levels of juvenile hormone compared to foragers. These stores allow them to feed the larval stages in the spring before pollen and nectar sources being sufficiently available. The colony however cannot rear all the brood using only nutrients stored in body reserves alone and it requires good stores of uncontaminated and nutritionally complete bee bread or early tree and other pollens such as crocus.

Foraging workers have low levels of vitellogenin and are fed just enough by the nurse bees to enable them to continue to forage for nectar and pollen.

Wintering bees reserves of vitellogenin and other proteins in their haemolymph decline over the period from late autumn to the end of the winter. They have enough fat / protein stores to survive several months on carbohydrates alone.

The fat body controls social foraging behaviour. The fat bodies of foragers in spring and the summer are less well developed than the young hive bees whilst in the overwintering worker bee the enlarged fat body is more complex. In the enlarged hypopharyngeal glands and the haemolymph of the overwintering bees there is much more protein, (mainly vitellogenin), arylphorin, and other jelly proteins.

21. The honey bee gut (micro) biome – an introduction

Microbial organisms associated with bees are collectively referred to as the 'bee microbiome'. These microbial species have varying types of interrelationships with the honey bee. They may be:

- Symbionts (the relationship between two different species of organisms that are interdependent, with each gaining benefits from the other species.
- Mutualistic (as per symbiosis).
- Commensal (either of two different species living in close association but not interdependent).
- Parasitic (an organism that lives on or in another organism and obtains nourishment from its host without benefitting the host or killing it).

The range of microorganisms associated with the honey bee *(Apis mellifera)* include viruses, bacteria, fungi and protists. Some of them are significant pathogens of bees, however for many the precise nature and significance of their effects on honey bee health and colony fitness remain unclear and the host bees seem asymptomatic.

This section of the book will look at one aspect of the microbiome that of the honey bee gut microbiome with respect to nutrition and honey bee health.

It is the subject of much research and many research publications and so we can only introduce the reader to the topic and emphasise its growing importance and significance to beekeeping as more knowledge and understanding is acquired. (Engel et al. 2016).

The bacteria present have the capacity to breakdown plant macromolecules found in the cell walls of pollen grains probably providing a source of extra energy in addition to that obtained from the sugars contained in nectar. They are also capable of utilising sugars such as mannose, arabinose and xylose which have been shown to be toxic to honey bees, as well as sucrose, glucose and fructose.

Studies have shown that whilst there are only about eight species of bacteria found in the honey bee gut microbiome each species occurs in the form of many strains with a lot of genetic variation together with activities involving many different metabolic pathways.

They include three major phyla, or groups, of bacteria (*Firmicutes, Proteobacteria and Actinobacteria and Bifidobacterium*). Within these groups there are several honey bee specific families and genera. These genus and species designations are grouped into clades which have a common ancestor and a lineal line of descendants.

In the adult worker few bacteria are found in the front and midgut but this changes closer to the pylorus, which forms the connection between the midgut and the hindgut. The microbial colonisation of the gut increases in the ileum with the rectum having the highest level of microbial colonisation.

The pupa does not acquire additional flora during the pupal stage. In newly emerged honey bee workers their guts are essentially free of microbiota.

Initially there is variability in the microbiota from one bee to another but by days three to four after emergence the basic core microbiota of the adult worker honey bee gut is established.

By day three the microbial species composition in the worker honey bee gut is about 107 species of bacteria reaching a plateau of 109 by the eight day. These then remain stable through the various changes in role in the colony as the worker honey bee ages

The new adult bees seem not to acquire their gut microbiome by oral trophallaxis (communication and food transfer). Newly emerged workers that have eaten their way out through the cell capping of their cells ingest residual gut symbionts on the surfaces of the comb and frame. Other routes may be through faeces and contact with other bees in the colony.

Workers in the same colony often have differing proportions of the various bacteria in their microbiome and this may reflect the demography (age-related) of the colony and / or seasonal changes in the proportions of the bacterial species.

In the case of the queen, knowledge about their microbial communities and their effect on health is important to understanding the queen's proper functioning and the future of the colony. Supersedure is a natural process of queen replacement actively managed and supported by the more enlightened beekeeper with the new virgin queen being raised and fed by her worker nurse-aged bee sisters and including attending her emergence and mating and subsequent maturation. Both queen larvae and adults will receive food and other materials from many different individual bees.

This contrasts the situation where the beekeeper intervenes in this natural process and replaces it with queens which have been reared and raised by worker bees which may not be related to her.

Queens have a big change in their bacterial communities as they age especially the period from larvae to adulthood. The queen's microbiota is dominated by honey-bee specific alphaproteobacterial. It does not have the core microbiome associated with adult honey bee workers. The honey-bee specific alphaproteobacteria are found in the worker hypopharyngeal glands which produces the food which is fed to the queen larvae.

During development the honey bee queen gut microbiome changes as she is exposed more and more to workers.

The gut microbiome of drones is similar to, but not the same as the workers honey bees whose microbiome contains bacteria which can ferment the components of nectar and pollen, whilst drones and the queen do not have this function in the colony.

22. Bee health, vitality and nutrition

22.1   What is meant by health and vitality?

As humans we talk about our health and vitality but when it comes to defining these terms with respect to eusocial honey bees it is a challenge.

True sociality (termed eusociality) is defined by three features.

Firstly, there is co-operative brood-care so that it is not each one (bee) caring for their own offspring.

Secondly, there is an overlapping of generations so that the group (the colony in the case of honey bees) will sustain others for a while allowing offspring to assist parents during their life.

Thirdly, there is a reproductive division of labour. Not every individual reproduces equally in the group (colony), as is the case in most insects. There are one or a few reproductive individuals (queens and drones in the case of honey bees) and the workers which are more or less sterile.

Dictionary definitions of health and vitality are usually anthropocentric.

'Health' is usually defined as a state of physical, mental and social well-being in which disease and infirmity are absent.

'Vitality' is defined as the capacity to live and develop, with some definitions including an energetic and activity aspect as well.

In this book we will use the terms health and healthy to encompass the scope and intentions of the definitions of both health and vitality at the individual honey bee and colony level.

We have already described the honey bee colony as a super-organism. It has physical, chemical and behavioural defences in place which can act either on the individual cell, the individual honey bee, the full colony and more widely in the local population of colonies.

Whilst consuming much of the nectar and pollen as soon as it is collected some is processed and kept in a form as a store for use in times of need. Nectar and honey contain carbohydrates and provide the energy for all stages in the individual honey bee life cycle and castes. Pollen and bee bread provide proteins and the nutrients required in the physiological processes involved in brood rearing, growth and immunity.

22.2 The honey bee immune system

Immunity is defined as the balanced status of having biological defences to fight infection, disease or other unwanted biological invasion whilst having adequate tolerance to avoid allergy and auto-immune disease and is a key mechanism used by organisms to protect themselves.

There are a number of mechanisms, behavioural, anatomical and physiological which the individual honey bees can deploy to stop pathogens getting inside the bees' body and mechanisms which can act at the colony level.

Honey bees have an innate immune system which is the dominant of the two immune systems it possess; the second being its adaptive immune system.

The innate immune system works because the individual bee's body is able to recognise chemicals, organisms such as pathogens as being foreign compared to those which belong to itself. The system recognises specific proteins which are present on the surface of each type of pathogen (e.g., bacterial or fungal). This is achieved through the presence of haemocytes and enzymes which move around the haemolymph and when recognizing foreign invaders (for example bacteria or fungi) they signal the enzyme phenoloxidase to begin a series of chemical responses. These either engulf or encapsulate the foreign material and kill it with a range of biologically active chemicals such as quinines, phenols and reactive oxygen species. The haemocytes which have responded to the invader may then kill themselves (apoptosis) and become coated (together with the foreign material) and encapsulated with dark melanin polymers. These can be observed as dark areas on white dead pupae. The dead pupae are removed from the colony.

At the same time that this humoral immunity mediated by macromolecules found in the extracellular fluids such as the haemolymph takes place, a humoral induced response is started (it takes 12-48 hours to reach peak levels) which results in the production and release of antimicrobial peptides. The effects of this response can last for weeks and that the peptides can be passed to other members of the colony to help them in their ability to prevent becoming infected.

Vitellogenin hitherto referred to in its protein storage capacity can also bind directly to multiple pathogens such as gram-positive and gram-negative bacteria, macrophages and fungi.

In the case of immune responses to viruses there is an intracellular response in the form of RNA interference which defends cells against foreign nucleotide (organic molecules that are the building blocks of DNA and RNA) sequences found in viruses and transposons. Transposons are small fragments of DNA which can jump from one chromosome to another.

They are important in mutations, resistance to antibiotics and disease.

The saliva of honey bees contains enzymes, such as glucose oxidase, which facilitate the generation of hydrogen peroxide. In this instance the honey bee is self-medicating with the generation of hydrogen peroxide, but the substance is also functioning at the colony level (social immunity).

The honey bee gut microbiome may be an important part of the bee immune system with some bacterial species having a defensive function including pathogen-specific interactions.

The immune function may be affected by a restriction in calories in the diet and this may result in the metabolic processes associated with innate humoral and cellular immune reactions and can also provide secondary plant metabolites that have antimicrobial properties (Erler, Denner, Bobis, Forsgren and Moritz 2014).

Honey bees are able to prime the immune systems of the next generation (transgenerational immune priming) of honey bees in the colony through binding of the immune - provoking surface molecules of the pathogen to vitellogenin which are then transported to the developing eggs in the queen's ovaries and improve the chances of the offspring surviving the pathogens present in the colony.

22.3   Environmental stressors

Nutritional stress and disease are two of the main concerns in honey bee health but increasingly the importance of environmental stressors on the health and well-being of honey bee is now being recognised.

Environmental stressors induce:

- Primary effects- at the molecular level where damage can be repaired.
- Secondary effects – through damage to structure or processes (for example in membranes, DNA, cell respiration). Cells remain viable after an intermediate level of damage.
- Tertiary effects – causing the collapse, loss of integrity of the cell structure. Cells undergo apoptosis (programmed and controlled cell death) or necrosis (a form of traumatic cell death).

Much of the challenge in understanding colony failures is the fact that the interaction of stressors on individual bees, colony dynamics and environmental factors such as season and resource availability are complex, dynamic and layered.

The role of nutrition in the honey bee immune response systems is the subject of much research and study. The immune system is key to the defence of the honey bee against being colonised by potential pathogens. Humoral immunity (immunity mediated the presence of macromolecules found in the extracellular fluids) involves the synthesis of antimicrobial peptides in response to infection by bacteria, fungi or parasites. The energy and structural components to make these peptides are found in the bee's diet so without adequate nutrition and the microflora required to digest it a bee is unable to generate an effective immune response to infection. Colonies are commonly exposed to fungicides and sometimes to antibiotics and these might destroy or damage the activities of the intestinal microflora.

For many years honey bees and other pollinators have been faced with a continual, sometimes rapidly changing, loss of abundant, diverse and nutrient-rich forage. This has probably meant that bees and consequently their colonies have had to expend more resources in increased flying activity looking for and obtaining adequate quantities and variety of forage.

Weakened brood rearing capability has led to the cannibalism of larvae. Increased swarming leads to the weakening of colonies. To some extent the colony can adapt its behaviour. The same is not the case for starvation, especially in the winter period when the bees can make little or no changes in their behaviour. The result of which can be the death of the colony.

As a consequence, beekeepers are faced with having to decide on whether to give their bees supplementary feeding, as well as be aware of the nutritionally related aspects of some honey bee diseases. Traditional beekeeping teaching suggests that the need for giving extra feed or 'topping up the stores' is par for the course, yet in reality it is very much the case of a craft being conducted in often adverse conditions and even unsuitable locations for the honey bees in our care. Often there is little understanding and appreciation by beekeepers for the reasons why they are losing colonies, or why their colonies are performing poorly.

The lack of immune-competence achieved as a result of some deficient nutrition, especially the variety and diversity of proteins and amino acids, can be mitigated to some extent however attempts to increase dietary protein intake *per se* does not necessarily improve it.

Any factor that negatively affects either larval survival or adult worker survival can make a huge difference to eventual colony size and honey production, (Oliver 2015).

The strength of a honey bee colony is dependent upon the protein vitellogenin and the protein reserves of other nurse bees.

Poor honey bee nutrition can exacerbate the negative impacts of infectious viral, bacterial and fungal diseases and conversely common honey bee parasites can adversely affect bee nutritional physiology. These can lead to feedback mechanisms which impact the health of the bee and the colony, (Dolezal and Toth 2018).

Improving the '*nutritional resilience*' of honey bees is important and can help the colonies to resist environmental stressors such as disease.

22.4   Carbohydrates

The wide variety of carbohydrates which the honey bee can acquire during their foraging include inulin. This is a fructan, one of a group of polymeric sugars which are found in nature as plant storage carbohydrates. Others include oligofructoses which are a range of fructoligosaccharides. These substances when present in the bee's gut promote the intestinal health through the selective stimulation of the growth and metabolic activities of beneficial lactic acid bacteria such as *Bifidosporum* spp. and *Lactobacillus* spp.

Starvation can arise during long non-foraging periods, the insufficient or untimely feeding of carbohydrates and /or the excess removal or robbing (by other honey bee colonies, beekeepers or other pests) of honey from the colony.

Beekeepers are encouraged to examine brood combs and to inspect them to check whether there are signs of brood diseases present but other important things should also be checked. Shortage, or lack of pollen, or the collection of low nutritional value status pollen will impact the colony and the first signs of this can be seen with the appearance of areas of scattered brood across the brood area of the brood comb.

Starvation influences the behavioural development of honey bees. Food shortage brings the early onset of foraging in young workers which affects the lifespan of the foraging population and the demography of the colony.

Nectar flows stimulate hygienic behaviour in the brood nest so that dead and dying adults and brood are removed more quickly, compared to when there is no stimulus. Dead or dying brood release E-β-ocimene (EBO), a terpene with a sweet, fragrant smell and oleic acid (Omega-9-Fatty acid). Colonies have differing tolerance thresholds for the presence of diseased and damaged brood.

Fermented honey is very detrimental to the health of the colony and will lead to the early death of the bees. Leaving fermenting stores or feeding fermented honey back to the bees is a poor practice.

Carbohydrates in the form of sugars act as humectants in the hive in that they absorb or retain moisture. By binding up the water, say in food, this can reduce microbial growth, inhibiting chemical reactions that can cause food deterioration.

Parasites which exert nutritional stress on honey bee colonies can intensify the effect of starvation.

To achieve well-fed and healthy colonies, it is essential that colonies can obtain well-balanced nutrition especially if colonies are going to be located in unfavourable environments, or are used for pollination purposes. In both situations there may be little or other forage to enable the colonies receive a varied intake of pollen and nectar thereby reducing the possible shortages of essential nutrients.

Better fed bees with high levels of body protein are stronger and live longer that other bees with lower body protein levels.

Honey bee immunity improves when fed a plant species-rich diet.

Feeding bees with multi-floral pollen inherently rich in various nutrients (e.g., amino acids) was found to strengthen some immune factors of worker bees, e.g. the production and secretion of the enzyme glucose oxidase, whose activities produces hydrogen peroxide which enables the honey bees to sterilise the colony, honey and brood food (Alaux, Ducloz, Crauser and Le Conte, 2010).

Feeding correctly formulated supplemental diets to colonies in addition to their foraged pollen and nectar can help minimise the risk of unhealthy colonies. However, colonies maintained entirely, or substantially, on artificial diets do not do as well as those which have access to even low quality (expressed in terms of crude protein and essential amino acids) natural pollen.

In general, in the British Isles it seems that there a few situations where colonies only have access to a single floral source and one which is known to be deficient in one or more essential amino acids.

Take for example the situation with dandelion *(Taraxacum officinale)* which has a low protein content ranging between 9.2 and 19.2%. The reason for this is the low content of the essential amino acids, arginine, valine, leucine and isoleucine. The nutritional content of the pollen cannot be fully utilised by the honey bee because the pollen is not fully digested. This issue is particularly important for brood rearing as the shortage of isoleucine causes serious disturbances in honey bee development. Foraging bees on the other hand seem less vulnerable. Foraging bees avidly collect dandelion pollen because it contains large amounts of fat (7.3%). During the time dandelion is in flower foraging bees will probably also be visiting other plant species and crops such as oil seed rape and such foraging behaviour will go some way to rectifying these deficiencies.

Some varieties of sunflowers *(Helianthus spp.)* can be deficient in tryptophan and if honey bees only have access to sunflowers, e.g. when they are being used to provide pollination services to pollinate a sunflower crop, the foragers will suffer decreased longevity and need alternative floral or nutritional supplements to enrich their diets and maintain health. Colonies moved back after they have provided pollination services in sunflower crops require time (some months) to recover their colony protein status and colony balance.

The keeping of colonies in areas where there is the large scale growing of maize (e.g. sweetcorn, *(Zea mays convar.saccharata var rugosa)* unless the bees have access to other forage plants will result in cannibalism of eggs and larvae because of the histidine deficiency of maize pollen. However even in the garden environment in years when there is a pollen dearth honey bees will collect maize pollen.

22.5   Nutrition and Immunity

The immune function may be affected by a restriction in the availability of resources yielding calories to serve the energy requirements of the individual bee and the colony as well as innate humoral and cellular immune reactions.

Dietary protein (pollen) provides essential amino acids that are required for the synthesis of peptides in immune biochemical pathways including the components of AMP (antimicrobial peptides) such as abaecin, apidaecin, hynoptaecin and defensin. These AMP can be found in the haemolymph of honey bees when a microbial infection takes place.

Social immunity is the collective defence of a colony against parasites and pathogens. This can be achieved by honey bees carrying out small tasks on a collective basis which then have a colony-wide impact on the pathogens and parasites. For example, workers removal of dead bees and larvae from the colony (undertaking or necrophoric behaviours) or the removal of parasitized larvae and bees (hygienic behaviours).

The collection and use of propolis (plant resins) which are used to create a water and airtight antimicrobial and antiviral envelope around the nest. Compounds in propolis such as p-coumaric acid up-regulate immunity genes. Upregulation (also referred to as modulation) is an increase in the activity (such as gene expression) which makes the cells more responsive to stimuli. For example, the flavonol quercitin is known increase the expression of gene encoding proteins responsible for detoxification enzymes.

Immunity against pathogens at the individual level consists of several lines of defence and these include physical and chemical barriers and the cellular and humoral immune responses.

If bees are parasitized by Varroa mite there is a decrease in protein metabolism, the inhibition of certain immunity genes and increased virus levels that cannot be reversed by pollen feeding.

Diets that are optimal for growth are not necessarily optimal for colony immunity. For example, colonies that are building in the spring may require nutrients geared towards growth in contrast to the time in the autumn when brood rearing is reduced, and the colonies are preparing for the winter and inevitable confinement for long periods.

22.6   The Balance of Carbohydrates and Proteins

Proteins and carbohydrates are critically important macronutrients that influence growth, performance and survival.

Honey bee nutritional requirements are dependent on the age and the behavioural role of adult bees. Bees receiving high amino acid containing diets had poor longevity. Bees fed with a lower protein to carbohydrate diet ratio survived longer with those fed with carbohydrate alone survived the longest.

22.7   Pollen and Longevity

The major factors affecting lifespan are:

- the amount of pollen consumed.
- the protein concentration in the pollen and
- especially the total amount of pollen protein consumed.

Pollens / spores which significantly reduce the life expectancy of adult worker honey bees include Uromyces sp. (a rust spore), bulrush (*Typha latifolia*), Ragweed (*Ambrosia artemisifolia*).

- Pollens which have been found to induce only a small increase in life span include Tree Groundsel (Baccharis halimifolia) and Taraxacum spp.
- Pollens which gave the greatest increase in life span included Blackberry (Rubus fruticosus), California Poplar (Populus trichocarpa) and the Heather family (Ericaceae)

Worker bees with a crude protein content of over 60% are strong and long-lived, whereas worker bees with a body content of <30% live only a short time, suffer from diseases and are poor honey producers.

22.8   Propolis and honey bee health

Simone-Finstrom and Spivak (2010) published a review article on propolis and bee health covering the natural history and significance of resin use by honey bees. Lipinski (2018) believes that propolis should be viewed more appropriately as a complex natural resource for the control of microorganisms, rather than as a source of potent antimicrobials.

Materials collected from plants (from surfaces of leaf buds and wounds) such as antimicrobial resins are manipulated by a worker bee who specialises in the collection of such resins. These are then used as collected or mixed with beeswax and saliva to produce a material collectively known as propolis. They are used to create an antimicrobial barrier around surfaces of the colony especially the entrance, the inner surfaces of the cavity, the faces of combs and in sealing holes and crevices.

Honey bee longevity has been found to be increased (6.6%) in colonies which produce and utilise more propolis in the hive, (Nicodemo et al. 2014).

Propolis contains many substances and polyphenolic compounds which have therapeutic properties for honey bees including:

- Flavonoids which have detoxicant, immunostimulatory, anti-fungal, anti-bacterial and anti-viral effects
- Caffeic acid phenylethyl ester with anti-biotic, anti-fungal, anti-viral and anti-tumour activity

These substances contribute to making the honey bee colony more resistant to poisoning and infections.

Simone, Evans and Spivak (2009) demonstrated that the honey bee nest environment affects the expression of immune-genes. Previous work had shown that a chronically high activation of the immune system at the individual level could lead to decreased productivity at the colony level. On the other hand, factors which reduce investment in immune related activities could lead to increased productivity. Simone et al. (2009) were able to demonstrate that resin -enriched colonies in the field were able to invest less energy on the immune functions for two divergent immune-related genes and that this was possibly due to decreased bacterial loads. In other words, the use of resins by honey bees may have implications for colony health and productivity.

Sugar syrups and protein supplemental foods containing propolis are considered to have a positive impact on the strength and health of the whole of the colony.

The selective breeding for the reduced tendency to forage for propolis may lead to impoverished colonies with a lack of important micronutrients (Fearnley 2001).

22.9   Cannibalism

When a colony suffers from an extended poor sugar – protein diet young hungry worker bees will eat and remove the remains of the youngest larvae thus prolonging the lives of the worker bees.

23. Bee diseases, pests, other stresssors and nutrition

23.1 Honey bee Colony Number Fluctuations

Honey bee colony losses and survivability over the past decades in the British Isles have been reported by the bee health authorities and annual surveys carried out by beekeeping associations such as the British Beekeepers Association (BBKA).

Honey bees have become a focus of wider attention on the decline and the current status of all pollinators. In spite of this recent heightened concern, it is important to recognise that honey bee colony numbers have fluctuated over the years, the causes of which are many and varied especially in the combinations in which they may manifest themselves.

Colony starvation, disease, pests, queen quality and other stressors all feature in the surveys which seek to monitor and understand the status of honey bee colonies in the British Isles. Variations occur throughout the British Isles and there is no single, simple, consistent explanation to account for the changes in colony numbers.

However, merely reporting honey bee colony losses is only one part of the story to be told about the state of honey bee health in the British Isles. Yes, there are overwinter losses, but colony numbers are replenished each season by the beekeeper performing colony splits, through queen rearing and the production of new colonies including through swarm management (or lack of).

23.2 Computer based Modelling

Honey bee research at the colony level in the field in real time scenarios is expensive and very time-consuming and requires access of the research scientists to competent beekeeping skills and experience to help them in their research. This becomes even more problematical when attempting to conduct field research and there are interactions between the various factors which are affecting a colony.

These interactions may dampen or accentuate the effects of these factors making the experimental design and interpretation of results difficult. In honey bee biology there are a number of feedback mechanisms which are a problem to interpret from empirical research. Empirical studies to test hypotheses rely on practical experience, experimentation and observations rather than from theories or belief. Thus, empirical relies solely on evidence obtained through observation or scientific data collection methods.

A complementary approach to research work in the field and the laboratory is to develop computer models which are designed to describe in modelling terms the structure and functioning of a honey bee colony. Then into these models can be introduced varying combinations of external factors which could influence and decide on the future prospects for the individual honey bee, as well as to the colony as a whole.

Such an example of modelling can be found in the development and publication of the BEEHAVE model described in the paper by Becher et al. (2014).

The model and associated manuals and guidance can be downloaded free on http://beehave-model.net

The purpose of BEEHAVE is to be able to explore how various stressors, including varroa mites, virus infections such as deformed wing virus (DWV) and Acute Paralysis Virus (APV), impaired foraging behaviour, changes in landscape structure and dynamics, and pesticides affect, in isolation and combination, the performance and possible decline and failure of single managed colonies of honey bees. Also included in the model are a number of beekeeping practices.

Becher et al. (2014) demonstrate in their paper that the predictions of colony dynamics made by the model fit within the range of experimentally derived data.

The availability of pollen and nectar both within the hive and within the landscape are key drivers in colony dynamics. The level of pollen and nectar stores in the colony affects the age at which workers forage with a consequent effect on the colony size as the mortality of foraging bees is higher than for in-hive bees.

Survival of the brood is affected by the number of in-hive bees available for nursing brood and the protein content of the jelly fed by nurse bees. The survival of foragers is mainly determined by their activity level (total foraging time).

23.3   Beekeeper responsibilities

Beekeepers are responsible for the health of their honey bees and should:

- Practice good honey bee husbandry, remembering that the beekeeper can be a significant source of stress to honey bee colonies.
- Carry out pest and disease surveillance and integrated pest management.
- Ensure colonies have sufficient stores to maintain strong colonies.
- Be aware of declines and apparent changes in colony health including looking for signs of new or exotic pests and pathogens.
- Manage colonies to minimize the population of *Varroa* mites and the spread of the deformed wing virus (DWV) as this will significantly improve colony health.

- Avoid high colony densities in apiaries, as this promotes pest and pathogen transmission and may also mean inadequate colony nutrition because of limited forage availability and quality.

23.4 The Challenges to the Colony

Honey bee colonies are able to exist because of the coordination of activities among members of the colony. Essential tasks such as brood nest thermoregulation, brood rearing and resource gathering are efficiently carried out due to the architecture and organisation of the nest and the spatial proximity amongst the individual bees.

Resistance to pathogen infection (intensity and prevalence) is dependent upon nutrition and the nutritional state of the colony and individual honey bees.

Conditions within a hive such as warm temperatures, high humidity, high carbon dioxide levels, high concentrations of resources e.g., carbohydrates, and periods of confinement in the hive give ideal conditions for pest and pathogen invasion and for their transmission throughout the colony.

Pathogen infection and susceptibility is affected by the nutrition of the host but can also contribute to the malnourishment itself. This may be through:

- Infection affecting physiological nutrition through the digestion process.
- Effects on honey bee behaviour that impact the level of nutrition in the hive.

Interactions between multiple stressors such as nutrition and pests and diseases can lead to a severe disruption and imbalances to the patterns of division of labour especially in the worker bees.

Individual honey bees have their own innate immune defence systems and they have also developed a number of behavioural defences. This is a form of 'social immunity' behaviour and involves the foraging for nectar which contains secondary plant metabolites which are known to have antimicrobial properties. These substances are highly plant species-specific and their effects on honey bee health depends on which plant species have been visited.

Honey acts as a self-medicating agent not only for the adult itself but also for the larvae and other members of the colony. Polyfloral honeys have been shown to be more effective in their antimicrobial activity compared to monofloral honeys in their activity against American Foul Brood (AFB) and European Foul Brood (EFB), (Erler et al. 2014).

Sugar has an inhibitory effect on the growth of bee pathogens but as the solution strength becomes weaker the inhibitory effect becomes more variable. Polyfloral honeys retain their inhibitory effects on AFB and EFB at lower sugar concentrations and inhibit secondary bacterial invasion pathogens associated with EFB such as *Enterococcus faecalis, Paenibacillus alvei, Brevibacillus laterosporus, Bacillus pumilis, and Achromobacter euridice.*

23.5   Important threats to honey bee colonies

Observant beekeepers will look at their colonies for signs of diseases and pests and try and identify those which are causing problems before seeking to use methods to manage, control or eradicate them or to seek further advice.

The Annual Bee Husbandry Survey findings described in the Healthy Bee Plan Review of November 2020 contains a table detailing the proportion (%age) of respondents to each year's survey reporting specific problems which beekeepers where reporting occurring in their bees.

Condition	2009	2010	2011	2012	2013	2014	2015	2016	2017	2018
Acarapisosis	1.87	0.65	0.6	1.34	0.1	0.04	0.11	0.06	0.26	0
AFB	n/a	n/a	0.7	1.92	0.1	0	0	0	0	0
Ants	n/a	n/a	8.77	6.4	9.39	8.29	8.76	10.06	8.63	8.69
Chalkbrood	82.03	48.28	33.2	43.53	34.34	32.17	31.82	27.9	29.02	28.04
Deformed Wings	n/a	26.42	29.41	35.72	23.33	20.31	34.04	29.94	28.37	33.95
EFB	n/a	n/a	2.09	6.4	1.72	1.38	1.22	1.08	0.65	0.58
Failing queen	n/a	34.34	26.72	29.19	38.28	24.45	32.04	39.88	37.91	36.96
Mice	n/a	n/a	11.67	21.64	11.82	7.72	12.2	9.46	8.24	12.17
Nosemosis	17.09	11.45	14.36	19.72	10.71	6.82	9.65	8.5	7.45	8.23
Paralysis	n/a	3.1	1.4	1.66	1.52	0.81	1.11	2.63	3.53	2.9
Rats	N/A	N/A	1.6	1.28	1.41	0.65	0.89	1.56	1.31	1.04
Sacbrood	3.33	2.88	1.6	1.92	1.92	1.42	2.38	3.53	3.2	3.07
Vandalism / theft	n/a	n/a	4.59	6.79	3.23	2.11	2.55	2.04	1.57	2.09
Varroosis	n/a	39.16	50.55	49.55	37.07	35.58	43.46	32.93	30.98	36.73
Wasps	n/a	35.63	43.07	46.61	22.63	41.27	26.72	42.99	33.33	28.85
Woodpeckers	n/a	n/a	19.44	19.85	10.3	6.66	5.76	4.55	5.49	5.68
Other	n/a	n/a	10.07	7.17	10.3	8.61	7.21	7.54	8.37	6.6

Note there is no mention of starvation or nutrition.

23.5.1 American Foul Brood (*Paenibacillus larvae larvae*) (AFB)

This is a contagious bacterium which attacks developing larvae and pupa. The larvae become infected by ingesting bacterial spores from larval food. The spores germinate on reaching the larval midgut and proliferate. The larva dies from bacteremia (presence of the bacterium in the haemolymph) and lysis (breaking down) of its organs often before pupation, but after the cell is sealed. It produces highly resistant spores and can only effectively be destroyed in beekeeping operations through the burning of infected colonies and frames / comb and the sterilization of hive parts by blow torch.

23.5.2 European Foul Brood (*Melissococcus plutonius*) (EFB)

This is also a contagious bacterium that attacks developing larvae and pupae. It usually affects unsealed brood but can be found in some older sealed larvae. It does not produce spores.

Some infected larvae can survive and develop into adults if they receive enough nourishment from nurse bees.

23.5.3 Nosema (*Nosema ceranae* and *Nosema apis*)

Infections of these microsporidian, a small, unicellular, spore-forming parasitic fungal species can adversely affect the health status of a honey bee colony and can result in the demise and death of the colony. Infection with *Nosema* spp. in the autumn leads to poor overwintering ability and subsequent colony performance in the following spring. Queens can be superseded soon after becoming infected with *Nosema*.

Nosema spp. are sensitive to temperature and their biological cycle is temperature dependent (Martin- Hernandez et al. 2009). Their research showed that *N. ceranae* infected more cells and produced more spores sooner than *N. apis* at relatively low temperatures (25^0C). Both species showed similar growth at 33^0C, but at 37^0C growth of *N. apis* had stopped, whilst *N. ceranae* only showed a few parasitized cells.

N. ceranae seems better adapted to complete its endogenous (within the host honey bee tissues) cycle at different temperatures which may also be a contributory factor in its increasing prevalence.

The dormant stage of *N. apis* is a long-lived spore which is resistant to temperature extremes and cannot be killed by freezing the contaminated comb.

Di Pasquale et al. (2016) tested the influence of pollen diet quality (different monofloral pollens) and diversity (polyfloral pollen diet) on the physiology of young nurse bees which have a distinct nutritional physiology (e.g. hypopharyngeal gland development and vitellogenin level), and on the tolerance to *Nosema ceranae* by measuring bee survival and the activity of different enzymes potentially involved in bee health and defence response (glutathione—transferase (detoxification), phenoloxidase (immunity), and alkaline phosphatase (metabolism). They found that both nurse bee physiology and tolerance to the *Nosema* were affected by pollen quality. Pollen diet diversity had no effect on the nurse bee physiology and the survival of healthy bees. In the case of bees infected with *N. ceranae* those fed with a polyfloral diet lived longer than those fed with a monofloral pollen diet, with the exception of the monofloral diet which contained the highest protein content. They also found that survival was positively correlated to alkaline phosphatase activity (metabolism) in healthy bees and to phenoloxidase activity (immunity) in infected bees.

In other words, both pollen quality and diversity influenced the ability of individual worker honey bee to respond to the parasite *N. ceranae*. The authors note these results help the understanding the influence of agriculture and land-use intensification on bee nutrition and health.

In a study carried out by Pettis et al. (2013) investigating the potential interaction between pesticide exposure and pathogens and the effects on managed honey bee colonies during crop pollination (blueberry, cranberry, cucumber, pumpkin and water melon). The pollens collected by the colonies were almost exclusively from weeds and wildflowers. They found that there was a detrimental interaction when honey bees are exposed to both pesticides and *Nosema*. There was a significant increase in *Nosema* infection following exposure to fungicides which were found in pollen.

Worker honey bees infected with *Nosema ceranae* have elevated hunger levels and changes in feeding behaviour (trophallactic). This lower energetic state caused by the parasite reduces the survival of bees in starving colonies and may play a role in the reported disappearance of foragers from *Nosema ceranae* infected colonies.

When Nosema is present in a colony the longevity of the adult bees is significantly reduced and the loss of foraging bees can exceed that of the birth rate. This is mainly apparent in spring and is known as spring dwindle. The presence of *N. ceranae* in the mid gut disrupts protein metabolism and causes energetic stress as it obtains energy from the bees' midgut cells and impairs midgut epithelial cells and midgut development during replication. This impacts on the midgut proteolytic enzyme activity.

Nosema ceranae may also affect hypopharyngeal gland structure and function and this is important because the hypopharyngeal glands produce and secrete major components of royal jelly and larval brood food as well as the synthesis of enzymes (e.g., invertase) involved in the conversion of sucrose to glucose and fructose and in nectar conversion to honey.

Infected bees demonstrate signs of hunger and are less likely to share food with other colony members via trophallaxis which will further influence colony health.

The effects of the *Nosema* can be reduced when there are good conditions for brood rearing, namely a nectar flow and good supplies of good quality pollen so that an increase in the brood area takes place and there is rapid replacement of the infected adult bees that are lost in the field.

Diets which have a higher pollen quantity increase the intensity of spores of *Nosema ceranae* in the bee gut but also enhance the survival and longevity of the bees, (Cameron et al. 2016).

Bee bread increases honey bee haemolymph protein content and assists in the better survival of honey bees infected with *Nosema ceranae* in spite of causing higher levels of the *N. ceranae* in the honey bees, (Basualdo et al. 2014).

23.5.4   Amoeba (*Malpighamoeba mellificae*)

This is a single-celled parasite which affects the Malpighian tubules (excretory organs) producing cysts and whose effect is not clearly known but in combination with *Nosema* spp is likely to damage the health of the honey bee.

It spreads less easily than *Nosema apis,* (Bailey and Ball 1991) and usually only through severe dysentery.

The development of the cysts is retarded at low temperatures and accelerates as the temperature in the colony cluster rises as is the case in the transition from winter to spring when brood rearing begins. This behaviour mimics the situation found in *Nosema* spp.

See Precautions to take in feeding sugar syrup.

23.5.5   Chalkbrood (*Ascosphaera apis*)

Chalkbrood is usually found in situations where the colony is under stress, and in particular when the ratio of nurse bees to unsealed developing brood is out of balance especially at the beginning of the active season as colonies begin to expand.

Chalkbrood development in honey bee colonies is stopped if the brood is kept at a temperature of 35^0C. If chalkbrood spores are ingested with food by larvae which have been chilled at 30^0C for a few hours, a situation which can arise if there are an inadequate number of adult worker bees to cover and warm the developing brood adequately, they can germinate in the gut of the larva and begin to grow mycelia.

Chalk brood spore germination increases in higher levels of carbon dioxide in the atmosphere inside a hive. Poor ventilation in winter can result in increased levels of CO_2 and in the overwintering colony the level is minimal at 10^0C.

Ascosphaera apis is temperature sensitive and experiments reported in Starks et al. (2000) indicated that the brood temperature was increased (on average by 0.56^0C) by the bees in response to the colony being inoculated with chalk brood mummy extracts. This appeared to have a preventative effect on the chalk brood developing.

Small colonies are at the greatest risk of becoming chilled because of their high surface area to volume ratio and cannot generate or retain sufficient heat.

If the colony has low or insufficient carbohydrate stores and is in a poor condition it will also be unable to generate heat further increasing the risk of chalk brood developing and spreading quickly through the brood. (Starks et al. 2000).

23.5.6   Acarine (*Acarapsis woodi*)

This mite infests the tracheae of the honey bee where it lays its eggs. On hatching the larvae begin to feed on the host bee's haemolymph.

The lifespan of over-wintering bees is reduced which leads to the winter bees dying off. The expanding brood in the spring therefore cannot be adequately cared for and so the colony dwindles.

23.5.7 Tropilaelaps (*Tropilaelaps claraeae* and *T. mercedeseae*)

This pest is an external parasite of the honey bee *(Apis mellifera)* and feeds on the haemolymph of the drone and worker bee pupae as well as reproducing on the honey bee brood. It produces multiple entry points in the larval skin to feed (*c.f. Varroa* which makes well-defined feeding holes).

It is unable to survive in brood-less colonies so is unlikely to survive in temperate conditions such as those found in the British Isles, although this is open to some debate, especially in the light of climate change. The effects of climate and weather pattern changes may mean that honey bee colonies in some parts of the British Isles may have low levels of brood rearing throughout the winter months and so could potentially sustain an infestation of Tropilaelaps.

Troplilaelaps is a statutory, notifiable pest under UK legislation, but at the time of writing is not yet found in the UK.

23.5.8 Small Hive Beetle (*Aethina tumida*)

The young beetle larvae and adults prefer to eat bee eggs and brood, but they will also eat honey and pollen causing much damage to the combs and food reserves and by destroying the brood nest of the colony. The queen stops laying and this, as well as the mess made of the brood nest results in the colony's ultimate death. Heavy infestations can generate enough heat inside the hive to cause the combs to collapse releasing honey and the colony may abscond.

One consequence of feeding colonies with a pollen supplement would be to give the beetle an additional food source which could further increase its reproductive capacity in the colony.

Beekeeper awareness and vigilance are crucially important to prevent this pest becoming established in the British Isles. Small Hive Beetle is a statutory notifiable pest under UK legislation, but at the time of writing is not yet found in the UK.

23.5.9 Varroa (*Varroa destructor*)

Varroa (*Varroa destructor*) is an obligate ectoparasitic mite.

Varroa parasitism can have a dramatic effect on the nutritional status of a honey bee colony which in turn impacts the development and future prospects for the honey bee colony.

Varroa is associated with the impaired development of immature bees, decreased lipid synthesis, reduced protein levels, desiccation, impaired metabolic function, inability to replace lost protein, precocious foraging, heightened winter mortality, impaired immune function, decreased longevity and reduced pesticide tolerance.

In the bee larva and early-stage pupa the tissue of the fat body is distributed throughout the haemocoel and varroa mites seem to have no preference where they fed. This changes as the fat body changes and in the case of the adult honey bee the fat body tissue is found primarily in the inner dorsal and ventral surfaces of the abdomen (termed the metasoma in Hymenoptera).

Ramsey et al. (2019) showed that the primary host tissue being consumed by the mite was the fat body and that the mite extracts the pre-digested body cells through the removal of haemolymph. A fully functioning fat body is essential for the development and general health of the individual honey bee which in turn defines the fate of the colony.

The mite will access anywhere on the capped larvae and early- stage pupae as the fat body is distributed throughout the haemocoel and insert the gnathosoma (mouthparts) and the salivary stylets so that the salivary fluid mixes with the host tissue. When the mite parasitizes the adult bee to feed it is usually found ventrally (in the front) in the abdomen under the sternite or tergite of the third metasomal segment which is the closest point to access the fat body. This preference for the third metasomal segment may be because it is the longest segment thereby giving the external parasite the opportunity to feed and the same time reduce the ability of any grooming behaviour of its host bee to be effective.

They observed that around 95% of varroa observed on the adult worker body were located below the metasoma, below the sternite, with a statistically relevant preference for the left -hand side of the honey bee. Varroa stay on the adult bees between 3-14 days, preferably nurse bees because they have the largest fat bodies and therefore are most nutritious for the varroa. The less nutritious the fat body the longer the varroa mite needs to feed to get the necessary nutrition. A poor nutritional condition of the colony has a direct negative effect on the development of the varroa and vice versa. A colony in the best development and nutritional state increases the reproductive success of the mite population.

There is a relationship between the effectiveness of social and individual immunity and the nutritional state of the colony. In the case of Varroa they reduce nutrient levels, suppress individual bee immune function and transmit viruses (DeGrandi-Hoffman and Chen 2015).

Varroa destructor has been shown to reduce the vitellogenin storing capacity of infested workers and has a severe impact on their overwintering capabilities, impair pupal development, decreased lipid synthesis, desiccation, precocious foraging, reduced longevity, heightened winter mortality, diminished immune function, inability to replace lost protein and to reduced tolerance to pesticides. Adult bees parasitized as larvae are unable to store protein obtained from the pollen they eat and the synthesis of fat is inhibited.

Recognition of the substantial damage done to the adult fat body and its consequences would explain the inability to produce antimicrobial peptides, lipophorin and wax precursors. Lipophorin and the wax precursors are key components in maintaining the water-proof seal around the bee's body and preventing the evaporation of water and the desiccation of the bee. The fat body is also involved in the process of metamorphosis, regulating metabolism and is important in thermoregulation.

It is now thought that simply the damage to the pupae caused by the varroa is sufficient to impact the capacity of the honey bee's immune system to be effective.

For many years it was considered that the damage to the honey bee through the Varroa feeding on the haemolymph of the capped brood and on young bees deprived the bees of nutrients, especially the low weight molecular proteins which had been held in the fat bodies.

Such deprived bees do not thrive and if they are destined to be the overwintering population of the colony and even though they may have access to some bee bread (thereby eating the overwinter reserves) they get progressively weaker. This can result in a very weakened or even dead colony as the winter progresses into spring.

Improved nutrition can help colony growth and the immune response to a virus but the presence of the Varroa might undermine these efforts to counteract those of varroa. Abundant resources can stimulate brood rearing and population growth throughout the spring and summer however any growth in the colony will also see an increase in the Varroa population as well.

The wounds caused to the bee by the mite's feeding activities have been seen to be colonised by bacteria some of which have been shown to be pathogenic to the bee, even though in other circumstances where varroa is not present no pathogenic activity has been reported (Budge et al. 2016).

Varroa damage to adult worker fat body tissue occurs very soon after they are parasitized, and recognition of this fact is important to beekeepers because when there are brood-less periods especially in late summer the mites have to migrate onto the bodies of the adult bees where their feeding behaviours damages essential tissue and transmits viruses.

Healthy fat body tissue is essential for the successful survival of overwintering bees and so effective control of varroa is necessary to ensure the fat body tissue of these bees is not damaged. Simple reduction in varroa loads in colonies may not be sufficient as there may still be enough mites present in the colony to continue damaging honey bees as they prepare for the winter. Vitellogenin stored in the fat body reduces oxidative stress and significantly extends the lifespan of the overwintering bee and the subsequent build up in the spring.

As well as overwintering longevity and spring build up, varroa feeding on larvae and pupae below the capped cell will also affect metamorphosis and this could affect the eventual size and health of the adult honey bee.

In the autumn when there is little or no brood and Varroa is not being actively managed there is a risk that most of the adult bees will have been parasitized during their development and have viral infections. These colonies have high overwintering mortality rates.

The original native host for *Varroa destructor* is the Asian Honey Bee *(Apis cerana)* and parasitized bees of *A. cerana* can produce antimicrobial peptides that can recognize and eliminate foreign agents. They can also produce immunity-related enzymes (e.g., phenoloxidase and lysozymes) that will promote wound healing of the feeding holes made by the mite. The varroa mite can suppress the production of some of these defence mechanisms. In contrast *Apis mellifera* does not respond to the parasitism of the varroa with anti-parasitic proteins as does *Apis cerana* probably because of the immunosuppressive effects of the mite saliva injected during the mites feeding events using its muscularized pharyngeal pump. The pump operates at a mean repetition rate of 4.5 cycles/s to ingest host fluids with each event lasting around 10 seconds and separated by about 2 minutes. The feeding behaviour events may number more than 30 each 24 hours.

There may also be biochemical differences exhibited by the different regions of the fat body tissue which the varroa senses and exploits (Li et al. 2019).

Improved nutrition can help colony growth and the immune response to a virus but the presence of the Varroa might undermine such efforts. Abundant resources can stimulate brood rearing and population growth throughout the spring and summer however any growth in the colony will also see an increase in the Varroa population as well unless it is subjected to monitoring and control.

It is therefore essential that the beekeeper remains vigilant to the varroa status of colonies and practices effective control and eradication of the varroa as a high priority of their beekeeping. Beekeepers should remember when we are checking for varroa our eyes are drawn to the presence of the mite riding on the back of the adult forager and this will usually prompt us to think varroa! When the bees are moving across the comb it is difficult to see the majority of the mites, which will be on the ventral surface (underneath) the abdomen. The use of varroa screens and monitoring the mite fall is an essential part of effective varroa management and control.

Drone culling and / or the insertion of frames of drone comb or foundation are techniques used to attract the varroa to preferentially infest the developing drone larvae which can then be removed from the colony and disposed of. The development of the drone requires the consumption of more protein and hence pollen than is required for worker bees and there is a significant risk that continual drone culling can create a shortage of protein available for the growth and development of larvae and adults of queens, workers and drones. This is especially the case when there is little pollen and poor diversity and quality available for the foraging bees to collect.

When varroa (infected with Deformed Wing Virus (DWV) containing colonies are modelled using the BEEHAVE model and no varroa management controls are used the colonies begin to suffer in the third year severely damaging the colony in the fourth year leading to its death in the late fourth or fifth year. The timing depends on the size of the colony which defines the critical colony size for survival and if this value is set at 4000 bees then most of the virally infected varroa infested colonies will die in the spring of the third year. This agrees with empirical survival data for untreated varroa infested colonies and other varroa models. The model has also been used including hypothetical treatments with acaricides and predicting the effect on the colony population and survival.

If bees are parasitized by varroa there is a decrease in protein metabolism, inhibition of certain immunity genes and increased virus levels that cannot be reversed by pollen feeding. In other words, there are limitations on the benefits of diet on the immune function in bees affected by varroa. Clearly the key message is to ensure exercising management and control over the varroa population in a colony.

BBKA Member Surveys in 1990- 2020 revealed that some 25.5% of beekeepers did not carry out a varroa treatment with an approved varroacide in the period the beginning of August to the end of September and 37.8% did not treat with an approved varroacide in the overwinter period of 1st October to 1st April. Some 7.9% reported varroa being present in their colonies with 3.3% stating that Deformed Wing Virus (DWV) was present in their colonies.

23.5.10 Greater Wax moth (*Galleria mellonella*)

Greater Wax moth and its larvae do not kill honey bees or their colonies, but are often found destroying unoccupied drawn comb in colonies that have failing queens, have been damaged by pesticides, are weak, queen-less, starving or diseased. Greater Wax Moth also have the potential to destroy or damage stored comb. Honey bees will also avoid areas where the either the larvae or the adult moths have been active and destructive.

23.5.11 Lesser Wax moth (*Achroia galleria*)

Though larvae of the Lesser Wax moth consume honey, pollen and wax they are not found in comb occupied by bees and do not damage hive components as do the larvae of the Greater Wax moth.

Cooling combs and combs containing honey to between 0-17^{0}C for several hours effectively kills all stages of both wax moth species without the need to use fumigant chemicals such as acetic acid, which in any event does not always kill the moth is it is protected within the debris of the comb destroyed by the moth.

23.5.12 Mice

House *(Mus musculus)*, Wood Mouse *(Apodemus sylvaticus)* and Field Mice *(Apodemus* spp*)* will enter hives and cause extensive damage to comb and brood. Often entrance is gained before the onset of winter and the honey bee colony is clustered or too weak to defend itself. The mice may then hibernate in hive.

We do not use mouse guards but use entrance blocks which have low entrance heights. Our reason for not regularly using mouse guards is that the sharp edges around the holes in mouse guards can easily strip off pollen loads from foraging bees which fall outside the hive or onto the floor inside the hive and are lost to the colony. Such pollen may have been collected at a very risky time of the year and many foragers may have been lost trying to secure adequate pollen to meet the protein demands of the colony.

23.5.13 Hornets

Hornets are large insects and apart from their size and colouration, the noise they make in flight is distinctive.

23.5.13.1 European Hornet (*Vespa crabro*)

The European Hornet is a long-established species in the British Isles. It will hunt many species of insects to feed their larvae and will predate on honey bees usually flying through apiaries and taking honey bees on the wing.

In general, they are not a problem to strong honey bee colonies. Hornets will however identify weak colonies and can subdue them. They will also seek nectar before their newly mated queens go into hibernation ready for the next year.

They are attracted to lights at night and will fly both in the day and at night.

23.5.13.2 Asian / Yellow- Legged Hornet (*Vespa velutina var nigrithorax*)

This is a non-native predatory invasive species originating in South East Asia, inadvertently introduced into France in 2004 and which has spread widely in Europe. The first verified UK sighting was in 2016. There continue to be isolated reports of the hornet being found in the British Isles and these have been successfully traced and destroyed. The hornet is a significant threat to honey bees and other native pollinator species. Beekeepers, bee health authorities and other responsible bodies in the British Isles continue to be on a high alert with the aim of finding and eradicating colonies before they become established and multiply in an area.

The hornet hawks outside honey bee colonies killing bees as they attempt to defend their hive and will readily subdue and destroy the honey bee colony bees, brood and food reserves.

Beekeepers should be aware of the characteristics of this hornet and its differences to the European hornet which poses less of a threat to a honey bee colony.

Possible sightings should be reported to alertnonnative@ceh.ac.uk or to the bee health authorities responsible in the devolved administrations of the UK and for the Irish Republic.

Beekeepers who find a suspected nest of the Asian hornet should not approach or attempt to destroy it and should warn bystanders or others at potential risk of exposure to keep away as the hornet is highly defensive of its nest at some distance away from it and approaching the location of the nest will provoke large numbers of the hornets to attack and sting.

Hornet traps can be inserted in hives but require careful attention in their setting up and use as they are also very attractive to bumble bees, honey bees and the European hornet. Proprietary wasp traps or traps made from jars with modified lids can also be used charged with bait solutions to attract the hornets, particularly the young queens. Beekeepers are advised to check on the latest guidance regarding which baits to use in the traps and when to put them out.

23.5.14 Wasps

Whilst for most of the year wasps can be seen as very beneficial, but in the late summer and autumn they can become a serious threat to honey bee colonies. Their need to make up for the shortage of the sweet substance secreted by the brood on which the worker wasps are fixated or the shortage of other insect larvae and caterpillars for feeding the young wasp brood means that they become attracted to honey bee colonies. Outside the hive they will take returning foraging worker bees or they will seek to gain entry into the hive to access the honey / nectar and if the honey bee colony is too weak to defend itself the wasps will remove brood from their cells. Poorly maintained hives with holes and gaps provide ready access to the colony which the honey bees may find difficult to deter.

Wasp traps containing a sweet liquid (not honey as this can attract honey bees) should be deployed around the apiary.

Great care should be taken when feeding the bees with syrup ensuring none is spilled or the feeders do not overflow and the syrup runs out from the hive as this will attract wasps and robber bees.

Of particular importance is to restrict the size of the entrance. In our beekeeping we use entrance blocks with small entrances all the year as this helps the colony defend itself against robbing bees and wasps, and they in use even during a nectar flow. The use of open mesh floors ensures good ventilation.

Wild honey bee colonies established in tree cavities have entrances with an area of 12.5 cm^2 which is quite a small area.

23.5.14.1 Yellow Jacket / Common Wasp (*Vespula vulgaris*)

The workers of this species have an anchor-shaped mark on their faces.

23.5.14.2 German Wasp (*Vespula germanica*)

The adult workers of this species have faces marked either with three dots or two dots and a line. This species is more bad-tempered (prone to sting) than *V. vulgaris*.

23.5.14.3 Yellow jackets (*Dolichovespula media, D. sylvestris, D. saxonica*)

There are three species of *Dolichovespula* in the UK with the most common being the Median Wasp *(D. media)* which arrived in England in the 1980s and has spread northwards and is becoming more common. They are bigger and have blacker abdomens than the *Vespula* spp. referred to above and are only seen in the early – mid summer.

23.6 Viruses found in honey bees

23.6.1 Viruses and their symptoms found in honey bees in the UK

Viruses can infect all developmental stages and castes of the honey bee and can often exist in colonies as asymptomatic infections.

Honey bee viruses can be transmitted horizontally, vertically or vectored by parasites such as the Varroa mite.

Transmission is the establishment of a new infection in a previously uninfected individual. This may be after acquiring virus directly from another infected individual, from a vector such as varroa, or directly from the environment e.g., from the contaminated parts of flowers.

Horizontal transmission is the transmission of infectious agents e.g., viruses, amongst individuals of the same generation. This may be achieved through contaminated food-borne transmission, and contact with faeces, venereal transmission from drones when the queen is on her mating flights or being artificially inseminated. Vector mediated transmission is also an example of horizontal transmission.

Vertical transmission is the transmission of viruses to the next generation e.g., from an infected queen to her eggs.

The intervention of humans in the life of the honey bee through honey hunting, beekeeping and pollination services has greatly expanded the range across which bee diseases have spread. As a result, outbreaks of Emerging Infectious Diseases (EIDs) have become more frequent. Viruses pose a particular risk because of their high evolutionary capacity, exponential replication rates and the ability to adapt to new hosts.

Viruses have been linked to a number of honey bee diseases and can act synergistically with other biotic (e.g. varroa) and other abiotic (pesticides) stressors.

Many of the viruses found in honey bees are also found in other insect species and it is important to put matters into the correct perspective in that whilst these viruses are found in honey bees their presence and causation of spread of the virus may be through other insect species, and the honey bee is merely one species at risk from such viruses.

Yañez et al. (2020) describe the routes of infection of bee viruses in Hymenoptera, including honey bees. Many of the viruses described as bee-associated viruses are found in other non-bee insects, for example in wasps, beetles (such as the Small Hive beetle), insects (Diptera), cockroaches, spiders (sampled from near apiaries). It is unclear whether some of these associations represent true infections (and thus host status) or were passively acquired through feeding on infected bees or contaminated materials, such as nectar in flowers. Wasps are a good example which predate honey bees and share access to flowers with honey bees. The direction of infection may be complicated. For example, in the case of DWV it can be found in honey bees, varroa and wasps.

Virus	Major strains	Transmission route					Seasonal incidence			
		Oral-faecal	Varroa	Contact	Sexual-vertical	Envir	Spring	Summer	Autumn	Winter
Acute Bee-Paralysis Virus	ABPV	+	+	-	+	+	+	+++	++	+
Black Queen Cell Virus	BQCV	+	-		+	+	+	+++	+	+
Chronic Bee Paralysis Virus	CBPV	+	-	+	?	+	++	++	+	?
Cloudy Wing Virus	CWV	?	-	unclear	?	?	+	+	+	?
Deformed Wing Virus A	+	+	-	+	+	+	++	+++	++
B	+	+	-	+	+	+	+	++	+
C	?	?	?	?	?	+	+	+++	+++
Sacbrood Virus	SBV	+	-	-	?	+	++	++	+	?
Slow Bee Paralysis Virus	SBPV	+	+	-	?	+	-	-	-	?

The following information has been abstracted from Table 1 in Beaurepaire et al. (2020) and is relevant to the honey bee in the British Isles.

Initially honey bee viruses were studied which showed symptoms which were either physical, developmental, behavioural, or demographic, however the development of high throughput sequencing (HTS) technologies has revealed a large number of viruses in honey bees which are asymptomatic. Symptoms often only appear when the honey bees have very high viral loads and in situations where the virus has been injected (as in Varroa and Deformed Wing Virus) otherwise the virus may be present, but the bee is asymptomatic. However, confinement, starvation, chemical residues, cold, humidity and other pathogens such as *Nosema* spp. may turn these viral infections into a symptomatic and overt (open and observable) infection.

The following table details the known symptoms of viral infections found in honey bee and the tissues where the viruses are found in the British Isles. The information in the following table has been abstracted from Table 2 in Beaurepaire et al. (2020).

Virus	Symptoms	Tissues in which the virus is found
Acute Bee Paralysis Virus complex (ABPV)	Trembling, inability to fly, gradual darkening and loss of hairs from the thorax and abdomen, crawling on the ground and upward on grass stems, rapid death for highly infected bees.	Nervous system, cytoplasm of fat body cells, brain and hypopharyngeal glands.
Black Queen Cell Virus (BQCV)	Yellowish queen larvae with sac-like appearance that resembles SBV and in time evolves to dark brown; infected pupae turn brown and die, dark brown to black coloured walls in queen cells, significantly shortened life span in adult bees.	Gut tissue
Chronic Bee Paralysis virus (CBPV)	Syndrome 1: trembling of the wings and bodies, bloated abdomen, inability to fly, crawling on the gras and upward on grass stems, gather in groups in the warmest areas of the nest, death in a few days. Syndrome 2: ('black robbers'); hairless (thus appearing to be smaller), darker, greasy in appearance, shiny, suffer nibbling attacks	Nervous system, alimentary tract, mandibular and hypopharyngeal glands.
Cloudy Wing Virus (CWV)	Opaque wings, shortened lifespan of adult bees	Tracheal tissues and thoracic muscles
Deformed Wing Virus (DWV)	Crumpled or aborted wings, shortened abdomens, paralysis, severely shortened adult life span for emerging worker ad drone bees, modified responsiveness to sucrose, impaired learning, impaired foraging behaviour	Whole body, including the queen ovaries, queen fat body, spermatheca, and drone seminal vesicles, tissues of wings, head, thorax, legs, haemolymph and gut.

Virus	Symptoms	Tissues in which the virus is found
Sacbrood Virus (SBV)	Pupation failure, 'sac' phenotypes: swollen larvae filled with ecdysial fluid full of viral particles, precocious foraging, reduction of adult life span and metabolic activities, impaired foraging activity	Hypopharyngeal glands of worker bees, cytoplasm of fat, muscle and tracheal-end cells of larvae
Slow Bee Paralysis Virus (SBPV)	Paralysis of the two anterior legs a day or two before death.	Paralysis of the two anterior legs a day or two before death.

Despite exposure for several weeks of combs to the sun, DWV remains infectious when contained in pollen and honey.

During the annual life cycle of the colony through the seasons the availability of pollen and nectar has an impact on the life cycle of the viruses.

Honey bee viruses can have a range of effects on their hosts. These include:

- The pathological impacts of bee viruses on their host.
- Cause molecular responses by honey bees to viral infections.
- Trigger antiviral responses.
- Affect metabolic pathways.
- Act at the transcriptional regulation level, affecting gene activity in the biomarkers of honey bee health.

Other factors which may affect the host-virus interactions include:

- The presence of phytochemicals.
- Agrochemicals in the forms of pesticides and fertilisers.
- Pesticides used by beekeepers, including bee medicines.
- The application of integrated pest management principles and techniques.

Viruses can be shared among members of a bee community that forage on common floral sources. Viruses originally identified in *Apis mellifera* have been found in other bee species as well as other insects including cockroaches, small hive beetle, ants, wasps and wax moths. This does not necessarily mean that the honey bee is the source of the viruses which infect other pollinators but may merely mean that because of the fact that the honey bee is one of the most studied insects the virus was first detected and identified in the honey bee.

Certain viruses will reproduce in varroa mites (Grozinger and Flenniken 2019).

Honey bees consuming higher protein containing diets have lower viral loads, (DeGrandi et al. 2010).

The control of varroa will suppress the viruses associated with varroa (DWV, BQCV) and if the colony collectively has a healthy immune system and a low varroa population the virus will remain suppressed until a point when controls are relaxed and the varroa population increases and the immune systems of the bees are overwhelmed and the signs of DWV become more severe.

23.6.2   Deformed Wing Virus (DWV)

In the case of DWV the transmission of the virus is through infected nurse bees offering food to uninfected healthy brood or the queen who may then lay infected eggs. Infected bees have accelerated behavioural maturation, with reduced foraging activity and reduced longevity. Worker bee larvae with light DWV levels when they become adults show reduced activity, increased mortality and never become foragers. If young worker bees forage precociously, this may result in there being insufficient adult bees left in the hive to rear and care for enough brood to replace them.

23.6.3   Chronic Bee Paralysis Virus (CBPV)

In the case of Chronic Bee Paralysis Virus (CBPV) the colonies may be asymptomatic until the point is reached when overcrowding and increased confinement because of weather or shortage of forage means more contact between bees in the colony and the potential for viral transmission.

Bad weather can lead to confinement and the crowding of bees in the hive which leads to light abrasion of the bee's cuticle providing routes for the transmission of the virus into the bee's body bypassing its defences.

23.6.4   Black Queen Cell Virus (BQCV)

Black Queen Cell virus (BQCV) transmission is through varroa and even if the queens survive, mate and lay eggs these eggs are affected and so will be the brood and the resulting adults.

Some viruses penetrate their hosts at the gut interface indicating that the viruses themselves could disrupt gut physiology, digestion and / or nutrition acquisition.

Co-infection by *Nosema apis* in combination with BQCV is more harmful to the infected bee than *Nosema apis* alone.

23.6.5   Acute Bee Paralysis Virus (ABPV)

This virus is one form of a complex of several, closely associated viruses. The virus does not seem to cause symptoms in the bee larvae but there is quick mortality in the pupae and the adult bees, (Ball 1989). If the virus is ingested at low levels, it can cause a covert infection (no obvious symptoms) and if about one billion viral particles are ingested this will cause death, but if the virus is vectored by varroa and injected directly into the bee's haemolymph it requires only 100 viral particles to cause death.

23.6.6   Sacbrood virus (SBV)

Larvae infected with the sacbrood virus fail to pupate. Young worker bees ingest the virus-laden ecdysial fluid released when infected larvae are damaged during removal from the comb. The virus then multiplies in the hypopharyngeal glands of the nurse bee and mixes with the brood food produced in the glands when this is subsequently fed to the larvae.

Ongoing infection seems to be regulated as their hypopharyngeal glands degenerate. The nurse bees stop eating pollen and then soon cease to feed and look after the developing larvae. Infected workers show reduced pollen collection activity.

23.6.7   Can viruses be managed through beekeeping practices?

The transmission of the virus from adult bee to the larvae is most likely to occur when the colony is under stress, especially in terms of the age profile of the adult workers or during periods of prolonged lack of forage and no honey or pollen stores available to the colony.

Recent studies (DeGrandi-Hoffman and Chen 2015) indicate that RNA interference (RNAi) is the major anti-viral innate immune response in insects, including honey bees. Studies also indicate that the alterations in DNA methylation patterns in response to viral infections indicate that honey bee may possess parallel epigenomic (modifications of DNA) and transcriptomic (analysis of messenger RNA) mechanisms to respond to viral infections.

There are no specific treatments for bee viruses however it can be seen that varroa is a common aggravating factor for several of the viruses which affect honey bees in the British Isles. Controlling varroa will reduce the vectoring of viruses, suppress the viruses associated with varroa and it is worth reminding ourselves about such measures. They include:

- Keeping healthy strong colonies.
- Cull colonies which are weak.
- Monitor the mite population regularly and treat for varroa infestation.
- Especially treat in good time when varroa levels are low (i.e., before varroa damaged bees and infested brood are seen).
- Practice good apiary hygiene.
- Clean equipment (especially the hive tool, and gloves) for example with washing soda between each hive examination.
- Ensure the colony has stores it can readily utilise in times of dearth including during the summer months when it might be expected there would be adequate forage.
- Carry out good beekeeping management practices.

Beekeeping husbandry practices can accelerate the demise of colonies.

- Through poor apiary lay-out with hives too close to each other.
- Too many colonies for the forage likely to be available.
- Hives positioned in a straight line, which will cause foraging workers to drift from one colony to another.
- Having colonies of varying strengths in the same apiary can cause robbing of honey and nectar and drifting unless special attention is given to minimising this.
- Poor apiary hygiene when it comes to feeding syrup or leaving old combs lying about can cause robbing by other colony's bees or increase the risk of wasp attack.
- Both drifting and robbing will promote the transfer of viruses (and other pests and diseases) from one colony to another.
- Keep up to date with authoritative advice posted on the websites of the national bee health authorities, such as BeeBase (https://secure.fera.defra.gov.uk/beebase) and other extension services.

23.7   Other stressors

23.7.1   Xenobiotics

Xenobiotics are foreign chemical substances found in an organism that are not normally present or expected to be present. They are also substances which are believed not to have existed in nature before their synthesis by man.

Honey bees are unusual amongst insects in that they do not have a large number of genes whose expression is to detoxify toxins which the honey bees ingest or are exposed to. As a result they are especially sensitive to xenobiotics such as insecticides, fungicides and phytochemicals. Honey bees have evolved a different strategy for dealing with natural and synthetic toxins such as pesticides in that they exhibit metabolic resistance which are an array of inducible natural defences. Included in these is the existence of enzymes in the fat bodies to detoxify metabolites through microsomal oxidases.

Lipinski (2018) discusses these defence mechanisms in more detail.

23.7.2   Pesticide residues in nutrients

Bees foraging in agricultural, rural and urban landscapes are exposed to multiple pressures during both their development and adult lives.

There are many different pesticides used in the British Isles for a variety of purposes some of which may be mixed on site into combinations or in a spectrum of products used over the course of a year.

Pesticides at sublethal doses can interact with complex and even unpredictable, physiological effects that may not kill bees. However, they could impact on their performance and survival when foraging. Field exposure of bees to a wide range of pesticides, including fungicides, sprayed onto crops can also increase bees' susceptibility to *Nosema cerana* infection which leads to an increase in the number of bees failing to successfully return to the hive after foraging for pollen and nectar.

Honey bees (and many other pollinators) can be potentially exposed to pesticide products whether the honey bee colonies are kept in agricultural areas or in the built environment where they can be exposed to use of pesticides (e.g. herbicides, fungicides, insecticides) by gardeners, local authority and other personnel.

Ingestion of pesticides (including fungicides / insecticides / herbicides / molluscides) or their transfer to inside the hive may be through contaminated pollen and nectar or on the external surfaces of the foraging bee. Such toxins can potentially reach lethal or sublethal levels in the nectar, in bee bread fed to larvae and via residues retained in the wax of the combs. Such sublethal effects may involve behavioural changes which impact on honey bee communication, memory, foraging and navigating abilities. Exposure to pesticides affects foraging bees, shortens worker bee longevity, decreases queen survival, increases the chances of queen supersedure and can affect colony vitality.

Exposure to fungicides can also make honey bees more sensitive to acaricides.

The ventricle (mid gut) is the location in the bee's digestive system where most digestion and absorption of nutrients takes place and is vulnerable to pesticides. The Malpighian Tubules are also vulnerable as they remove toxic compounds and metabolic products from the haemolymph. Sammataro and Yoder (2012) Chapter 15 *Cellular response in Honey Bees to Non-Pathogenic effects of pesticides* gives a good introduction into the topic.

The use of herbicides to control and eradicate undesirable plant species (usually referred to as **weeds**) deprives all pollinators of the diversity and availability of forage which could be essential for or greatly contribute to their survival. The obsession with tidiness and being weed- free is endemic in our society and the unnecessary and unjustifiable use of herbicides is practiced in all areas of society and land use and land management, including the nation's infrastructure. There appears little appetite to enforce the regulations governing the licences for the use of these materials, including members of the public.

Traces of fungicides found in pollen reduce the amount of *Aspergillus* fungi in bee bread and this results in a higher incidence of chalkbrood symptoms in colonies (Yoder et al. 2013).

All pesticides, for example insecticides in order that they can be placed on the market are required to undergo extensive testing and evaluation. This includes:

- to confirm their effectiveness on the target organisms (e.g., insect and fungal pests and diseases).
- to evaluate whether they have any undesirable effects on other non-target organisms.
- Determine that risk to humans in their intended use (safety) is acceptable.
- Determine the risk to the environment in their intended use.

The onus is on the user to justify the reasons for their need to use the products, in the event that other prevention and control measures such as those set out in an Integrated Pest Management (IPM) approach and in the case of beekeeping an Integrated Bee Health Management (IBHM) approach would be considered to be ineffective.

23.7.3   Veterinary Medicines for Honey Bees

In UK legislation products intended to be used to manage and control pests and diseases in honey bee colonies are classed as Veterinary Medicines for Bees and are the responsibility of the Veterinary Medicines Directorate (VMD).

Products, such as miticides and insecticides, used in the management and control of bee pests and diseases, such as *Varroa* and Small Hive Beetle *(Aethina tumida)* can also contaminate honey, bee bread, propolis and royal jelly.

Substances used, often referred to as active ingredients include thymol, synthetic acaricides (bromopyrolate, amitraz, coumaphos, fluvalinate, flumethrin), organic acids (lactic, formic, oxalic acids), constituents of essential oils, and antibiotics used in brood disease control (tetracyclines, streptomycin, sulphonamides, chloramphenicol).

The use by beekeepers of many of these substances or products containing them is not authorised in the British Isles. They are under statutory control. The role and responsibility of the user must be recognised by beekeepers who use such products in their beekeeping practices.

Despite these regulatory regimes, there are improvements required to be made to the testing and evaluation regime, in particular any sub-lethal effects, and these are currently being addressed around the world.

Recently the class of insecticides called neonicotinoids have been the focus of concerns on their impact on pollinator species, in particular bumble bees as well as the honey bee. These compounds have been shown to affect the honey bee's metabolism and in the case of clothianidin, a neonicotinoid, it has been shown to weaken resistance and increase viral replication, including latent infections as well as other behavioural traits such as communication, homing and foraging. Godfray et al. (2014) published a restatement of the natural science evidence base concerning neonicotinoid insecticides and insect pollinators and is well worth reading as it sets out the issues needed to be addressed when considering the potential impacts of pesticides such as insecticides on insect pollinators.

Work published by Tosi et al. (2017) demonstrated that the neonicotinoid pesticides (clothianidin and thiamethoxam) and nutritional stress synergistically reduce survival in honey bees. Greater susceptibility to insecticides has been reported when bees are infected with parasites or diseases. Pesticide residues e.g., neonicotinoids, can compromise immune pathways and result in an increase in the Deformed Wing Virus load in the bee.

23.7.4 Effects of Antibiotics on the beneficial microflora of honey bees and their Immune systems

The composition of the microflora or microbiome of the honey bee gut is determined to some extent by the nutrients which have been acquired during foraging and that require processing. These microbes could affect the immune function by providing essential nutrients, inducing host immune responses or by reducing the growth of the pathogens.

The flavonol, quercitin, is known to upregulate detoxification enzymes in adult bees and larvae. Quercitin can reduce the toxicity of tau-fluvalinate (pyrethroid insecticide) against varroa.

Broad-spectrum antibiotics may also have an adverse effect on beneficial gut microflora and these may be replaced by other microbes as superinfections and are resistant to most antibiotics. A superinfection is a second infection superimposed on an earlier one, especially by a different microbial agent(s) (either exogenous or endogenous) and resistant to the treatment being used against the first infection.

The use of antibiotics in beekeeping is strictly regulated. and antibiotics should not be used by beekeepers. Authorised Bee Inspectors can, under special circumstances, apply them, for example, oxytetracycline (OTC) in the British Isles for the treatment and control of European Foulbrood (EFB).

Other antibiotics, including oxytetracycline, are used (legally and illegally depending on national regulations around the world) and if used frequently and indiscriminately they will increase the likelihood of a drug-resistant microflora developing which will eliminate the normal microbial flora and prevent the digestion and other physiological process taking place. Prophylactic (preventive) use of antibiotics such as tetracycline to control American Foulbrood is resulting in the increase of antibiotic resistance in the infected honey bee gut microbiome in countries such as the United States of America.

Worker honey bees have been found to have the following species of bacteria *Lactobacillus* sp., *Bifidobacterium* spp., *Batonella* spp., *Simonsiella* spp. mainly in the posterior part of their digestive tract (ventriculus) and posterior intestine including the rectum.

The honey stomach, or crop, also contains a microbial flora consisting of *Lactobacillius* spp. and *Bifidobacterium* spp.

Gut microflora are symbionts (defined as a close and long-term biological interaction between two different biological organisms) in the honey bee and the bees and the colony need them for food preservation and colony level metabolic functions.

Clean pollen sources can be converted to bee bread and help re-establish a healthy gut microflora and prevent the growth of pathogenic microbial organisms.

The precise role of microbes in the production of bee bread and its conversion into brood food is still to be determined. There are many other potential effects on bee behaviour if these microbial communities are damaged by the presence of antibiotic, (Forsgren et al. 2010).

23.7.5   Transgenic products

Transgenic engineering refers to the movement or insertion of a foreign gene (transgene) into an organism that does not normally have a copy of that gene. In the case of plants these genes may come from another plant of the same or different species, or a completely unrelated kind of organism such as bacteria or animals.

23.7.6   Genetically Modified Organisms (GMO) and crops

The development and cultivation of GMO crops has implications for honey bees.

The growing of Bt crops (e.g., sweet corn and cotton) appears to have some benefits because of the reduction in the frequency of pesticide applications to Bt protected crops. Bt crops are genetically engineered to produce the toxin (specific proteins called 'cry proteins') found naturally in the soil dwelling bacterium *Bacillus thuringiensis* (Bt) and are toxic to insects, thus protecting the crop plant.

The use of GE (genetically engineered) crops with herbicide resistance and the blanket application of herbicides (e.g., glyphosate) onto the cropped area, the field margins and run off into ditches and field drains has meant the loss of floral diversity and abundance thereby denying pollinator species like honey bees of vitally important nutrition both in quantity and variety.

It is hoped that the growth of uncropped areas and other landscape and land management practices will be put forward in forthcoming legislation and not only encouraged but demanded in response to the receipt of public funding. Persuasion of the benefits is far more powerful force than dictation by legislation and the authors hope that all involved will recognise the big improvements to biodiversity and greater human enjoyment of the wider environment, locally, regionally, nationally and globally which could be achieved with the reduced and more selective use of herbicides.

23.7.7 Minerals in Honeydew

The mineral content of honey dew honey has been implicated in health issues in the honey bee colony. High K (potassium) and / or P (phosphorus) and low Na (sodium) concentrations have been associated with the paralysis of adult bees in Germany. High mineral levels were also implicated in causing dysentery in adult bees.

23.8 Specific examples of nutritional impacts and stressors

23.8.1 Pollen and honey bee health

Some constituents of pollen may have an inhibitory role on bacteria. For example, linoleic acid from pollen inhibited the two foulbrood diseases European Foulbrood (*Melissococcus pluton*) and *American Foulbrood (Paenibacillus larvae subsp. larvae)* and Chalk Brood (*Ascosphaera apis*, (Frias et al. 2016).

Other fatty acids such as capric, lauric, myristic, linoleic and linolenic acids are also known to have antimicrobial properties.

In the case of varroa, pollen has been shown to improve the longevity of bees exposed to varroa. Different solvents have been used to remove either the polar or the apolar components of pollen and such modified pollens have been fed to varroa infested bees. The results clearly showed that when the apolar components were removed from the pollen the mortality of the infested bees was significantly higher than when fed with the complete pollen. The apolar fraction contains lipids that lipids may have a role in the beneficial effects of pollen which had been observed in laboratory and small field scale experiments on honey bees. In the case of the polar component removed pollen there was a higher (but not significant) mortality compared with complete pollen (Annoscia et al. 2017).

23.8.2 Bee bread and fungicides and chalkbrood symptoms

Traces of fungicides found in pollen impact on the abundance of *Aspergillus* in bee bread and this has been shown to increase the incidence of chalkbrood *Ascosphaera apis*). Competition amongst the fungi and bacteria help keep the incidence of the bee pathogens at a low level. (Yoder et al. 2013).

Aflatoxins and ochratoxins are produced by *Aspergillus* spp. fungi (e.g., *Aspergillus flavus and A. fumigatus)* which in the right conditions will grow on the stored pollen and bee bread. Aflatoxins can kill larvae and adult bees and the resulting dead bees and larvae found in the combs of infected colonies are referred to as Stone Brood.

23.8.3 Essential amino acids and their role in survival of infection of Deformed Wing Virus (DWV)

Newly emerged adult worker honey bee under a 'no choice' feeding regime given a diet containing increasing doses of essential amino acids showed that the nurse bees had a significantly shorter survival rate and a marked higher level of DWV when fed with the diet containing the highest levels of amino acids. These results indicate that unbalanced nutrition characterised by an excess of amino acids can enhance bee susceptibility to viral pathogens and is associated with higher mortality.

23.8.4   Nectar and honey bee health

A nectar flow or the feeding of bees with sugar syrup will increase the cleaning / hygienic behaviour of the worker bees, especially the older bees and this will generally result in the reduction (but not elimination) of brood diseases.

Workers when foraging can utilize polysaccharides for flight metabolism.

However, mannose / galactose / arabinose / xylose / melibiose / raffinose / stachyose and lactose are toxic to bees. Typically, these sugars can be found in pollens, nectars of the plant families Tiliaceae and Theaceae, and the exudates of tulips. About 40% of the sugars found in soybeans (used as pollen substitutes) are toxic to bees. In typical exposure conditions this toxicity can be reduced when bees can collect nectar from a diversity of sources.

Some plants including those that are recommended as suitable for honey crops produce substances that are poisonous to honey bees. Poisonings commonly called May disease can occur from the protoanemonin present in flowers in the Ranunculaceae (Buttercup) family.

A toxic saponin is found in the nectar and pollen of Horse-Chestnut (*Aesculus* spp.), (Family Hippocastanaceae), (Barker 1990).

Nectar from sunflowers is relatively rich in raffinose which is toxic to honey bees, (Barker 1977).

23.8.5   Non-Infectious diarrhoea or dysentery

The term dysentery describes the condition where there is heavy faecal soiling of hive parts or combs by adult bees. It is not caused by an infection but by other factors including poor nutrition.

Overfeeding of bees, especially in the spring, with food which has too much water can result in the appearance of non-infectious diarrhoea (dysentery). This can also occur in bees that have wintered too long and are overloaded with faeces. The faeces absorb water and as a result the bees can spread liquid faeces the colour of milk coffee on the combs and interior hive parts. Healthy bees will produce faeces in the form of discreet brown spots.

The prevention of creating the conditions in which dysentery can develop are key bee health management practices and the following actions will help reduce the incidence of the condition.

- Allow good time for the colony to take down the sugar feed from the feeder and reduce its moisture content before capping it to a safe level at which the risk of fermentation is low. For example, completion of autumn feeding by the first week in October is desirable in most years.
- Do not feed fermented honey or sugars of uncertain origin and chemical constitution.
- Use only refined white sucrose, table sugar or commercially prepared syrups with declared compositions.
- Ensure the bees can exit and re-enter the hive for voiding flights.
- Check there is not an accumulation of dead bees blocking the hive entrance (with or without a mouse guard).
- Remove dead bees to allow access and prevent further stress caused by the presence of decaying honey bee bodies on the floor of the hive.

23.8.5   Propolis and honey bee health

Terpenes in propolis inhibit the growth of American Foulbrood (*Paenibacillus larvae v larvae*). Colonies supplemented with propolis show decreased infection of Chalk Brood (*Ascosphaera apis*). There is also evidence for colonies increasing plant resin collection for propolis in response to elevated levels of chalkbrood disease in their colonies (Simone-Finstrom and Spivak 2012).

It is possible that the selective breeding of bees with little tendency to collect propolis could lead to colonies not obtaining important micronutrients (Fearnley 2001).

24. Phenology & Pollination

24.1   Phenology – definition

Phenology is the study of seasonal changes in plants and animals from year to year, especially their timing in relation to the weather and the climate.

The practice of recording phenological events in many parts of the world has a long history.

24.1.1   Recent findings in the British Isles

In the British Isles today, there is the Nature's Calendar scheme run by the Woodland Trust in conjunction with the Centre for Ecology and Hydrology (CEH). Large quantities of observational data collected by members of the public and enthusiastic volunteers are collated and analysed. These include the time of first flowering of a number of plant species, some of which are of relevance to honey bees and beekeepers.

The following data are abstracted from Table 1 of a paper by Sparks et al. (2020) and describe the annual mean dates of first flowering of several species covering two time periods 1891-1947 and 1999-2018. In the third column the means of the records of these two periods are compared using a t-test, the *t-statistic* and *p* value. Statistically significant results are shown in bold.

Species	1891-1947	1999-2018	T-and p values
Coltsfoot (*Tussilago farfara*)	Mar 4	Mar 20	-5.52 <0.001
Snowdrop (*Galanthus nivalis*)	Feb 2	Jan 22	1.22 0.247
Lesser Celandine (*Ficaria verna*)	Mar 8	Mar 1	1.41 0.178
Cuckooflower (*Cardamine pratense*)	Apr 15	Apr 6	2.01 0.085
Blackthorn (*Prunus spinosa*)	Apr 3	Mar 17	5.83 <0,001
Hawthorn (*Crateagus monogyna*)	May 9	Apr 26	6.00 <0.001
Horse Chestnut (*Aesculus hippocastanum*)	May 5	Apr 27	3.96 0.001
Ivy (*Hedera helix*)	Sep 27	Sep 20	4.84 <0.001

More recent information can be found posted on the Woodland Trust website (https://www.woodlandtrust.org.uk) which compared the first flowering dates for 2017, 2018 and 2019 with those for those of 2001. This year is used as the benchmark date as the weather conditions closely reflected the 30 year average for weather (1961-90).

Negative numbers represent days earlier than the benchmark date of 2001.

2017	2018	2019
-15.8	-8.5	-17.9

In the case of the snowdrop (*Galanthus nivalis*) overwinter in 2019 it flowered 18 days earlier than the benchmark and around 7% of the records received described snowdrops flowering in November and December. Blackthorn (*Prunus spinosa*) was reported as flowering 27 days and hazel (*Corylus avellana*) 30 days earlier.

Such changes will affect pollinating insect species especially those that survive overwinter as pupae. They may not complete their life cycle in time to emerge and feed on these earlier flowers. Provided the weather conditions at the time of flowering are suitable for honey bees to forage and that the flowers have nectar and or pollen available to collect such earlier flowering may benefit honey bee colonies.

24.2 Pollination – definition and introduction

Pollination is the movement of pollen from the male (anthers) to the female (stigma) parts of flowers and is essential for successful crop production (seed and edible) and wildflower reproduction.

Most flowering species in the British Isles rely on flower-visiting insects to enable pollination to take place. There are large numbers of species of wild bees (solitary and bumblebees), flies (e.g., hoverflies), butterflies and beetles and not forgetting managed and unmanaged honey bees. Moths, bats and birds are also important pollinators.

Many other forms of plants use the wind to carry the pollen from one plant to another and this enables the pollination of grasses, sedges, rushes, nettles, docks and many tree species.

A clear description part played by honey bees in pollination was first published in 1750 by Arthur Dobbs who lived in Co. Antrim, Ireland, (Walker and Crane 2000).

Proctor and Yeo's book on the ***Pollination of Flowers***, first published in 1973 in the New Naturalist series contains much information on the subject. This book was updated and republished in 1996 by Proctor, Yeo and Lack under the title ***The Natural History of Pollination***. Both volumes contain so much information; so much more than we have included in our book. Our purpose is to introduce the subject to the beekeeper or interested person for them to pursue.

More recent and specialist books are those by Glover (2007) on ***Flowers and Flowering***', and Willmer (2011) on ***Pollination and Floral Ecology***. In our book we are focussing on the western honey bee in the British Isles and the forage resources it can seek to utilise and only passing reference has been made to the many other pollinating organisms which exist and also interact with the floral resources exploited by the honey bee.

Willmer (2011) contains a good chapter on the Global Pollination Crisis and introduces the factors which have and continue to contribute to pollination disruption, namely:

- Assessing Pollination Services and Pollinator Declines.
- Habitat Degradation and Destruction.
- Habitat Fragmentation.
- Intensive Agriculture.
- Monocultures and the Loss of Hedgerows.
- Agrochemical Effects, because of the use of Herbicides and Insecticides.
- Overgrazing.
- Selective Harvesting.
- Genetic engineering of Crops.
- Increasing Fire.
- Introduced Animal Species and Pollinators.
- Invasive Plant Species.
- Diseases and Other Natural Threats to Key Pollinators.
- Climate Change.

There is a tendency in some circles to denigrate the role of the honey bee in pollination and to accentuate the positives for some other bee species which are specialist in the range of flowers in which they can effect pollination. As we shall see the honey bee benefits from its generalist behaviour, communication skills and plasticity of behaviour enabling a colony to respond to and exploit a wide variety of plant species.

It is important to review the characteristics of honey bees and their contribution to providing pollination activities for a wide range of plant species including those which form crops for humankind and for the continuance of biodiversity. Honey bees are valued pollinators because they:

- Are perennial colonies and able to readily respond to pollination requirements whenever there are forage resources available for it to exploit and hence carry out its pollination activities without the need to wait for the numbers of solitary bees or bumblebees to increase to be as effective as the numbers of bees in a honey bee colony.
- Nectar and pollen are their only food resources and so bees are very attracted to flowers presenting them.
- Anatomically honey bees have plumose body hairs which are very effective at removing and holding pollen from flowers and the through visiting other flowers and flower constancy behaviour effecting pollen transfer and pollination from flower to flower. of the same species thereby promoting pollination.
- They can be managed, and pollinator populations manipulated to coincide with times of maximum pollination requirement.

Honey bees are polylectic. They have an ability to collect pollen from the flowers of a variety of unrelated plant species. Honey bees are generalist in nature and have the capacity to visit many flowers doing so in an organised way which not only optimises the use of the resources expended on foraging but also benefits the plant species being visited and pollinated.

During the evolutionary process flowers of different plant species have developed certain characteristics or morphologies and reward (e.g., pollen and nectar) patterns exploiting the abilities and preferences of particular kinds of visitors (potential pollinators). The recognition of this led to the concept of pollination syndromes and in the case of bees the pollination syndrome is called Melittophily. The following information on melittophily has been abstracted from Table 11.1 in Willmer (2011).

Characteristic	Experience of bees
Timing of anthesis	Dawn, day
Main flower colours	Pink / purple/ blue / white / yellow
Nectar guides	Yes
Floral Scent	Moderate, usually sweet
Flower shape	Bilateral or radial, Exposed or short / medium tube
Nectar site (location)	Exposed or concealed
Nectar volume and concentration	Mid- volume & mid-concentration
Pollen amount	Mid
Pollen deposited	Head, dorsal, or ventral body

Honey bee colonies consist of thousands of individuals and even in the early spring when the colony population is probably at its lowest there are relatively large numbers of bees providing pollination services in the landscape area where there are beehives compared to other insect pollinators. Such comparatively large numbers will far outnumber other species whose life cycles may mean their populations may not have reached significant numbers.

Pollination is a mutualism with plants benefitting from and often requiring the pollinators for sexual reproduction.

Conversely pollinators typically depend on floral sources of pollen and nectar for their own nutrition, energy and reproduction.

It is now well accepted that climate change with its associated changes in weather patterns is affecting the phenology (timing of periodic biological phenomenon in response to climate change) of flowering times and the flying times of honey bees. In the case of other pollinator species, they may be less able to adapt to changing flowering periods.

As the growing season expands in many places the flowering period now begins earlier and where at one time different plant species might flower at or around the same time these are changing to the extent there may be floral gaps and consequent scarcity of forage for the honey bee.

If pollinator services are to be sustained there may need to look for new interactions or to try to simultaneously change plant and pollinators.

The honey bee provides nearly half of all crop pollination services worldwide.

Much is known about honey bees and the fact they can be managed and provide a significant pollination capability, their role in the provision of ecosystem services becomes even more important. Potential collateral impacts on other pollinator species do need to be recognised and need to be taken into account.

We have followed the general view that honey bees are pollinators, however like many things in biology all is not what it seems and there are considerable intricacies in the biology and ecology of honey bees and the flowers they visit.

Honey bees may visit flowers and be ineffective (thieves) pollinators, for example if there is no transfer of pollen from plant to plant in dioecious species, a honey bee visitor merely collecting and removing pollen from the male plant. In some plant species the honey bee may use force to gain access to the floral parts and in doing so may cause damage to the floral parts or in trying to access the floral parts and not effect pollen transfer in an effective, consistent and reliable way. Such bees might be called robbers. There is also a third possibility where the interaction of the bee with the flower has no positive or negative effect.

It is therefore more accurate to refer to honey bees as a 'flower visitor,' rather than 'pollinator' until a pollinating effect has been observed and demonstrated for each plant species visited. The plant's breeding system needs to be understood and knowledge of the actual movement of bees in and between flowers and plants is necessary to determine their true role.

24.3   Crop pollination by honey bees

J.B Free's book *Insect Pollination of Crops* first published in 1970 with a second edition in 1993 is still a good source of reference and information, as is Proctor, Yeo, and Lack, *The Natural History of Pollination* (1996).

Delaplane and Meyer (2000) published a more recent title *Crop Pollination by Bees* and this book considers other types of bees as well as honey bees.

Pollinators make a significant contribution to crop production in the British Isles.

Insect pollinators benefit not only the yield but also the quality of many crops. These cover some 37% of the UK food crops including top fruits like apples and cherries; soft fruits including strawberries and raspberries; and field crops such as beans and oilseed rape. Field beans are more effectively pollinated by bumble bees and the beans will give higher yields, but honey bees will also pollinate them especially in the absence of bumblebees.

In addition, another 37% of UK crops such as carrots and onions need insect pollination for seed production.

The level of dependence on insect pollination is crop and variety specific and the contribution they make to productivity can vary. For example, around 85% of apple production is dependent on insect pollination, but only 15% of oilseed rape.

Honey bees are used widely to supplement wild insect pollinators (bumblebees, hoverflies, solitary bees and flies) whose abundance is affected by climate, weather and land management.

A diverse pollinator community should provide some degree of 'insurance' against pollination failure, but here again the fact that honey bees can be managed and moved to carry out pollination services is very important to recognise. The population dynamics of the honey bee are beginning to be quantified and the role of the honey bee not just for honey or beeswax production but its current and future potential to help provide effective pollination services to meet perceived pollination requirement is being recognised not just by beekeepers, growers and the public but also by policy makers.

Due to our changing climate and market forces the types of crops being grown in the British Isles is changing. Modern agricultural practices have changed large areas of the landscape to become cropped and grazed land with few areas of natural or wild habitat left.

The following table reproduced from Williams, Carreck and Little (1993) lists the crop plants grown in the British Isles which required or would benefit from insect pollination, in particular bees. Please note the crop may be the seed.

UK Crops dependent on, or benefitting from, pollination by bees		
Largely self-incompatible, monoecious, dioecious or dichogamous	Partially self-incompatible, heterostylous or protandrous.	Self-fertile but not auto self-pollinating.
Bees essential	Bees important	Bees beneficial
Crops grown for their fruit or seed		
Apple	Black Mustard	Aubergine
Blackcurrant	Broad Bean	Blackberry
Caraway	Buckwheat	Dewberry
Celery	Field Bean	Gooseberry
Coriander	Plum	Grapevine
Fennel	Raspberry	Loganberry
Gherkin	Runner bean	Morello cherry
Kiwifruit	Sunflower	Okra
Marrow	Turnip rape	Peach
Melon	White mustard	Peppers
Pear		Red currant
Pumpkin		Strawberry
Sweet Cherry		Swede rape
Sweet Fennel		Tomato
		White currant

UK crops dependent on, or benefitting from, pollination by bees		
Largely self-incompatible, monoecious, dioecious or dichogamous	Partially self-incompatible, heterostylous or protandrous.	Self-fertile but not auto self-pollinating.
Bees essential	**Bees important**	**Bees beneficial**
Crops that need bees for propagative seed production		
Vegetable crops	Vegetable crops (cont)	Green Manure / forage crops
Asparagus	Pakchoi	Alsike clover
Basil	Parsley	Lucerne (alfalfa)
Beetroot	Peppermint	Sainfoin
Cabbage	Petsai	Vetch
Carrot	Radish	Red clover
Cauliflower	Rocket	White clover
Chervil	Rosemary	Essential oil crops
Chicory	Salsify	Borage
Chives	Spearmint	Evening primrose
Endive	Sugar beet	Lavender
Marjoram	Turnip	
Onion		

The following crops grown in the British Isles require pollination by or are visited by **honey bees.**

Crop name	Species name
Oilseed Rape / Rapeseed (winter and spring oilseed	*Brassica napus* ssp. *napus*
Vegetable and fodder rape)	*Brassica napus*
White mustard	*Brassica alba*
Chinese Cabbage / pak-choi	*Brassica chinensis*
Turnip (also known as Swede, neep and snagger), green manure	*Brassica rapa* ssp. *rapa*
Radish	*Raphanus sativus*

Crop name	Species name
Cabbage (wild and domesticated forms)	*Brassica oleraceae*
Horse-Radish	*Armoracia rusticana*
Kiwi fruit (Chinese Gooseberry)	*Actidinia arguta*
Aubergine / Egg plant	*Solanum melongena*
Grapevine – Muscadine Grape 1	*Vitis rotundifolia*
Alfalfa / Lucerne	*Medicago sativa*
Red Clover	*Trifolium pratense*
White Clover	*Trifolium repens*
Ladino Clover	*Trifolium repens* (Giant variety)
Crimson Clover	*Trifolium incarnatum*
Alsike Clover	*Trifolium hybridum*
Field Bean	*Vicia faba*
Runner Bean / Scarlet Runner Bean	*Phaseolus multiflorus*
Common, French, Kidney, haricot Bean	*Phaseolus vulgaris*
Sweet Clover	*Melilotus alba*
Sainfoin	*Onobrychis viciifolia*
Field or Garden Pea	*Pisum sativum*
Blackcurrant	*Ribes nigrum*
Gooseberry	*Ribes grossularia*
Red Currant	*Ribes rubrum*
Loganberry	*Rubus ursinus x R. idaeus*
Cucumber	*Cucumis sativus*
Courgette	*Cucurbita pepo*
Carrot	*Daucus carota*

Crop name	Species name
Parsnip	*Pastinaca sativa*
Sunflower	*Helianthus annuus*
Lettuce	*Lactuca sativa*
Bilberry (European blueberry)	*Vaccinium myrtillus*
Capsicum	*Capsicum annuum*
Sugar Beet, Garden beet, spinach, fodder beet and mangold	*Beta vulgaris* ssp. *vulgaris*
Swiss Chard /Spinach Beet / foliage beet	*Beta vulgaris* ssp. *cicla*
Apple	*Pyrus malus*
Pear	*Pyrus communis*
Plum	*Prunus domestica*
Sweet Cherry	*Prunus avium*
Almond	*Prunus amygdalis*
Cultivated Strawberry	*Fragaria x ananassa*
Raspberry	*Rubus idaeus*
Onion, Common and Shallot	*Allium cepa*
Leek	*Allium ampeloprasum*
Garlic	*Allium sativum*
Welsh Onion	*Allium fistulosum*
Chive	*Allium schoenoprasum*
Asparagus	*Asparagus officinalis*
Medicinal, opium or oil poppy	*Papaver somniferum*

Superscript 1 The extent to which selfing, wind and insects contribute to fruit set and quality in muscadines and other grape species (*V. vinifera* and *V. rotundifolia*) is poorly understood. Sampson et al. (2001. Muscadine grape flowers are visited by honey bees where they collect nectar and either voluntarily or involuntarily collect pollen as the calyptras eject the pollen.

Strong colonies are required if they are to be deployed in pollination services because they can provide more field / foraging bees. Strong colonies will also forage at lower temperatures and in stronger winds than will a weak one.

To be most effective in providing a pollinator service for the crop in question the aim is to get strong and populous colonies ready for the projected time when the crop is due to flower.

Colonies with an expanding brood pattern are more effective in pollination because the presence of brood stimulates the collection of pollen.

Taking brood taken from other colonies and given to ones which will be deployed for pollination services will stimulate the colonies to be used as pollinators to increase their pollen collection, however this practice requires care so there is no transfer of pests and disease between the colonies and that the colonies used to provide the brood will not suffer a set- back. It requires good knowledge of the health status of the donor colony.

The position of the brood nest relative to the cavity opening / entrance to the hive with a single brood box influences the rate of pollen collecting because foraging bees found close to the brood nest are likely to be pollen foragers. This closeness of the brood nest to the entrance reinforces the stimulus for the foraging bees to go and forage for pollen.

24.4   Historical aspects of beekeeping, crop pollination and honey production

Commercial pollination has for many years been a key revenue generating activity of bee farmers who derive much or all of their livelihood from the ownership, management and deployment of honey bee colonies (usually in substantial numbers) in supplying contract-based crop pollination services, honey production and the production and sale of colonies of honey bees, queens and beekeeping equipment.

Hobbyist beekeepers usually keep fewer colonies and may move colonies to benefit from key crops, but they do so through seeking permission from the landowner and do not usually receive payment for the services of their bees. Having said this the dispersed nature of the colonies managed by hobbyist beekeepers across the landscape means they provide a significant pollinator resource for all plants requiring pollination and to which the honey bee can gain access and benefit be it nectar and / or pollen.

In 1994 a survey (reported in Carreck et al. 1997) of commercial bee farmers found that 82% and 67% of them moved colonies for crop pollination and honey production respectively. The crops reportedly pollinated by colonies supplied Bee Farmers are listed below and included in the table is a ranking of the need for insect pollination:

Crop	Need for insect pollination+	Crop	Need for insect pollination+
Field crops		**Seed crops**	
Oilseed rape	*	Clover	***
Field Beans	*	Field Bean	*
Borage	*	Oilseed rape	*
Linseed	0	Cauliflower	***
Phacelia	?	Kale	***
Tree fruit		Cabbage	***
Apple	***	Carrot	***
Pear	**	Parsnip	**
Plum	**	Swede	*
Cherry	***	Onion	***
Soft fruit		Turnip	*
Strawberry	*	**Glasshouse crops**	
Currants	***	Strawberry	*
Blackberry	**	Runner bean	**
Raspberry	*	Raspberry	*
Gooseberry	*	Tomato	*
		Peach	**
+ Key for classification: ? unknown, 0 none, *moderate, ** beneficial, *** essential			

A 1994 survey (also reported in Carreck 1997) of members of the British Beekeepers Association (largely an association of hobbyist scale beekeepers) showed 5% of members moved their colonies for crop pollination purposes and 17% for honey production. The results of a similar survey conducted in 1985 also showed 5% of members moved colonies for pollination and 21% for honey production.

Carreck (1997) includes a table which lists the 10 most important nectar sources declared in a BBKA Survey 1984-86, 1994 and a Bee Farmer's Association survey which reported the percentages of their colonies moved to these nectar sources.

	BBKA Survey	% of Respondents	BBKA Survey	% of Respondents	BFA Survey	% of Colonies
	1984-86		1994		1994	
1	Oilseed rape	34	Oilseed rape	55	Oilseed rape	44
2	Lime trees	16	Blackberry	20	Field beans	24
3	Apple, pear, plum, cherry	14	Apple, pear, plum, cherry	19	Apples	24
4	Blackberry	13	Lime trees	18	Pears	5
5	Heather	8	Field beans	14	Clover	5
6	Clover	7				
7	Field bean	7	Clover	11	Strawberries	3
8	Sycamore	5	Sycamore	7	Currants	3
9	Chestnut	4	Chestnut	7	Cherries	3
10	Hawthorn	4	Hawthorn	5	Borage	3

Earlier a detailed study of an area in Ashtead, Surrey made by Dorothy Hodges during the period between 1940 and 1957, (Crane 1980), listed the flowering times of 75 plant species recorded. In total 105 species were worked by honey bees, of which 18 were considered to be major honey plants. These are listed in the following table.

Month	Plant species
May	Dandelion, sycamore, hawthorn
June	Sainfoin, bean, raspberry
June / July	White clover, blackberry, second-cut red clover
July / August	Willowherb, bell heather, lime
August / September	Ling heather, mustard

25. Plant and flower structure

Honey bees are often termed supergeneralists and will work most flowers in most habitats they have access to. Honey bees are thought not to be particularly attracted to zygomorphic (a flower with bilateral symmetry, for example flowers of the pea family) shaped flowers. They seem to prefer a more open or radial symmetry flower and favouring flower colours of white / yellow / orange. They will also visit flowers of some species which are considered to be wind-pollinated and this is sometimes seen in the British Isles when honey bees will collect pollen from the tassels of sweet corn (*Zea mays convar.saccharata var.rugosa*) and maize (*Zea mays* ssp. *mays*) especially in periods when there are no other sources of pollen available.

Aston and Bucknall in their book ***Plants and Honey Bees – their relationships*** (2004) introduced the beekeeper to flower structure and how it relates to honey bees and their activities.

25.1 The availability of pollen – timing of pollen presentation

25.1.1 Flower Opening and Closure

The structure of flowers in the Angiosperms varies widely and the mechanisms of flower opening are equally as varied. The time of flower opening begins the time when pollinators will be attracted to visit the flower initiating the transfer of pollen from male and bisexual flowers by pollination leading to fertilization and seed set in female and bisexual flowers.

In many species the flowers are open permanently until pollinated with the flower closing by a closure movement or is finished by petal withering or abscission. In some species the periods of the flower being open are alternated with periods of closure.

In 2003 Van Doorn and Meeteren published a review on flower opening and closing and the physiological mechanisms responsible. The paper considers the following topics:

- The variations of flower opening and closure.
- The Mechanism of opening and closure.
- Carbohydrate metabolism.
- Cell wall expansion.
- Water relations.
- Hormonal regulation.
- Regulation of opening and closure by external clues.
- Is flower opening due to changes in humidity, temperature, or light?
- Effects of temperature and light on flower closure.
- Regulation of opening and closure by duration of darkness and light.
- Repeated opening and closure, role of endogenous rhythms.
- Control mechanisms: the molecular basis.
- Strategies of flower opening and closure.
  The review concludes that the co-ordination of processes culminating in synchronized flower opening is, in many species, highly intricate through the interaction of endogenous and exogenous factors.

So why include this topic in a book on bee nutrition?

As we have noted a honey bee colony requires considerable quantities of nectar and pollen to enable the colony to thrive, reproduce and to hoard sufficient stores to be able to survive periods when there is a shortage of nectar and / or pollen.

The availability of the forage is not only determined by the variety and quantity of the plant forage species and pollen /nectar but also weather conditions and the amount of pollen and nectar produced by the flowers that are available for the bees to forage from.

The honey bee is adapted to be able to work a wide range of plant species and is able to benefit from habitats where there is a wide diversity of plant species.

Beekeepers could usefully keep observing as they travel around the area in which their bees might forage and note the diversity of forage available as well as any large areas of crops or other monocultures which their honey bees might visit. It is the nature of the total entity of the forage available that will determine whether colonies do more than just survive and are healthy, have a high nutritional status, and can thrive. The honey harvest for human consumption should be viewed as a bonus.

Diurnal chasmogamic (flowers which are commonly open and showy with open petals encircling exposed reproductive parts, namely anthers and stigmas) flowers of the British Isles normally release their pollen from the anthers in the period 7am to 5pm. The peak of pollen presentation (the release of the pollen from the anther sac making it available for contact with visiting insects) varies widely.

Flower opening can take place very quickly, e.g., in the Evening Primrose (*Oenothera biennis*) full flower opening takes 20 minutes and in the case of Ivy (*Hedera helix*) it occurs in about five minutes.

The following data are selected from Percival (1965) and Percival (1955).

These data were reported from observations made in the 1950's and it may be that some species adapted to the changing weather conditions in part brought about by climate change. They do however illustrate the wide variation in flowering timings. In addition, there are the impacts of weather conditions on the availability of pollen and nectar in response to relative humidity, temperature range and whether rain limits anthesis or flower opening.

Early Morning availability	
Large Bindweed	*Calystegia sylvatica*
Long-headed Poppy	*Papaver dubium*
Broom (Common or Scotch)	*Sarothamnus (Cytisus) scoparius*
Charlock	*Sinapsis arvensis*
Scotch rose	*Roas spinosissima*
Common Rock rose	*Helianthemum chamaecistus (H. mummularium)*
Traveller's-Joy	*Clematis vitalba*
Ribwort Plantain	*Plantago lanceolata*
Chiefly Morning availability	
Bulbous Buttercup	*Ranaunculus bulbosus*
Creeping Buttercup	*Ranunculus repens*
Dandelion	*Taraxacum officinale*
Wood Anemone	*Anemone nemorosa*
Blue Anemone	*Anemone apennina*
Lesser Celandine	*Ranunculus ficaria*
Aubretia	*Aubretia deltoides*
Gorse	*Ulex gallii*
Mid-Day availability	
Water plantain	*Alsima plantago-aquatica*
Crocus	*Crocus aureus*

All Day availability	
Wallflower	*Chieranthus cheiri*
Bramble	*Rubus fruticosus*
Nasturtium	*Tropaeolum majus*
Foxglove	*Digitalis purpurea*
Phacelia	*Phacelia tanacetifolia*
Chiefly Afternoon availability	
Apple	*Pyrus malus (Malus domestica)*
Pear	*Pyrus communis*
Afternoon availability	
Broad bean	*Vicia faba*

Anthesis is variously defined as a time period during which a flower is fully open and functional; or the onset of that period; or the opening of the flower bud to the setting of the fruit.

The rate at which dehiscence (splitting of the anther sacs to release the pollen grains) takes place is very variable. It can be all the anthers in one flower dehisce simultaneously and may even be simultaneous over the whole plant or even over a whole population of plants. It may be sequential between flowers on the same plant, or sequential for different anthers within one flower as is the case in many hellebore flowers where the anthers ripen and dehisce one at a time. Some flowers may have two or more whorls of anthers which mature and dehisce then shrink back against the petals before the next whorl repeats the same process several hours or even days later. Examples of this can be found in the common cranesbills (*Geranium* spp.), the stitchworts (*Stellaria* spp.) and some other members of the Caryophyllaceae (Pinks and Carnations family).

The timing of the anthers releasing their pollen is generally correlated with the behaviour of the pollinator and this is a good example of how bees and plants interact. In the case of the Cistus, Sun Rose or Rock Roses e.g., *Cistus salvifolius (Sage -leaved rock rose)* grown in gardens a close time relationship has been observed between the time of honey bee activity and pollen release with a time lag of only 28-60 minutes over a 10 day period.

The factors controlling dehiscence are still not understood as some can be seen to be dependent on the state of development of the flower and some linked to the environment. For example, the daily cycles of declining relative humidity and rising ambient temperature increase evaporative water loss resulting in structural changes in the plant tissues.

25.1.2 Presentation of the pollen and its attachment to the honey bee

Plants have different ways in which to make available (present) the pollen grains for transfer from one flower to another via the pollinator. With respect to insect pollinators such as the short -tongued honey bee they may be released as single pollen grains and this particularly found in plant species where nectar is the main reward for the pollinator.

In other species the pollens may be aggregated together also known as 'pollen cementing'.

The pollen may be inherently sticky with the sticky pollen coat material pollenkitt as is found in the Compositae (Asteraceae) and Labiatae (Lamiaceae), in fact in most families of angiosperm plants pollinated by animals. The other sort of sticky pollen coat material is tryphine which seems to be only found in the Brassicaceae.

Pollenkitt enables the cohesion of single pollen grains after dehiscence and the attachment of these aggregations of pollen to flower-visiting insects. The pollenkitt also acts as a store for the recognition substances involved in the pollination process.

Pollenkitt is a complex mixture which includes lipid and viscous substances (including carotenoids).

In some species, such as members of the Ericaceae (Ling, heather *Calluna vulgaris*) pollen grains may be cohered (stuck together) by means of common walls loosely held by lipid deposits or exine bridges into tetrads or polyads. Acacias commonly known as mimosa and species such as *Acacia dealbata* are grown by gardeners in the British Isles and these produce polyads.

Another example of pollen aggregation is found in the rosebay willowherb (*Chamerion angustifolium*) (Family Onagraceae) and here the pollen may also have aggregations of the pollen held together by viscin threads which are non-sticky and non-viscous. They are attached to the exine of the pollen grain. The pollen grains are also linked by pollenkitt.

Viscin threads contain sporopollenin.

The threads fasten the aggregations or pollen grains (e.g., tetrads) by entangling with the insect's hairs or bristles and even the smoother surfaces of the bee's body (Hesse 1981).

Some flowers only produce pollen and are worked by honey bees. These include poppies (*Papaver* spp.), *the rock roses (some Cistus* spp., most *Potentilla* spp. *and Helianthemum* spp.) and Meadow-rue (*e.g., Thalictrum delavayi).*

Plants with distinctive pollen colours		
Near Black / purplish	Field Poppy	June-July
Deep Blue	Scilla	Mar-Apr
Brick-red	Horse Chestnut	May-June
Pale blue	Rosebay Willowherb	July-Aug
White	Bluebell	Apr-June
Off-white	Himalayan Balsam	Aug-Sept

25.2 The availability of nectar and nectaries

25.2.1 Floral and extrafloral nectaries

Nectaries are locations from where nectar is secreted onto the outer surfaces of the plant, but their locations and structures, sometimes not visually obvious, very diverse and their structures, sometimes very complex.

They may be **floral**, i.e., found in parts of the flower; or **extrafloral**, located outside the flower.

Nectaries can be connected to the phloem, the xylem, to both or have no direct connection to the plant's vascular system. The nectar may exit the plants through modified stomata (stomata are pores in the epidermis of the leaves, stems and other plant organs that facilitate gas exchange) which remain open or though specialised trichomes (e.g., hairs, glandular hairs, scales and papillae).

Heil (2011) is a good review paper and introduction to nectar and a more recent review can be found in Roy et al. (2017).

In this book we have restricted to ourselves to consider plant species which grow in the British Isles and which are of potential use principally to honey bees, but also where relevant as a matter of interest to other pollinators.

25.2.2 Nectar as an attractant

As we have seen nectar has a wide range of components and short-tongued bees (which include honey bees) are said to favour floral nectar rich in hexoses. (monosaccharides). The amino acids present in floral nectar have been shown significantly increase the attraction of the nectar to different groups of pollinators, but there appears no specific reference in the literature to the honey bee and the flora of the British Isles.

The levels of amino acids in nectar are two to three orders of magnitude less than they are in pollen which would therefore be a better source of nutrients, however the presence of amino acids in nectar may affect its taste. Honey bees prefer essential amino acids compared to non-essential amino acids. Phenylalanine and proline (both phagostimulatory) are two of the most abundant amino acids present in the nectar of bee-visited flowers and bees are attracted to them, whereas they are deterred by glycine (the simplest amino acid).

VOCs (Volatile Organic Compounds) and scented petals are known attractants to honey bees, however the odour of the nectar is also an attractant and relevant signal for a pollinator.

25.2.3   Nectar as a protectant

Nectar proteins have been known about for a number of decades and presumed to be part of the food source of organic nitrogen, but it appears they also have a protectant function as well. These proteins are referred to as nectarins (e.g., secreted in *Allium* sp. (Leek) flowers) and appear to be involved in the protection of the nectar against microbial attack (Park and Thornberg 2009). They also protect the nectaries from invasion by plant pathogens through the stomata and nectiferous tissue.

Some plant families contain species whose nectar is toxic due to the presence of secondary metabolites including non-protein amino acids, phenolics and alkaloids, but it is unclear as to what the advantage is to the plants to produce toxic nectar. It is possible they could be mechanisms by which plants manipulate insect behaviour or have health promoting effects on bee gut parasites.

25.2.4   The availability of nectar

Plants control nectar production and floral nectaries produce and secrete nectar in response to consumption rates. Plants with floral nectaries can also reabsorb non-secreted nectar, but this appears not to be the case with extra-floral ones.

The floral and extra-floral nectaries of many temperate plant species produce their nectar overnight or in the early morning in time for their diurnal (during daytime) rhythms to attract pollinating visitors and in response to consumption and affects caused by the temperature conditions. Such plant species are said to have unimodal patterns of nectar secretion.

In extra-floral nectaries (e.g., *Prunus avium*, variously called Wild cherry, Sweet cherry, Gean) experiments using simulated herbivory has been shown to induce extrafloral nectary production (Pulice and Packer 2008).

Nectar secretion and flow involve the hormones called jasmonates which promote flower and fruit development. Other hormones such as auxins and gibberellins also play a part.

More than one third of plant species examined for a connection between the plant's vascular system and its nectiferous tissue showed no direct connection (vascularisation). However, it does seem that the plant's vascular system does supply the secreted phloem sap to secretory parenchyma cells which form the nectiferous tissue.

The mechanisms of nectar production and secretion are still not fully understood, but Heil (2011) contains the following hypothesis. The hypothesis is that prenectar is formed, metabolised, and transported in vesicles (small sac like structures in this case containing processed prenectar) within the nectiferous tissue. The high density of plasmodesmata (microscopic channels which traverse the cell walls of plant cells enabling transport of chemical substances and communication between the cells) is common in nectiferous tissue and these allows the vesicles to move rapidly amongst the cells. In the case of floral nectaries these may be able to reach high peak sugar secretion rates through a combination of carbohydrates stored in amyloplasts (small structures found in plant cells which can synthesise and store starch and can also convert it to glucose) through the breakdown of starch and at the same time the loading of sucrose from the phloem. The sucrose may then be converted to hexose sugars by invertase enzymes contained in the cell walls. The nectar is then held in the extra-cellular space before passing out through the stoma / stomata (of the nectary).

The energy obtained from nectar depends on its sugar content and this can vary considerably with nectar sugar concentrations ranging from 7-70% by weight.

Once the nectar is secreted and potentially exposed to varying weather conditions the nectar may undergo post secretory changes

We discuss elsewhere the ways in which worker foraging bees access the pollen and nectar in the various shapes of flowers which can be found in the flora of the British Isles, but it is interesting to know that the nectar sugar composition may be a consequence of the flower morphology and the microclimate that exists in and around the key floral parts. Sucrose-rich nectars are more common in protected tubular flowers which are more usually visited by specialist pollinator species, whilst open exposed flowers usually have hexose (fructose and / or glucose) sugar-containing nectars and are visited by generalist pollinator species of which the honey bee is one. Hexose solutions have higher osmotic concentrations than sucrose solutions and therefore their evaporation rate is lower.

Nectar secretion is influenced by the ambient temperature and plant species vary in their response. Crane (1951) contains a table which includes a range of plant species considered important to bees and provides the minimum shade temperature which initiated nectar secretion. The data contained in the are the work of Dugat (1949).

Common Name	Botanical Name	Min. Shade Temp. for nectar secretion ^{0}C
Rape	Brassica napus oleifera	12
Mustard	Brassica sinapsis	12
Red Horse-Chestnut	Aesculus carnea	10
Lime	Tilia spp	20
Acacia	Acacia spp	17
Bird's Foot trefoil	Lotus corniculatus	15
Lucerne / Alfalfa	Medicago sativa	15

Common Name	Botanical Name	Min. Shade Temp. for nectar secretion ^{0}C
Sweet Clover	*Melilotus alba*	16
Sainfoin	*Onobrychis sativa*	15
Sophora	*Sophora* spp	20
White Clover	*Trifolium repens*	15
Apple	*Malus pumila*	15
Plum	*Prunus domestica*	10
Cherry	*Prunus cerasus*	16
Hogweed	*Heracleum spondylium*	18
Ivy	*Hedera helix*	10
Dandelion	*Taraxacum officinale*	12
Golden Rod	*Solidago* spp	20
Heather / Ling	*Calluna vulgaris*	15
Lavendar	*Lavandula* spp	22
Rosemary	*Rosmarinus officinale*	13
Savory	*Satureia* spp	18
Thyme	*Thymus serpyllum*	20
Buckwheat	*Fagopyrum esculentum*	12
Sweet Chestnut	*Castanea sativa*	19
Phacelia	*Phacelia tanacetifolia*	12
	HONEYDEW	
Norway Spruce	*Picea excelsa*	27

The speed of evaporation depends on the floral morphology and the microclimates in and around the flowers as well as the ability of the plant to reabsorb / secrete nectar as a means of controlling nectar concentration. Enclosure of the nectar inside corolla tubes or under 'hoods' formed by the petals, the presence of hairs just above the nectary are all mechanisms which can help the plant to maintain the nectar sugar concentration and physical form, so that it continues to be attracting and accessible to the pollinating insect.

The pendant, or hanging down drooping aspect of flowers, such as those found in the inflorescences of Lime (*Tilia* spp.), help protect the nectar from dilution by rainfall. On the other hand, sugar concentration may increase due to the flower's exposure to the direct sun and in the case of lower- flowered species of plant whose flowers have this drooping habit these may trap warm moist air rising from the ground under the flower.

High sugar concentrations in nectar mean the nectar is increasingly viscous. A 60% sugar solution is about 28 times more viscous compared to a 20% solution. In the case of the honey bee the optimal sugar concentration for energy intake and ease of licking up the nectar is about 60%. Sugar concentrations greater than 75% means that the nectar becomes crystalline. Ivy *(Hedera helix)* nectar being a good example of this. However, honey bees will collect nectar of varying sugar concentrations (preferably 30-50% concentration) depending on what is available which in terms depends on the environmental conditions, the colony status and the effectiveness in communication of forage through the language of the bees dances.

Honey bees increase the concentration of the sugar in the nectar as it is processed into honey or consumed soon after gaining it by oral regurgitation and evaporation. Changes in the nectar concentration and volume secreted often vary throughout the day and this may mean different visitors during the course of the day, much depending on the accessibility of the nectar and the relative reward received and perceived by the flower visitor, e.g., the adult foraging honey bee.

25.2.5   Types of Floral Nectaries

Floral nectaries are found associated with the various parts of a flower and we have selected examples of each type of floral nectary present in flowers of the flora of the British Isles.

25.2.5.1   Sepal nectaries

These are found on a sepal or its equivalent on a calyx.

The Malvaceae (Mallow family) contains several genera which include *Abutilon, Pavonia* and *Lavatera* which have sepal nectaries and the nectar is secreted from tissues on the sepals.

The flowers of some species of lime *(Tilia* spp.) have sepal nectaries which appear as tiny slits of secreting tissue which are invisible to the naked eye. The boat-shaped sepals are lined with hairs whose surface tension assists in retaining the drop of nectar as it forms in the hollow or inner surface of the sepal and is important because the individual flowers of lime often hang down.

25.2.5.2   Petal nectaries

Here the nectary is formed from part or the whole (when they are called nectar leaves) of a petal.

These occur throughout the Ranaunculaceae. In the *Ranunculus genus* (Buttercups) itself the nectary is situated under a tiny flap of tissue near the base of the petal. In Hellebore species, e.g., *Helleborus orientalis* the petals which have become modified into tubes, become filled from half to two thirds of their length with nectar consisting of 26-41% sucrose. The coloured parts of the flower are modified sepals.

In *Trollius* spp. (Globe Flowers) there are special spoon-shaped petals within the flower that are light yellow in colour and contain the secreted nectar.

25.2.5.3   Staminal nectaries

The Violas have staminal nectars in the form of two slender tongues of green tissue arising from the filaments of two of the stamens. These project back into the spurred petal which receives their secretion. The spur tissue does not secrete. In *Anemone pulsatilla* the anthers of a few of the outermost stamens secrete nectar from the base of the stamen instead of producing pollen. The Wood Anemone *(Anemone nemorosa)* is nectar-less.

25.2.5.4   Carpel or Gynoecial nectaries

There are three types of carpel nectaries.

The 'valve' nectary is usually a groove of secreting tissue on the flanks and towards the base of the young carpel, e.g., in the Marsh Marigold or Kingcup *(Caltha palustris)*.

The 'septal' nectary mainly occurs in the tricarpellary ovary of the Plant Order Liliales and their derivatives. The young ovary has three vertical grooves marking the position of the septa. Nectar is secreted in the form of beads along these lines and runs down the groove and collect around the base of the ovary in the cup of the perianth. In the Bluebell *(Hycanthoides non-scriptus)* the honey bee soon learns to insert its proboscis in between the base parts of the perianth which only overlap and can then access the nectar.

The 'disc' nectary is found in the case of inferior ovaries e.g. in the Umbelliferae (Apiaceae) family which includes celery, carrot and parsley and in Ivy *(Hedera helix)*, a member of the family Araliaceae . The inferior ovary is surmounted by a thick pad or 'disc' of nectar secreting tissue.

25.2.5.5   Receptacle Nectaries

The receptacle cups of the flower are lined with nectar secreting tissue and those of the Raspberry, Blackberry, Plum (perigynous in the Rosaceae) secrete nectar abundantly, but only to a small extent in the Saxifragaceae (which include the genera *Saxifraga*, *Heuchera* and *Astilbe*).

25.2.5.6   Axial nectaries

Many flowers of plant species classified in the orders Lamiales, Personales and Solanales have nectaries in the form of a circlet of secreting tissue embracing and are more or less fused to the base of the ovary.

25.2.5.7   Extrafloral nectaries

The definition of extrafloral nectaries can include nectaries which are not involved in pollination (defined on an ecological function basis) or by their location, structure, or developmental origin *per se*. They have been found in nearly every non-floral aboveground plant structure, including leaves, stems, petioles, cotyledons, bud bracts, the outside of sepals, and fruit (Weber and Keeler 2013).

The website, www.extrafloralnectaries.org, can be accessed to find the world list of plants having extrafloral nectaries in both angiosperms (flowering plants) and ferns. The presence of extrafloral nectaries in gymnosperms (conifers) is controversial (Weber, Porturas, and Keeler 2015).

The morphology of extra-floral nectaries is variable and can include vascular and non-vascular structures, patches of indistinct secretary tissue, glands (concave and convex) and secretory hairs. Confusion can arise unless the secretions are tested for sugars as there are other extrafloral plant glands such as hydathodes and lactifers which are not nectaries.

Nectar secretion from extrafloral nectaries can be continuous, seasonal or associated with particular plant activities., for example. a flush of new leaves.

The composition of extrafloral nectar typically includes mono and disaccharide sugars and amino acids, all at varying concentrations.

The Cherry laurel *(Prunus laurocerasus)* has small areas of nectar-secreting tissue in the angle between the midrib and main veins on the back of the leaves.

Extrafloral nectaries are found in some species of the following botanical families, genera and species:

Fabaceae	Pea Family
Populus spp.	Poplars
Salicaceae	Willows
Malvaceae	Mallow family, including *Gossypium hirsutum* (cotton), a member of the Malvaceae

Non-flowering plants such as Bracken *(Pteridium aquilinum)* have a smooth, shiny nectary at the point where each pinna joins the rachis of the fern. Bracken can be an important spring source of nectar forage. Extrafloral nectaries attract ants in particular with the benefit to the plant that the ants deter herbivores. The extrafloral nectaries of bracken are described in Lawton and Heads (1984).

26. What we understand honey bees know about flowers

26.1   What do honey bees know about flowers?

Honey bees can obtain knowledge about flowers in three ways:

* Information which is inherited and embedded in the genome (**instinct**).
* Information which is gained from experience (**learning**).
* Information which is gained from other experienced members of the species (**communication**).

It is not intended that a comprehensive description is given of the state of knowledge of the relationship of honey bees and flowers, however sufficient is given to enable the reader to appreciate the importance of the subject and gain an insight into its importance in the care of our honey bees with respect to the places we keep them and the potential availability, quality and type of forage they might gain access to.

Tautz (2008) describes the activities which the foraging bees must be able to perform to exploit such floral resources. They include:

* Recognize the flowers as such.
* Distinguish between different kinds of flowers.
* Recognize the state of the floral parts e.g., is the stigmatic surface of the flower receptive?
* Know how to work the flower effectively with legs and mouthparts.
* Determine the geographic location of the flower in the landscape.
* Determine the daily time window in which various flowers produce the most nectar and pollen.
* Share the information with nest mates as a messenger in a communication system.
* Receive and understand such information oneself in this communication system of where to find flowers.

26.2   Honey bee vision

Kelber and Somanathan (2019) published a review article on spatial vision and visually guided behaviour in Apidae. Bees use a pair of large compound eyes and three small lens eyes, the ocelli, for visual tasks.

The compound eyes have a small dorsal region, called the dorsal rim, which is specialized for the perception of polarized light.

The compound eyes provide the ability for the bee to undertake behavioural tasks such as finding flowers or the detection of mates, both of which require the bee to have high spatial resolution.

The visual system of bees is such that each of their compound eyes is made up of many single 'eyes' or facets.

These single eyes are an ommatidium, each having a facet lens and 8 or 9 photoreceptors the output being equivalent to a pixel in the image.

The bee brain assimilates the information coming from each of the single optical units and synthesises a picture of their surroundings from the collation of many, quite large, separate single points. The downside of this type of eye compared to that of the human where a single large complete image is formed on the retina through a single lens is that the honey bee has poor optical acuity and they can only resolve the fine details of objects and flowers when they are a few centimetres away from them.

In addition, there are three ocelli located on the top of the head whose functions and interactions with the compound eyes are still not fully understood. The ocelli have large visual fields covering the entire dorsal and frontal part of the world.

Bee ocelli are thought to have several functions, including horizon detection and flight control by the dorsal retina of the ocellus and the evaluation of the polarized sky.

Kastberger (1990) determined that the ocelli help the bee to react photokinetically (change in the velocity of movement of a bee resulting from changes in light intensity) to photic stimuli in a much shorter time than do the compound eyes alone. The ocelli and the dorsal rim of the compound eye provide functions which require only low resolution, for example, phototaxis or the use of the polarization pattern of the sky as a compass for navigation.

The eyes and ocelli in honey bees are sexually dimorphic.

Sex / Caste	Maximal Facet Diameter µm	Number of Facets/ Eye	Median Ocellus Diameter (mm)
Queen	26.1	4460	0.30
Worker	25.2	5375	0.28
Drone	40.1	9993	0.34

In 2019 Horridge published a book titled *The Discovery of a Visual System. The Honeybee.*

This book sets out his understanding of the history of research and deductions made from the decades of research which gave rise to the general perception of how the honey bee visual system works. Namely:

- Honey bees detect some visual features such as edges and colours but there is no sign that they reconstruct patterns or put together features to form objects.
- Bees detect motion but have no perception of what it is that moves.
- They do not recognise objects or colours by their shapes.

The Ultraviolet (UV) light of the sky is poorly reflected from natural objects, so the UV comes almost entirely from above, i.e., from the Sun. A disturbed honey bee always heads towards the brightest UV part of the sky to escape a threat. Honey bees use the UV light of the sky to stay the correct way up when in flight.

Honey bee workers have eyes that can look or see to the side and downwards with a small overlap in front that sees the food. There is a small blind region below and behind the animal. The honey bee (*Apis mellifera*) has a 50-60^0 (25-30^0) binocular field in the region of the eye which looks forward in flight.

The dorsal rim of the eye has a line of specialised ommatidia for the detection of UV light that is polarised along particular axes from which the bee can be aware of the position of the Sun when it is behind clouds.

The frequent occurrence of nectar guides on flowers which attract pollinators such as the honey bee suggests that bees can see between their legs and feet.

The honey bees' natural visual system relies on rates of change of the optical inputs and the power and the total energy of the green modulation signal as well as the intensity of blue.

These observations question the traditional understanding about all the correlations made between flower colour, shape and behaviour.

According to Horridge (2019), honey bees only need blue light and the coincidences of a few cues that identify the place, and that they use the same cues at decision points along the flight path route.

He believes that the evidence today questions the established view that bees have something like human colour vision of hues and tints. They **can** distinguish a palette of colours, but in a very novel and extremely simple way. They distinguish different flowers as to whether the flowers are a more or a less blue than the background of foliage or soil (earth).

The visual system relies on a number of inputs working in parallel and which seem to have an order of preference to the bee. They readily locate and learn to go to blue and distinguish all colours by a measure of their blue emissions and blue contrast relative to the background. They confound (mix up) the blue content of emissions with aspects of pattern structure. The bees detect, or have a large field of view, that is useless for distinguishing pattern in the usual sense of the meaning of the word pattern. They also measure the intensity of blue.

Green channel detectors in the ommatidia detect radially or tangentially arranged edges in relation to a hub. Green modulation is confounded with edge length and aspects of the pattern structure.

The separate summation of blue and green detectors makes it impossible to reassemble a pattern - but it is the gross distribution of structure across the eyes that is somehow partially recoverable.

It is very hard for the non-specialist to understand this whole subject let alone try and give it the importance and relevance in a book such as this.

As well the science, Horridge observes that the matters of correct thought and intuition are crucial to the understanding of honey bee vision. Humans fall victim to errors of thought because of their habit of looking at, then comparing and measuring the performance (e.g., what the animal does; and then making guesses and judgements based on their own experience). Humans have the strong tendency to observe performance and infer an intuitive conclusion based on outwardly similar systems, or human performance.

In the case of honey bee vision:

- Bees distinguish some colours therefore they have colour vision:
- Bees (sort of) distinguish a colour from all shades of grey- therefore they have colour vision – *like humans.*
- Bees distinguish between many pairs of shapes, but that does not show that they use the shape in the task. They use small clues.
- Bee vision is not intuitive like that of humans.
- Whilst humans evolved in the same world as the honey bee we have almost no ability to imagine the unknown mechanisms of insect vision which must perform superficially like theirs.
- Bees learn asymmetry not symmetry.
- They learn modulation, not contrast.
- They distinguish colours by using blue content relative to the background.
- Each ommatidium has good resolution, but the information is summed.
- They distinguish patterns by structure of modulation, not by shape.

The implications of these findings for our understanding of the honey bee, how the colony functions and what this means in relation to the foraging activities of the honey bee are still being developed and at some point beekeepers in general will have to study and re-interpret their understanding and communication about honey bee vision.

The orthodox concept of trichromatic colour vision is no longer useful for the honey bee.

The bee ommatidium has receptors for UV, green and blue but nothing for grey, white or black.

It is possible to calculate the relative stimulus to each receptor type from the emission spectrum of any coloured paper in sunlight relative to that of white paper.

Bees reared in isolation have an innate preference for UV / blue, then for green, not for yellow irrespective of background, intensity or green contrast.

Bees have an innate preference for radial patterns that are easily learned.

Without green contrast at the boundaries, bees merge together areas of different colours (new average colour indistinguishable from the original pattern).

What do bees actually detect when they fly around? The visual input is combined with other senses to build up a total impression of the adjacent environment……as humans the way a bat or a ship guided by radar in a fog is aware of the surroundings. Bee behaviour is probably affected by cognitive behaviour, concepts or detection of abstract properties such as shape, similarity, symmetry, colour in general or texture.

26.3   Optic flow

Optic flow is measured as the amount of image motion in the lateral visual field of the bee's eyes while flying towards a goal. Honey bees use optic flow to measure distance flown on their way to and from food sources (Kelber and Sonanathan 2019). The bee's perception of the distance flown varies in relation to the landscape. Honey bees estimate flight distance using only the green receptor contrast of the patterns. Bees also use optic flow to control flight speed and flight height and to avoid flying to close to obstacles (Kelber and Sonanathan 2019).

26.4   Use of landmarks for Navigation and Homing

When a bee leaves a hive for the first time it performs a learning flight which allows her to find her way back to the hive (or nest in the case of wild / feral colonies). These flights are also referred to as orientation flights or turn back and look behaviour and is done in a series of successive arcs that are approximately centred on the hive or nest location. Each time on exiting the hive the bees turn to face the hive. The flights enable the forager to learn features around the nest site and update itself on any changes, as she moves progressively further away from the hive (Kelber and Sonananthan 2019).

26.5   Foraging and Flower Detection

The honey bee (*Apis mellifera*) can detect a flower when it fills 3^0 or 5^0 of their visual field, thus from a distance of 12-18 cm.

26.5.1   Foraging behaviour in relation to pollen quality

The current understanding of honey bee foraging behaviour and the collection of pollen is that honey bees do not preferentially collect high quality pollen (Roulston, Cane and Buchman. 2000).

It is said that honey bees cannot communicate about pollen so they cannot recognise whether a pollen they consume is a good pollen source or not. Having said that honey bees seem to recognise whether or not they are consuming a good protein source because they may prefer protein rich pollens and respond to the presence of phagostimulants in the pollen, or the fact that the plant species is a good source of nectar. Wind-pollinated plants such as alder, birch, maize and hazel are less well visited probably because the flowers are not scented and have no nectar.

It is believed bees have some sense of a protein deficiency if the collected pollen is not optimally balanced in terms of its essential amino acid content. This is said to happen when foraging on alfalfa, sunflower and blackberry. Honey bee workers whose bodies are well nourished with high quality proteins are able to forage more productively for nectar and produce more honey.

27. The hoarding instinct

The honey bee superorganism has developed ways of being independent of fluctuating sources of energy through the adaptation of *Apis mellifera* colonies to store large quantities of carbohydrates, more specifically and ideally in the form of honey which contains many other important substances and not just the source of the sugars, in order to enable the colony to survive dearth periods and low temperatures

They have also developed ways to self-produce food materials such as the adaptation to store protein (as pollen) in the brood nest, bee bread and as body proteins in fat bodies.

These adaptations have enabled the colony to become more independent of a fluctuating resource and changing weather and to enable it to create a living space, whether in a natural hollow cavity in a tree or a man-made beehive where it controls temperature, humidity and is surrounded by the varying forms of nourishment it requires to survive and hopefully thrive.

Honey bee colonies in their natural state do not expand at the same rate as those managed by beekeepers, nor do they collect large amounts of nectar and store ever larger stores of honey. Instead, the colonies are more likely to slowly increase in size and to swarm much less frequently than those managed by beekeepers. These colonies show little or no signs of stress. These should be compared with colonies being artificially stimulated to increase the colony size, produce more queens, develop colonies early for pollination contracts or to work crops for honey production. Any or all of colonies subject to these factors may show signs of stress and even disease, e.g., European Foul Brood (EFB).

The influence of the beekeeper interfering with or thwarting the natural behaviour of the colony should not be underestimated.

Usually, stronger colonies can obtain and store enough honey while weaker ones in the same apiary and with the same resource availability opportunities can suffer malnutrition and even starvation.

28. Honey bee foraging

Honey bees need to be able to access a wide range of plant species in order to satisfy their heterogenous nutritional requirements. The food collection behaviour of foragers reflects the nutritional needs of the colony but how this is accomplished is presently unknown, (Wright et al. 2018).

28.1   Scout bees and foraging

Scout bees have been described as adult foraging bees which look for new food sources by independent searching rather than having been stimulated and directed through following the waggle dances of their nestmates. These dances have been well described in the literature, including Seeley (1983). It has been estimated that between 5 - 35% of the foraging bees in the colony perform this scout duty at any one time. Scout bees tend to be the more experienced foragers, and the proportion of scout bees in relation to the total foraging force depends on the availability of the nectar and pollen at that time. Scout bees may be novice foragers that are being recruited and just beginning to start entering the final chapter of their lives as foragers or they may be experienced foragers who have finally finished working an area of forage. In most cases these unemployed foragers are recruited by their nest mates through the waggle dance to visit new sources, but some find new forage sites on their own and these are termed scouts.

If forage is short, the proportion of scout bees increases thereby improving the chances of finding new resources.

Foraging bees returning to the hive bring back information about the location of the forage and its value. Only foraging bees that have found highly valuable sources of forage perform waggle dances on their return to the hive. This information is collated in the colony which acts as an information exchange centre so that the foragers can concentrate on the highest quality and the closest resources (Winston 1987).

Differences in the forage due to chance discoveries by individual bees partly explains the differences in foraging behaviour of individual colonies in an apiary where one colony may be exploiting a different forage source to its neighbour. This can be observed by the differences in pollen load colours being taken into the colonies and placed in the combs of the colonies in the apiary at the same time.

28.2   Flower constancy

Apis mellifera is a polylectic species. That is a species visiting and gathering pollen or feeding on the flowers of a variety of unrelated species of plants. In contrast other bee species (e.g., bumble bees or solitary bees) may be oligolectic and visit a narrow range of flowers of several species, typically of the same plant family or genus.

Although the honey bee species visits many species of flower an individual foraging bee on a daily basis will frequently keep to one flower species during a single foraging trip. This is referred to as flower constancy or fidelity (Free 1963). This has an advantage to the individual bee and the colony as a whole, because it makes for more efficient foraging as the bee does not have to constantly learn the locations and how to work the nectaries and anthers on each flower species it visits and enabling it to keep to a species which is yielding nectar and /or pollen. Foraging bees only tend to change to other species when the pollen of the species they had been visiting became unavailable for longer periods.

28.3   Forager worker honey bee ways of working different types of flowers

The foraging behaviour of honey bees seems to be both inherited and learned in that bees are able to forage on their initial flights to the resource but can improve their efficiency through experience with particular flowers.

Foraging bees employ different strategies to gain access to the pollen and nectar according to the type of flower.

28.3.1   In **Open** flowers, e.g., bramble and raspberries, she bites the anthers and uses her forelegs to pull the anthers towards her.

28.3.2   In **Tubular** flowers, e.g., ling heather and the disc florets of dandelion, she inserts her proboscis for nectar, and the pollen is incidentally caught on legs and mouthparts.

28.3.3   In the case of **Closed** flowers the foraging worker bee uses her forelegs to force petals apart, and gathers pollen on her mouthparts and forelegs. In field beans, garden broad beans, and scarlet runner beans the flower size and the degree of openness of the flower determine whether honey bees can enter and work it by entering at the front of the flower. Bumblebees bite holes in the base of the flower and access the nectar and honey bees will exploit this using the ready-made holes and access the nectar themselves. This technique is used with aquilegia where nectar collects in the base of the spurs of the flowers.

28.3.4   For **Spike** and **Catkin** flowers, e.g., hazel, the worker bee runs along the catkin brushing the pollen onto her body hairs.

28.3.5 In **Presentation** flowers, e.g., salvias, the abdomen is pressed against the flower forcing the pollen mass outside the flower and onto the bee's body.

Bees are capable of remembering the time and location where a particular resource (flower species) was found for up to 24 hours and the location of the flower species can be communicated to other foraging bees in a colony.

Normally the colony population has bees exhibiting a balanced division of labour and there is a balanced relationship between bees attending to brood nest related activities and those which are engaged in foraging. This balance is maintained by a pheromone mediated social inhibition whereby the presence of older worker bees in the colony inhibits younger worker bees becoming foragers.

The use of empty drawn comb, whether newly drawn or previously used stimulates the worker honey bees to store honey.

The presence of brood stimulates the collection of pollen.

If this balance is disturbed because the colony loses its foraging bees, the lack or lower presence of the forager bees in the colony means that younger adult worker bees are recruited to the foraging force. These so-called 'precocious' foragers may arise because of individual or colony starvation, pollen deprivation, diseases, pesticide damage and wax deprivation.

These precocious foragers have a shorter longevity, complete fewer foraging trips and were more likely to die during their first few flights outside the hive in comparison to adult bees which became foragers at the typical; age of more than two weeks after emergence.

Colonies which are short of protein reserves have workers which move from brood nest-based tasks to foraging at an earlier age than normal. Workers that delay the onset of foraging activities have longer lives. Colonies which have a food shortage will reduce the amount of brood they rear and because the workers are required to forage earlier their longevity is reduced.

The following information has been abstracted from Tautz (2008) and Wright et al. (2018).

28.4 Some facts about foraging

A single forager transports between 20 and 40mg of nectar in her crop.

A single forager completes between three and ten flights per day.

A single forager collects over a period of 10 to 20 days.

A single colony deploys between 100,000- 200,000 foragers per year.

A colony's foraging bees which obtains a minimum 20mg of nectar / flight x three flights per day for 10 days x 100,000 bees would produce 60 kg of nectar. This will have varying sugar concentration levels depending on plant species, weather and other factors.

A colony's foraging bees which obtains a maximum value 40 mg of nectar / flight x ten flights per day for 20 days x 200,000 bees would produce 1,600 kg of nectar. This will have varying sugar concentration levels depending on plant species, weather and other factors.

Assume 50% sugar concentration in the nectar then we could expect 30-800kg honey / year per colony.

Nectar loads are typically smaller than the full honey crop capacity (approximately 70μL) because of energetic consideration relating to flight distance, time and ambient temperature.

A medium-sized colony collects about 30 kg pollen / year.

The amount of resin brought into the hive is several hundred grams.

28.5   Basic extrapolations re nectar

Honey bee colony produces about 300 kg honey / year which take about 7.5 million excursions, assuming a full crop on return will mean almost 20 million km are flown.

25 flights are needed to fill a cell with honey, based on special group of bees concentrating the nectar of 40% sugar into honey of 80% sugar.

28.6   Basic extrapolations re pollen

A payload of 15 mg pollen distributed between two pollen baskets.

A colony collects 20-30 kg pollen /year.

This requires about 20 million foraging flights to assemble these stores of pollen.

Lipinski (2018) notes that:

The most effective foraging is done within half a mile (0.8 km)^2 of the apiary which equates to an overall foraging area of about 1 square mile, or 2.6 km^2.

The rate of energy loss / usage sets the foraging range to a maximum of about 6 miles (10 kms). After about 500 miles (800 kms) of flight the individual forager's glycogen exchange system decays and the bee dies.

During flight up to 60% of the energy is released from metabolism is as heat.

28.7   Pollen gathering by honey bees

As the bees visit the flowers pollen is transferred onto the bee's body; and periodically, often after visiting each flower, the bee hovers in the vicinity of the flower before moving to the next flower. The foraging bee first carefully grooms the pollen grains along the plumose hairs onto brushes on the middle metatarsi where it is moistened with nectar from the honey sac. From here it is scraped into the comb-like metatarsal brushes on the hind legs and then into the pollen baskets (corbiculae) on the outer surfaces of the hind tibiae. It is then compressed by the 'pollen press' formed between the top of the metatarsus and the base of the tibia. As pollen is gathered and manipulated the accumulating sticky pollen load in the corbiculum is shaped into a kidney-shaped mass, held together around a central bristle. This forms the pollen load. Displacement sensors inform the bee on how full the corbiculae are (Ford et al.. 1981).

As a rough guide the weight of two pollen pellets from a pollen forager ranges from 7.7 to 8.6 mg. García-García et al. (2004) studied the variations in the weights of pollen loads collected by *Apis mellifera* L. in Spain and the paper is interesting as it discusses the various variables which affect when and what quantities of pollen can be collected by foraging honey bees. Some of the plant species studied can also be found in the flora of the British Isles and included field bean, hawthorn, rosemary, strawberry, Echium and Anchusa.

Their findings showed:

- There were differences in the mean load weights collected in the different periods of the day.
- The heaviest loads were recorded at midday or the beginning of the afternoon.
- Considerable differences were found in the mean load weights between the different plant species.
- The pollen loads contained considerable amounts of pollen from anemophilous plant species including poplar and whose pollen loads are dry, not very cohesive and are not moulded using nectar.
- The pollen grain size and the presence or absence of nectar in the pollen load were not determining factors of the mean load weight for each species.
- Each species showed a different pattern of variation in pollen load throughout the day. The predominating species were those in which the heaviest loads were collected at midday or at the beginning of the afternoon.
- The range of variation in mean load weight during the day was very different and depended on the pollen source.
- The mean load weight of each type varies through the day and depends significantly on the rate of collection of the pollen resource in question, i.e., when the collection is more intense, the loads seem to be greater.

Clearly an additional important factor was the timing of the presentation of the pollen and its duration for each species during the day.

Honey bees do not 'buzz pollinate' even though they have the well-developed ability to vibrate their indirect wing muscles.

28.8   Communication dances and signals

Honey bees, as social insects, have evolved a highly developed ability to communicate with each other. This is used to exchange information about the status of forage sources and their qualities, and how to find them in an effective and efficient way. However, although a bee can remember shapes and the appearance of objects which she cannot communicate this information to other bees. The pioneer of the work on bee communication was Karl von Frisch and his students; their work is described in his book ***Dance Language and Orientation of Bees*** published in 1967. Bee dances and their role in honey bee colonies have been well described in Winston (1987), Seeley (1995) and more recently in Schneider and Lewis (2003). Horridge (2019) sets Von Frisch's work in a socio-political context as well as offering a critique of the conclusions from some of his work.

A number of dances are now recognised, including the round dance, the waggle dance, and a dorso-ventral abdominal vibrating dance (DVAV), now more often called vibration signals. Other dances involving trembling, jostling, buzzing, and shaking have been described by researchers. Knowledge of the significance of these activities in honey bee colonies continues to increase as a result of research and observation. More information and understanding is now available about the role of the shaking signal in the honey bees colon, (Koenig et al. 2020).

Shaking signals do not lead the recipient to change its behaviour in a specific way but instead lead to a general increase in the activity of the receiver bee (worker, queen or drone) and an increased probability that they will behave in certain ways, depending on their identity.

Worker honey bees of all ages can initiate the shaking signal but the majority of the shaking signals are initiated by the older workers. The shaking signal involves the worker bee grabbing hold of the recipient (worker, drone or queen) and vibrating the recipient bee at 16.3 +/- 5.8 Hz for 1.2 +/- 0.3 seconds. It is a vibro-tactile signal, but there are additional cues associated with it.

When workers are shaken, they increase their movements and change them in the brood nest. Those worker bees of a foraging age move to the 'dance floor'; that area of the brood nest where successful foragers perform their waggle dance to advertise foraging sites. Shaken drones increase their movement are fed and groomed more than those who are not shaken.

When a virgin queen is shaken, she is more successful at eliminating her rivals and more likely to become the queen of the colony than the virgin queen which has not been shaken.

When mated queens are shaken, they prepare to fly away with a swarm of workers and found a new colony.

Because the shaking signal is so frequently used it is unlikely to be directed towards specific workers but is a general signalling that means 'increase your activity'. It appears to play a part in organising colony-scale processes and may be involved in the regulation of the division of labour within the colony and a mechanism for older workers to progress younger workers along their life cycle in response to the changing colony needs.

The round dance is the simplest dance and serves to inform other workers in the colony that there is a useful resource near to the colony, usually within 15 metres. The dance does not communicate the precise distance or the direction of the resource from the hive.

In the round dance the worker bee makes small circles and every 2 –3 rotations she reverses and goes in the opposite direction. As she performs the dance she is closely followed and antennated (antennal contact made) by other worker bees.

The waggle dance communicates distance in terms of energy expenditure needed to reach the location, as well as direction and the quality of the forage at distances of more than 100 metres from the colony. This is achieved in its characteristic figure of eight, the number of waggles of the abdomen per vertical run across the comb face, and the direction of the runs on the comb in relation to the angle of the sun are interpreted by attendant worker bees which are recruited to forage. A food source 100 metres away from the colony is communicated by a waggle run lasting 1.35 seconds, whereas a food source located 4500 metres from the colony is shown by a waggle run of about 4 seconds.

Vibrations signal behaviour is used to regulate the foraging and other activities of workers by enhancing the performance of many different tasks. For example, the vibration signal increases the likelihood of older bees to engage in foraging, whereas younger workers will respond by increasing the time they spend on tasks inside the nest, such as food processing and brood care. In performing the vibration signal behaviour, a worker vibrates her body, particularly the abdomen, dorso-ventrally using her legs to grasp another worker.

Tremble dances are performed by workers that are nectar foragers. These dances stimulate and recruit additional bees to function as receivers and food storers when the colony's rate of nectar collection has risen above its capacity for nectar processing. These dances also inhibit the production of waggle dances, and thus reduce the nectar collection rate, until a time when the rate is in balance with the colony's nectar processing capacity. In the tremble dance the worker holds her front legs in the air, whilst moving her body from side to side and up and down in a trembling movement. She then turns her body in a different, apparently random direction and continues her trembling movement. The worker bee may continue this behaviour for more than an hour as she moves across large areas of the comb thus recruiting more receiver bees.

28.9   Food odour density

Odour signals which a foraging bee may return with from a foraging trip may override the spatial information which the bee may try to communicate in her dances and recruited experienced foragers will commonly forage for a known source of the odour rather than seek the new site encoded in the dance. It may well be that when a bee approaches a flower, she responds to the flower odour before alighting on it.

28.10 Optimisation of colony foraging efficiency and effectiveness

Seeley (1995) noted that only bees which were working the better sources of forage performed waggle dances and so account must be taken of other patches of forage which might be visited foraging and scout bees.

Honeybee foraging efficiency has been said to be most efficient at 3km radius from the hive (Visscher and Seeley 1982). They reported the most common distance was between 600-800 m from the hive with a mean of 2.3 km and 95% of foraging activity taking place within a circular foraging area with a radius of 6km. However foraging bees are capable of foraging up to 10km from the hive (Seeley1986).

Beekman and Ratnieks (2000) in their studies on the foraging for heather by honey bees from a hive set up in Sheffield (UK) found that the median distance foraged during their travelling was 6.1 km and the mean 5.5km. Only 10% of the bees foraged within 0.5km of the hive, 50% foraged at more than 6km, 25% more than 7.5km and 10% more than 9.5km from the hive location.

Bees have several instinctive behaviours which combine to help them utilise resources at close to the optimal limit. The better the quality and quantity of the nectar in the flower they are currently foraging in then the shorter the distance they will fly when they leave the hive. On the other hand, after visiting a poor source they will fly longer distances.

Honey bees in larger colonies are more effective in their foraging efforts than those in smaller colonies. This is due to the greater information-gathering capacity of larger colonies and their ability to communicate resource location via waggle dances. They achieve this by

- Foraging earlier in the day.
- Forage at greater distances than weaker colonies.
- Their foragers return with more nectar per trip.

Foraging behaviours are flexible and opportunistic in looking for new pollen and nectar sources, and when a good source of forage is found bees may remain faithful to it until it becomes depleted or less attractive for more than a day or so. After this they will sample other sources and return to the original source only if it starts to be productive again.

Foraging involves the expenditure of considerable amounts of energy and influences the activities of the foraging bee as to how she moves between patches of forage and between flowers.

A worker honey bee with a body mass of 100 mg uses about 800 J $g^{-1}h^{-1}$ of energy when flying, and around 1-3 J $g^{-1}h^{-1}$ when resting. Pollen foragers have a metabolic rate 10% higher than for nectar foragers and higher thoracic temperatures. Hovering over flowers is very energy intensive.

Foraging workers can only fly a limited distance over their lifetimes as metabolic functions associated with flight degenerate over distance. Consequently, foragers when collecting nectar may only partially fill their honey crops but the result of this behaviour may mean that there is an increase in the amount of information which the colony may receive and be able to respond to the changes in resource availability.

Research has shown that the majority of honey bee foragers do not find flowers by themselves; they are directed there by means of the waggle dance carried out by other foragers acting as long distance scouts that are already working those flowers. The scout bee that finds a worthwhile resource returns to her colony and informs other worker bees of the location and worth of the resource through the waggle dance, recruiting them to go out and exploit it.

The success of the foraging bees in their collection of pollen has a direct effect on the amount of brood that can be reared. Feeding the brood is a core activity for the nurse bees, and they adjust their brood rearing behaviour in relation to the amount of pollen stored in the brood combs. If an insufficient amount of pollen is being brought into the hive depletes the pollen stores, the foraging bees will detect it as they inspect the comb when depositing their pollen load, and more workers will be recruited to forage for pollen.

Honey bees do not forage randomly and their behaviour is modulated by learning which plants have an 'optimum nutritive' value (Hendriksma and Shafir 2016).

Foraging preferences are also influenced by the addictive qualities and visual attractiveness of plants, which are not always linked to nutritional rewards (Nicholls and de Ibarra 2017), (Thomson, Draguleasa and Tan 2015).

It is important to consider both the species composition of the food stores (bee bread) and their nutritional content. Although flora sources may be sufficiently abundant and attractive, they can only support a healthy bee population when they contain sufficient nutritional macronutrients (protein, carbohydrate, lipids etc) and micronutrients (amino acids, vitamins and minerals, etc) (Hendriksma et al. 2014).

The factors which drive temporal trends in foraging dynamics are numerous and complex and include plant resource availability, landscape diversity and climate change on the broad scale and weather patterns on the local scale.

Bees are capable of choosing the better of two forage patches according to the rate at which they can collect nectar and pollen from them. A practical consequence of this is that honey bee colonies used for crop pollination should be placed at the edge or within the target crop.

Bee visitation rates increase proportionately in flower patches that have increasing numbers of nectar-bearing flowers, nectar volume, and increasing sugar concentration.

If the nectar output is relatively high, bees pollinate more efficiently because they will visit more flowers in a given period of time, and this is important when honey bees are being used to pollinate a crop. Having encountered a patch of profitable flowers bees tend to forage in a straight line from the colony, and this has implications for the way hives are sited in crops.

Foraging bees returning from the sources do not transfer their nectar directly into the comb, but instead to a number of receiver bees that, before placing it into the comb, manipulate the nectar with their mouthparts, starting the process of reducing the moisture content of the nectar. More invertase is added to continue the conversion of the sucrose to glucose and fructose. During the transfer of the nectar the receiver bee extends her tongue and sucks up the nectar that the forager bee has regurgitated. The forager's tongue is not extended. The transferring of nectar to several receiver bees results in efficient use of time, and the faster multiple transfer of information through the colony.

The foraging bee is able todecide whether to recruit more foragers, or to recruit more receivers and these interactions have a big impact on the efficiency of exploitation of the nectar resource. If the forager perceives that she is experiencing a long wait for a receiver bee she moves away from the area of the comb where the main nectar transfer activity is being carried out and will often make a tremble dance to recruit more bees to become receivers. If the forager experiences only a short delay she often makes the waggle dance, which serves not only to direct other workers to a specific nectar source, but also to recruit more bees to foraging activities.

The bee colony operates on a decentralised basis with each forager making her own decision as to whether to recruit more foragers or receivers by making waggle and tremble dances. In this way the colony can cope with the nectar flows being brought into it. In summary, the waggle dance recruits more foragers, and the tremble dance recruits more receivers.

Honey bees from different colonies compete with each other for nectar, they are not aggressive to each other, each just tries to gets its share; however aggression is seen when robbing takes place between colonies. Some strains of honey bee are robbers and will actively seek out other colonies to rob them of their stores of honey and nectar if they can gain entrance and not be evicted or killed by the guard bees.

28.11 Foraging for water

Bees can collect water in extremely different environmental conditions. In temperate climates bees fly out and collect water at temperatures between 8-10^{0}C. The individual honey bee crop can contain around 25 mg of water.

Bees are recruited to a water source by communication dances similar to those used to recruit bees to sites rich in nectar and pollen. The bees being recruited are given droplets of water by the recruiting bees.

Water foraging bees hold their collected water in their crops, return to the hives where young bees are waiting for them on the combs. The foraging bee then transfers the water to the young bees, and this is further spread through the process of trophallaxis. There is a feedback mechanism operating so that if the water foragers are relieved of their water loads quickly, they perceive there is a demand for water and will continue foraging for water. If there appears to be a slow transfer of the water, then the water- foraging bees become more sluggish and reluctant to forage and there is slower collection of water. Water as such is stored in the crops of bees which remain motionless on the comb at the edge of the brood nest. These bees have greatly enlarged abdomens. When weather conditions prevent foraging for water this water acts as a reservoir.

In exceptionally warm and dry years bees will store water in the upper parts of combs in small, cell-like, compartments made from old wax and propolis.

28.12   Foraging for plant resins

Most resins collected by honey bees in the British Isles come from tree and poplars, in particular, if they are present locally.

The cues that honey bees used to find resinous plant sources remain unknown. One possibility is that there is a chemical released by the resinous plant which the foraging bee can detect.

Resin- foraging honey bees follow a diurnal pattern of behaviour and typically forage between 10 am and 1530 pm on sunny days, presumably because the resins are more pliable at these times.

Resin foragers can be found in colonies from May to November with the peak period from the end of June until late autumn this being related to a seasonal change in foraging.

Typically, a total of 5-15 foragers from a colony will continuously collect resin during a single day. Contrast this with around 150 nectar / pollen forgers returning to a colony every five minutes during a flow.

Resin foraging bees do not forage for pollen at the same time as collecting resin.

The way the resin forager honey bee collects and packs her corbiculae are as follows:
1.   Using her mandibles, she breaks off a particle of resin from the plant.
2.   The particle is worked by the mandibles before being taken by the forelegs.
3.   The particle is then transferred from the forelegs to the middle legs.
4.   From there it is transferred to the corbicula on the same side.
5.   Having made the particle transfer the bee might fly around for a few seconds above the resin source before landing again and adding more resin to each corbicula; the short flight possibly being a means to assess the weight of the corbicula load.
6.   Resin collection takes time and a full corbicular load make take from seven minutes up to one hour depending on the weather.
7.   She then returns to the colony.

What happens next is described in the section on its use by honey bees.

29. Landscape, land use and forage

29.1 Types of landscape and land-use patterns in the British Isles

In 1955 W.G. Hoskins published his book *The Making of the English Landscape* in which he described the evolution of the landscape combining geography with history, politics with economics, geology with botany woven into a tapestry of social and historical perspective. His book begins with the landscape before the English settlement and ends with the landscape of his day (1950's). In more recent years the ecologist Oliver Rackham was to publish several texts on woodlands and the history of the Countryside describing how their structure, composition and shape was the result of centuries of management and control by man. These included *The History of the Countryside* (1986).

Then, in 2016, Fiona Reynolds published her book *The Fight for Beauty- our path to a better future* in which she describes how our landscape and land use, nature, farming and urbanisation have been impacted by public policy decisions, economics and events. She reminds us that our sense of beauty, the landscape and nature have been subsumed into words like biodiversity, ecosystem services, natural capital and sustainable. Our landscape and land use patterns are driven by economics and increasingly attempts to retrofit these environmental descriptors into our decision making and lifestyle.

The British Isles, including Ireland, still have vegetation which reflects the factors which determine the habitats and plant species to be found in them. Michael Proctor's book *Vegetation of Britain and Ireland* (2013) was published in the New Naturalist Series in which he has traced and described the development of the landscape, habitats and the vegetation of Britain and Ireland.

All of these books, and there are others, help us as beekeepers to establish an understanding and environmental framework within which we might expect, or perhaps hope, to manage our honey bees.

In essence the British Isles contain a range of habitats with a wide diversity of plant species, some of which are visited by the honey bee.

These include:

- Agricultural land.
- Chalk downs and grasslands.
- Ponds, marshes and riversides.
- Woodland.
- Scottish pine forest and conifer plantations.
- Lowland heath.
- Moorland and mountain.
- Coasts.
- Gardens and Roadsides.

Agricultural habitats are important because this is where the pollinators live and survive. Such habitats in terms of some of the crops grown (such as oil seed rape) can provide significant amounts of pollen and nectar, but only over relatively short periods.

Agricultural intensification with its cultivation techniques, removal of competing weeds and maximisation of production (wall to wall farming) impacts habitat and consequently floral diversity at the local scale relative to the foraging capabilities of the honey bee.

Small numbers of wildflowers and trees e.g., *Acer* spp., *Prunus* spp. *and Salix* spp. are particularly attractive to honey bees and may be preferentially visited to crop flowers such as sunflower *(Helianthus annuus)* and oil seed rape *(Brassica napus)* (Requier et al. 2015).

Improved grasslands have been suggested to be of high nutritional value to pollinators because of the relative high abundance of white clover nectar, however they also represent a low floral diversity meaning that although these environments may supply additional supplies of nutrients when the white clover is in flower, they may lack the additional supplies of nutrients present in more diverse environments over a longer period.

Urban and especially suburban environments *per se* may have high local biodiversity providing more diversity and quantities of nectar and pollen than the surrounding 'natural environment'.

The habitats in which managed honey bee colonies exist in the British Isles continue to undergo rapid change. Intensively farmed areas of land provide little or no forage added to which the increasing growth of urban development and the changes these bring has further reduced the forage available for the honey bee (and other pollinating insects).

Local authorities, enlightened commercial and private developers and many of the owners and managers of our utilities and infrastructure manage significant areas of land associated with these activities and they have started to manage the land they are responsible for in a more pollinator - friendly way. Honey bees are one of the beneficiaries of these resources. As beekeepers we should recognise and encourage this.

The ability of honey bees to continue to be kept and provide their essential pollination activities inevitably means that bee health and productivity (honey and pollination) will be best served by targeted approaches to establishing and managing landscapes to improve the availability of nutritional resources to improve bee and colony level nutritional resilience to the effects of environmental stressors, including the beekeeper.

There are many locations where honey bees are kept in the British Isles where they have access to pollen from a variety of plant species and therefore have the potential to have a balanced and varied diet.

It sees however that although honey bees are effective pollinators and able to utilise many plant species observations have shown that honey bees may only be collecting pollen from a proportion (<50%) of the plant species available to them.

Where bees are kept in areas where there is intensive agriculture practised this can lead to an impoverished diet and this diet may not contain all the essential amino acids and other nutrients which the bees need to develop and sustain healthy colonies and additionally from the beekeeper's perspective, productive colonies.

Generally single plant species diets are inferior to those of mixed diets. Diverse pollen diets may compensate for essential nutrients in the pollen of one species (an example is arginine in dandelion). Simply eating more pollen from the same plant species does not make any difference and continues to cause nutritional stress.

In most places where honey bees are typically kept in the British Isles it is unusual for a colony to be only able to source pollen from one plant species at any one time even in areas of intensive agriculture and so it is unlikely that a colony will be forced to exist on a low protein-containing pollen which is lacking one or more of the essential amino acids.

A constant, moderate and long-lasting good nectar flow coupled with a low protein content pollen is highly significant for the ability of the honey bees to adapt to different environmental forage sources.

It is the total quantity of pollen available to a colony over the course of the active season which has the most impact on affecting the development, expansion and ultimate survival of a colony.

Even access to a wide variety of sources or a single high- quality pollen will not better a colony which has access to large amounts of pollen as this influences the amount of available food for the young adult bees and the developing larvae. It should also be noted that plant species that produce lots of nectar may not always provide pollen for the honey bees (Ellis et al. 2017).

Other authors have concluded that in many cases honey bees collected more pollen from plants poorer in protein, which were in total more beneficial to the colony compared to pollen collected which was rich in protein, but obtained in smaller quantities (Lilolios *et al* 2015).

29.2   Changes in forage with especial reference to honey bees as reported by beekeepers

Reference has already been made to Donald Sim's reflections on changes which had taken place during his 60 years of beekeeping experience and this section further details the changes which have taken place and been reported by beekeepers when they have participated in various surveys and other reporting over the years since 1945.

Since the end of the Second World War there have been considerable changes in the landscape and land use patterns throughout the British Isles.

Much land had been ploughed up for the war effort and after the war much of this land was repurposed and was covered in urban development in the rebuilding that took place. This was subsequently followed by the development of the New Towns and significant publicly funded house building, commercial development and then the growth of private house building. Much of this building took place and continues to take place on formerly prime agricultural land. Pasture-land, often found on poorer soils, ploughed up in the war was returned to grassland for use by the now less viable livestock farming sector and to an increase in forestry. The grubbing up of old orchards, woodland and hedgerows continued unabated as arable fields were increased in size and intensification of farming practices, bigger machinery and the use of pesticides (insecticides, fungicides, herbicides, molluscicides, acaracides) were deployed to control pests and increase yields, (Carreck 1994) describes these changes in more details.

In the times when Manley (1936) and the Herrod-Hempsall (1936) brothers practised beekeeping it seemed that clover and fruit trees were considered the most important bee forage plants.

In 1958, ASC Deans prepared a thesis for submission to the National Diploma in Beekeeping Board. It contained the results of the first large scale survey of honey sources in the British Isles. It involved the analysis of honey samples from 66 counties of the British Isles. Included in the thesis is a table (No. 15) setting out the most useful nectar yielding plants in England, Scotland, Wales and Ireland. (Deans 1958). This is reproduced below.

29.3 Most useful nectar-yielding plants in England compared with Scotland, Ireland, and Wales

Plant.	Percent of samples in which pollen occurs.			
	England	Scotland	Ireland	Wales
Trifolium repens.	91.2	93	100	93
Prunus/Pyrus spp.	77.4	47	76	73.3
Rubus spp.	54.1	45.1	36	79.7
Acer spp.	49.8	36.3	40	67
Castanea spp.	47.4	7.8	8	20
Tilia spp.	49.7	20.6	4	33.3
Brassica spp.	52.5	38.2	32	46.6
Ligustrum spp.	33.8	7.8	20	40
Vicia spp.	33.9	24.6	8	6.7
Trifolium pratense.	31.8	10.8	20	-

Plant.	Percent of samples in which pollen occurs.			
	England	Scotland	Ireland	Wales
Chamaenerion spp.	22.7	34.2	4	40
Cirsium spp.	23.2	25.5	8	40
Campanula spp.	21	10.8	4	13.3
Calluna vulgaris.	6.8	41.1	20	6.7
Heracleum type.	25.6	18.6	20	13.3
Taraxacum type.	25	25.5	52	33.3
Centaurea spp.	21	13.7	32	13.3
Aesculus spp.	15.9	3.9	-	13.3
Erica spp.	9.3	20.2	4	33.3

Source: Deans 1958 unpublished

Deans concluded that on the basis of his 1958 survey the most important honey producing plants in Britain were *Trifolium repens*, *Prunus/Pyrus* spp. Also included were *Crataegus* spp., *Rubus* spp., *Acer* spp., *Tilia* spp., *Castanea* spp., *Brassica* spp., *Ligustrum* spp., *Vicia* spp., *Trifolium pratense*, *Heracleum type*, *Aesculus* spp., *Chamaenerion* spp., *Cirsium* spp., *Campanula* spp., *Taraxacum* spp., *Centaurea* spp., *Erica* spp. and *Calluna vulgaris*.

He further concluded that individual analyses indicate that only two single source type honey are normally gathered in the British Isles. (1) white clover honey and (2) ling heather honey. Regions which offer a wide range of economic bee plants whose flowering time extends over the greater part of the summer season and which do not show marked topographical differences, tend to produce honey of a composite nature. In counties where it is possible to distinguish clearly defined natural regions such as high moorland and lowland, characteristic, single source honeys may be obtained.

Suburban type honeys tend to show a greater number of species in the pollen spectra and may contain, in some cases, an admixture of honey-dew.

In the BBKA surveys carried out in 1985 and 1994 members listed, in order of importance, the following species as being important forage plants, (Carreck 1997). The last column details the result of a survey with members of the Bee Farmers Association. More information on these surveys are included in a previous table.

1985 Survey by BBKA Members	1994 Survey by BBKA Members	1994 Survey by Bee Farmers Association
Oilseed rape	Oilseed rape	Oilseed rape
Lime trees	Blackberry	Field beans
Apple, pear, plum, cherry	Apple, pear. Plum, cherry	Apples
Blackberry	Lime trees	Pears
Heather	Field beans	Clover
Clover	Clover	Strawberries
Field beans	Sycamore	Currants
Sycamore	Chestnut	Cherries
Chestnut	Hawthorn	Borage
Hawthorn		

Carreck (1994) describes the changes which took place in agriculture and other landscape changes in the 50 years up to 1994.

Since then, the processes described by Carreck (1994) have continued and in spite of a number of farm- based initiatives. If anything, the process has accelerated. Later in the book we look consider the changes which will take place now that the UK has left the European Union (EU) and is intending to change from the requirements of the Common Agricultural Policy (CAP) to a policy which is intended to pay farmers for environmental 'public goods' rather than a subsidy system which was based on the amount of land tenure.

More recently Jones et al. (2020) reported the results of a nationwide UK project where beekeepers provided honey samples which were subjected to the techniques of DNA metabarcoding to enable identification of the plant species whose nectar had contributed to the honey samples.

Abundance & frequency within the 2017 samples	Common name of species	Botanical name
Most frequently found and abundant. All species have long-flowering periods, typically May to September	Brambles	*Rubus* spp.
	White Clover	*Trifolium repens*
	Oilseed rape & wild and cultivated species	*Brassica napus*
Next Group. Spring-flowering shrubs and trees	Hawthorn	*Crataegus monogyna*
	Apple	*Malus* spp.
	Cherries and plums	*Prunus* spp.
	Cotoneaster	*Cotoneaster* spp.
	Sycamores and Maples	*Acer* spp.
Towards end of season peaking in September	Heather / Ling Himalayan balsam (non-native invasive species)	*Calluna vulgaris* *Impatiens glandulifera*

The list of taxa found in more than 5% of the 441 honey samples obtained in 2017 are contained in the following list. The listing is from highest to lowest frequency and abundance. The paper contains more details of the frequency classes and the relative abundance for each taxon.

Botanical name	Common name
Rubus spp.	Currants
Trifolium repens	White Clover
Brassica spp.	Brassicas
Maleae (Crataegus/Malus/Cotoneaster)	Hawthorn/ Apples/ Cotoneaster
Acer spp.	Sycamore / maples
Quercus spp.	Oaks
Rosa spp.	Rose
Prunus spp.	Cherries
Asteraceae	Daisy
Aesculus hippocastanum	Horse-Chestnut

Botanical name	Common name
Vicia spp.	Vetch
Taraxacum officinale	Dandelion
Sambucus / Viburnum spp	Elder / Viburnums
Ligustrum spp.	Privets
Sorbus spp.	Whitebeams
Ulex spp.	Gorses
Myosotis spp.	Forget-me-nots
Fabaceae	Pea family
Salix spp.	Willows
Impatiens glandulifera	Himalayan Balsam
Filipendula spp.	Meadow-Sweet
Ilex spp.	Holly
Centaurea spp.	Knapweeds
Ceanothus spp.	Californian lilac
Camellia spp.	Camellia
Epilobium / Oenothera spp	Willowherbs / Evening Primrose
Phormium tenax	New Zealand flax
Allium spp.	Onions
Apiaceae	Carrot family
Tilia spp.	Limes
Eucalyptus spp.	Eucalypts
Calluna vulgaris	Ling
Hedera helix	Ivy

Botanical name	Common name
Hyacinthoides non-scripta	Bluebells
Pinus spp.	Pines
Lotus spp.	Bird's-foot-trefoils
Cirsium / Hypochaeris spp.	Thistles / Cat's Ears
Castanea sativa	Sweet Chestnut
Buddleja / Verbascum spp.	Butterfly-bush / Mulleins
Ailanthus altissima	Tree-of-Heaven
Trifolium pratense	Red Clover
Magnolia spp.	Magnolia
Borago officinalis	Borage

The paper reported there were no overall regional differences between England, Scotland and Wales in terms of the most frequently found plant taxa in the honey samples provided in 2017. Such a statement should perhaps be tempered by appreciation of samples taken from honeys produced from habitats such as heather moorlands which can be large areas.

Jones et al. (2020) compared the findings of Dean and his 1952 year's samples with those of the samples provided in 2017. Those reported by Deans (1958) who had used melissopalynological techniques (honey analysis using pollen grains present in the honey) whilst Jones et al. (2020) used the technique of DNA metabarcoding.

Common Name of Species	Botanical Name	% presence in honey samples		Change reported in Countryside Survey between 1978 and 2017
		Year 1952	Year 2017	
White Clover	*Trifolium repens*	74	31	-13%
Red Clover	*Trifolium pratense*	5	1	-27%
Brambles	*Rubus* spp.	5	31	+21%
Brassicas (including oilseed rape)	*Brassica* spp.	1	21	
Field Beans	*Vicia faba*			Production of field beans increased since 1945
Vetches(wild)	*Vicia* spp.			-26%
Himalayan Balsam	*Impatiens glandulifera*	1	6	

The UKCEH Countryside Survey has been monitoring the countryside of Great Britain since 1978, recording status and trends and compiling information which will inform decision-making. It is an unbiased sample of the soils and the common plants of the wider countryside https://countrysidesurvey.org.uk/

In December 2020 a free new app was published containing data on plant diversity and occurrence over 40 years showing the changes which have taken. https://shiny-apps.ceh.ac.uk/common_plant_change/

The UK Centre for Ecology and Hydrology is also conducting the CEH National Honey Monitoring Scheme which will, over a period of years, will analyse samples of honey supplied by beekeepers under a strict sampling protocol with one of the outcomes being the monitoring of plant species whose pollen is found in the honey samples, using DNA Barcoding techniques and examining the results in relation to the land use around the apiary sites from which the honey samples were taken. Other environmental factors will be included, such as pesticide residues. Progress and reports from the project should be able to be found on honey@ceh.ac.uk and https://honey-monitoring.ac.uk

30. The plant palette in the garden and pollinators

30.1 Where to Begin?

There are many informative gardening books, scientific publications and websites with an ever- increasing interest in and containing information and advice on supporting pollinator species in the garden.

Much information on which plants are attractive to pollinators, including honey bees, can also be found on the websites of a number of organisations including charities such as Royal Horticultural Society (RHS), the Woodland Trust, Plantlife, the British Beekeepers Association and other beekeeping associations, as well as from suppliers of plants to name but a few.

It is beyond the scope of this book to catalogue all plants which are available to the gardener and whether or not a plant species or its subspecies, varieties and cultivars are of benefit to honey bees.

The tables set out in this book contains compiled information from a range of sources, including the internet. It is intended to stimulate interest in the variety of plants of value to the honey bee and to encourage gardeners and growers to understand the benefits their efforts bring to helping pollinators survive and thrive and to recognise the contribution and importance of pollinators, especially the honey bee in the case of this book.

We have also included plants found outside the garden and in the natural or semi-natural habitats found throughout the country. These are a vital resource and require appreciation, careful husbandry, and further encouragement.

Just to remind readers there are a number of books include a wealth of information of particular relevance to honey bees and the reader is encouraged to take the opportunity to sample them. They include, Maclean (2015); Hooper and Taylor (1988); Kirk and Howes (2012); Howes (1979).

Gardens can be large or small, needing high or low maintenance, floristic, or devoted to growing crops and soft and top fruit. There is always space for plants which will benefit pollinators.

Starting from a general perspective the next table contains a non- exhaustive list of plant species which can be grown in gardens and will attract and benefit many species of pollinating insects (including bees, butterflies, moths and hoverflies).

Ideally a garden which is intended to be beneficial to pollinating insects, including honey bees, should be designed and managed to provide a continual flow of nectar and pollen all through spring, summer, autumn and in warmer climes, in winter.

The table also contains information on whether the species provides a source of nectar and or pollen to the honey bee.

Trees are listed in a separate section.

30.2 Garden plants rich in nectar and / or pollen which attract 'pollinating insects' over the course of a year in a garden

Common name	Botanical name	Visited by honey bees, for either nectar / pollen, or both	
		Nectar	Pollen
Winter			
Winter flowering Clematis	*Clematis cirrhosa*		
Crocus	*Crocus aureus*	+	+
Winter Aconite	*Eranthis hymalis*	+	+
Snowdrop (single flowered)	*Galanthus nivalis*	+	+
Hellebores	*Helleborus* spp. *e.g., Stinking hellebore (Helleborus foetidus)*	+	+
Mahonia (Oregon Grape)	*Mahonia aquifolium, M. japonica*	+	+
Sweet Box	*Sarcocca* spp. e.g., *S. confusa*	+	+
Winter-flowering heather	*Erica carnea, E. arborea*	+	+

Spring			
Aubretia	*Aubretia deltoidea*	+	+
Berberis	*Berberis* spp., e.g., *B.darwinii,*	+	+
Marsh Marigold	*Caltha palustris*	+	+
Japanese Quince	*Chaenomeles speciosa*	+	+
Quince	*Cydonia oblonga*	+	+
Wallflower	*Cheiranthus cheiri*	+	+
Hawthorn	*Crataegus monogyna*	+	+
Crocus	*Crocus aureus*	+	+
Spring-flowered heathers	*Erica* spp.	+	+
Candytuft	*Iberis saxatilis, I. umbellata*	+	+
Mahonia	*Mahonia aquifolium*	+	+
Dessert and Culinary Apples	*Malus domestica*	+	+
Crab Apples	*Malus floribunda*	+	+
Cherry (single flowered)	*Prunus cerasus*	+	+
Lungwort	*Pulmonaria officinalis*	-	+
Pear	*Pyrus communis var sativa*		
Flowering currant	*Ribes sangineum*	+	+
Skimmia	*Skimmia japonica*	+	+
Wood Anemone	*Anemone nemorosa*	-	+
Summer			
Yarrow	*Achillea millefolium, A. filipendulina*	+	-
Anchusa	*Anchusa officinalis*	+	-
Anthemis	*Anthemis tinctorial, A. cupaniana*	+	+
Snapdragon	*Antirrhinum majus*	+	-
Columbine	*Aquilegia vulgaris*	-	-
Majorwort	*Astrantia major / A. maxima*		+
Borage	*Borago officinalis*	+	+

Buddleia	Buddleja alternifolia ; B.davidii, B. globosa	+	-
Catmint, Catnip	Nepeta catarea, N. gigantea, N. racemosa	+	+
Field Marigold	Calendula arvensis		
Bell flowers Harebill, Clustered bell flower, Carpathian harebill	Campanula spp. Campanula rotundifolia C. glomerata C. carpatica	+	+
Clematis Traveller's Joy	Clematis spp. (single-flowered) Clematis vitalbi Clematis almondii Clematis montana	+ + +	+ + +
Tickseed	Coreopsis grandiflora / C. vertillata	+	+
Cosmos	Cosmos bipinnatus	+	+
Cotoneaster	Cotoneaster conspicuus	+	+
Bell Heather	Erica cinerea	+	+
Ling (Heather)	Calluna vulgaris	+	+
Cross-leaved Heath	Erica tetralix	+	+
Cornish Heath	Erica vagrans	+	+
Irish Heath	Erica erigena	+	+
Dahlia	Dahlia spp (single-flowered)	+	+
Foxglove	Digitalis purpurea	-	-
Globe thistle	Echinops humilis	+	+
Viper's bugloss	Echium vulgare	+	+
Fleabane	Erigeron spp.	+	+
Escallonia	Escallonia macrantha	+	+
Californian Poppy	Escholtzia california	+	-
Fuschia	Fuschia magellanica	+	+
Blanket Flower	Gaillardia aristata	+	+

Crane's Bill	*Geranium (single-flowered) various including G. sanguineum, G. cinereum,*	+	+
Water Avens	*Geum rivale*	+	+
Sneezeweed	*Helenium autumnale*	+	+
Hydrangea	*Hydrangea anomala petiolaris*	+	+
Hyssop	*Hyssopus officinalis*	+	+
Scabious	*Scabiosa atropurpurea*	+	+
Field Scabious	*Knautia arvensis*	+	+
Lavender	*Lavandula angustifolia; L. stoechas*	+	+
Ox-eye daisy	*Leucanthemum vulgare*	+	
Pot Marigold	*Calendula officinalis*	+	+
Ribbed Melilot	*Melilotus officinalis*	+	+
Kniphofia	*Kniphofia caulescens*	+	+
Privet	*Ligustrum sinense*	+	+
Garden mint	*Mentha spicata*	+	+
Forget-me-not	*Myosotis spp., e.g., M. arvensis*	+	+
Catmint	*Nepeta gigantea*	+	+
Catnip	*Nepeta cataria*	+	+
Evening Primrose	*Oenothera biennis*	-	+
Wild Marjoram / Origanum	*Origanum vulgare*	+	+
Poached Egg Plant	*Limnanthes douglasii*	+	+
Penstemon	*Penstemon spp./ cultivars*	-	+
Phacelia	*Phacelia tanacetifolia*	+	+
Jacob's Ladder	*Polemonium caeruleum*	+	+
Firethorn	*Pyracantha coccinea*	+	+
Dog Rose	*Rosa canina*	+	+
Rosemary	*Rosmarinus officinalis*	+	+
Coneflower	*Rudbeckia* spp. (single open-centred); e.g., *R. laciniata*	+	-

Sage	*Salvia officinalis*	+	+
Stonecrop	*Sedum spectabile*	+	-
Golden Rod	*Solidago virgaurea*	+	+
Lime	*Tilia* spp.	+	+
Verbascum / Mullein	*Verbascum chiaxii, V.thapsis, V.olympicum*	+	+
Verbena / Vervain	*Verbena bonariensis*	+	-
Broad Bean	*Vicia faba*	+	+
Weigela	*Weigela florida*	+	-
Autumn			
Monk's Hood	*Aconitum napellus*	-	-
Windflower	*Anemone blanda;* Japanese Anemone, *A. japonica*	-	+
Aster / Michaelmas daisy	*Aster novi-belgii*	+	+
Dahlia	Dahlia (single-flowered)		
Giant Sunflower	*Helianthus annuus*	+	+
Salvia	*Salvia officinalis;* Clary Sage *S. viridis;* Wild Sage *S. verbanaca;* Meadow Sage *S. pratense*	+	+
Pasque flower	*Potentilla vulgaris*		+

Please Note that <u>not all</u> of the species listed are visited by honey bees.

Bees of all kinds do not benefit from 'double' flowered varieties of the following species paeonies / marigolds / dahlias / roses / clematis / busy lizzies /daisy / chrysanthemums / camellias / begonia / zinnia / hellebores as they usually do not produce pollen and / or nectar.

31. Hedgerows and boundaries and screening

31.1 The importance of hedgerows, boundaries and screening

Then there is the use of tree and shrub species in the creation of hedges, boundaries and screening. And, of course, the protection and maintenance of our existing hedgerows.

Pollard, Hooper and Moore's book titled **Hedges** published in the New Naturalist Series in 1974 is still well worth reading and comparing their descriptions with what we see in today's landscape.

In 1987 Muir and Muir published their book **Hedgerows – *Their History and Wildlife*** which is full of information and reading it makes one appreciate what is left of our hedgerow legacy and yearn for a more enlightened understanding of their beauty, enhancement of the landscape as well as being a resource of practical and ecological importance.

More recently Wright (2017) published his book on the natural and cultural history of hedges (as well as ditches, dykes and dry stone walls), from the arrival of the first settlers in the British isles to the modern day.

Sadly, the traditional skills of hedge-laying by hand, cutting and shaping using hand tools have all but disappeared from mainstream land and boundary management practices but there is an increasing interest in reviving them and re-introducing them into land management, including agricultural field boundaries by farmers and land managers who see the contribution and enhancement they can make both aesthetically, practically, and ecologically.

Hedgerows can be managed to provide good sources of pollen and nectar from spring to the autumn. Then after pollination and ripening the nuts, fruits and berries can be harvested either by humans foraging, or by birds and mammals.

In early spring pussy willow catkins and blackthorn flowers are especially important. Other shrubs and trees include hawthorn, crab apple, wild cherry and wild plum flower later in the spring.

In summer it is the flowers of the hedgerow margins which are important and in periods of drought or extended periods with no rain and hot weather becoming an increasing tendency to occur, flowers growing in the hedges along ditches are especially important.

Later in the autumn ivy can produce a lot of nectar and pollen.

Management techniques such as mechanical flailing or hedge-cutters significantly reduce the ability of wildlife to benefit from the fruit, berries and nuts which could be produced. Such management also significantly reduce the ability of the plants to produce flowers and so deny pollinating insects valuable forage resources. However, such species when left or are managed intentionally to become trees are of value to honey bees and other pollinating insects. This species included in this table are those which will benefit honeybees.

Hedgerows in a landscape can be managed so that they can provide nectar and pollen sources from spring to autumn.

Postponing the hedge cutting until February or March can increase the amounts of berries available to overwintering birds.

For example, allow the hedge shrub and tree species to flower by cutting them on a three- year rotation basis – cutting no more than a third of the hedges in any year. Cutting only once every three years results in 2.5 times more hawthorn and blackthorn flowers than cutting every year.

Alternatively, if cutting annually, raise the cutting height by 10 cm (4 inches) each time. Also leaving a few uncut bramble shoots and rose outgrowths will be beneficial.

Hedges can be 'gapped' up with mixed species including hawthorn, blackthorn, buckthorn, wild cherry, wild privet, guelder rose and crab apple.

Perennial wildflowers in hedgerows, banked walls and ditch banks can provide nesting placers for pollinator species as well as a source of pollen and nectar throughout the majority of the year, excluding winter. These flowering times will vary depending on season and location and the timing of the cutting regime employed by the persons responsible for their management.

31.2   Plants species commonly found by hedgerows, on ditch banks and banked walls visited by honey bees

There are herbaceous plant species associated with the hedgerows, ditch banks and banked walls which are visited by honey bees.

The following non-exhaustive list identifies species commonly found by hedgerows and on ditch-side banks and bank walls.

Hedgerow / ditch banks / banked walls		
Plant Name	**Species Name**	**Typical Flowering Period**
Blackberry	*Rubus fruticosa*	July-August
Blackthorn (Sloe)	*Prunus spinosa*	March-May
Bugle	*Ajuga reptans*	May – late July
Common Agrimony	*Agrimonia eupatoria*	June-July
Common bluebell	*Hyacinthoides non-scriptus*	April – June
Common fleabane	*Pulicaria dysenterica*	July to September
Common gorse	*Ulex europaeus*	Throughout the year
Common knapweed / hardheads /scabious	*Centaurea* spp.	June to September
Common vetch	*Vicia sativa*	April to September
Field Marigold	*Calendula arvensis*	May to September
Cowslip	*Primula veris*	April to May
Dandelion	*Taraxacum officinale*	May-June
Devil's-bit scabious	*Succisia pratense*	July-August
Dog rose	*Rosa canina*	June-July
Field scabious	*Knautia arvensis*	July - August
Foxglove	*Digitalis purpurea*	June - September
Garlic mustard	*Alaria petiolata*	April - June
Goat willow	*Salix caprea*	February-March
Ground ivy	*Hedera helix*	September - November
Hawthorn	*Crataegus monogyna*	May - June
Hedge bedstraw	*Galium mollugo*	June - September

Hedgerow / ditch banks / banked walls

Plant Name	Species Name	Typical Flowering Period
Hedge bindweed	*Calystegia septum*	June – September
Hedge woundwort	*Stachys sylvatica*	June - September
Holly	*Ilex aquilifolium*	April - May
Ivy	*Hedera helix*	September -October
Lesser burdock	*Arctium minus*	July - September
Musk mallow	*Malva moschata*	July - August
Ox-eye daisy	*Leucanthemum vulgare*	May - September
Perforate St John's wort	*Hypericum perforatum*	May - August
Primrose	*Primula vulgaris*	December - March
Red campion	*Silene dioica*	May - June
Red dead nettle	*Lamium purpurea*	February - November
Rosebay willowherb	*Chamerion angustifolium*	July-August
Ragged robin	*Lychnis flos-cucili*	May - August
Selfheal	*Prunella vulgaris*	June - November
Spear thistle	*Cirsium vulgare*	July - September
Tufted vetch	*Vicia cracca*	May - September
White dead nettle	*Lamium alba*	February - November
Wild teasel	*Dipsacus fullonum*	July - August
Wood avens	*Anemone nemorosa*	March - May
Wood sage	*Teucrium scorodonia*	July - August
Yarrow	*Achillea millefolium*	July - September

32. Roadside Verge Management

32.1 An important resource

The last few decades have seen changes in the flora of our roadside verges. These have been driven mainly by the way highway and local authorities, the public and farmers cut and manage verge habitats and the increasing fertility of roadside soils through the increase in the deposition of nitrogen.

Muir and Muir (1987) include a chapter titled *A Case for the Roadside Verge*, observing that many roadside verges have ancient origins marking routes often used since prehistoric times, but of course many have been artificially created as the result of road schemes.

Roadside verges may abut hedges, banked walls and ditch banks and so many of the species listed in the table may also be found in the roadside verge.

Mowing frequencies vary from location to location, seasonally in response to weather conditions, with the policy of the local authorities and the decisions made by individuals also varying. The type of mowing machines and the effective scalping of the vegetation used often reduce the vegetation and the soil to dust and this effect is exacerbated during long periods of dry weather.

The use of herbicides along road edges and around the bases of poles, utility manhole covers, road signs have also added to the factors causing a reduced road-side flora. Health and safety on the highway and the obsession with tidiness are two key drivers to these practices. Grass cuttings, if left where they are cut leave nutrients which further encourage the grass and smother the wildflowers.

Some species have done well and are spreading, while others have suffered, and have been lost. With the move to earlier and earlier cutting in spring, summer flowering species are being lost from our verges. Only plants that flower early have a chance to set seed before the mowers and the sprayers arrive. As a result, some spring flowers are thriving and spreading, but summer flowering plants – many of which can be found in our meadows are disappearing.

The ability of plants to reflower and in what abundance after mowing determines whether the plant is of any further use to pollinating insects. Such reflowering might result in a phenological mismatch between plants and their pollinators and this particularly true where a plant species is dependent on a particular pollinator species or type which have a limited lifespan.

This isn't just bad news for flowers, it is bad news for the bees (including honey bees), but especially bumble bee and solitary bee species, beetles, butterflies and birds that rely on the plants for food and have a limited foraging range. In addition, vigorous perennial plants and some invasive introductions are better at surviving the tough roadside conditions created by management practices, so these are increasing at the expense of more delicate species.

Many bee species are oligolectic which is a term in pollination ecology used to describe bee species that have a narrow, specialized preference for pollen sources, typically to a single genus, multiple genera in a single plant family or even more restricted to a single plant species. The honey bee is a polylectic species which forages from a wide range of plant species and also forages relatively widely across the landscape.

Plantlife (www.plantlife.org.uk), is a charity that speaks up for our wild plants, lichens and fungi and aims to raise their profile. Their guide *The Good Verge Guide - a different approach to managing our wayside verges* was published in June 2016. It states that more than 700 species of wildflowers (>45% of the UK flora) grow on verges somewhere in the UK.

32.2   Top 10 plant species for roadside verges

The top 10 road verge flowers that support the highest numbers of invertebrates are as follows with those visited by honey bees marked with an asterisk:

Bird's-foot trefoil	*Lotus corniculatus*	160 insect species supported*
Yarrow	*Achillea millefolium*	141 species
Red clover	*Trifolium pratense*	115 species
Dandelion	*Taraxacum officinale*	107 species*
Ragwort	*Senecio jacobaea*	107 species*
Lady's bedstraw	*Galium verum*	101 species
Hedge bedstraw	*Galium mollugo*	100 species
White clover	*Trifolium repens*	98 species*
Meadowsweet	*Filipendula ulmaria*	91 species*
Ox-eye daisy	*Leucanthemum vulgare*	85 species*

33. Gardeners and their grass ('lawns')

33.1 Useful flowering plant species which can be grown in grass and the effects of different management techniques

Cutting the grass is seen by many as a chore with the result that the grass is cut as short as it can be and hopefully survive and regrow minimising the amount of time, effort and expense needed in the future. For many other gardeners the lawn receives the attention of scarification, fertilising, drainage, weed eradication using weed killers and the immaculate close-cut finish with stripes.

There is a third way and that is to think of the grass as a surface onto which you add colour and texture achievable by the creative use of the mowing machine. Patterns can be cut into the grass and the grass managed at various heights, creating pathways of shorter grass and islands of longer grass. Data collated by Plantlife has shown that short or low growing species are adapted to areas that are cut, every one to four weeks. These species grow closely to the soil surface, often spread by vegetative growth, and tend to produce a lot of flowers on short stems which produce seed between the cutting events.

Such species include:

- Daisy (*Bellis perennis*)
- Bird's foot trefoil (*Lotus corniculatus*)
- Dove's foot cranesbill (*Geranium molle*)
- Selfheal (*Prunella vulgaris*)
- Mouse-ear hawkweed (*Hieracium pilosella*)
- Thyme- leaved speedwell (*Veronica serpyllifolia*)
- White Clover (*Trifolium repens*)

Species which are adapted to growing in taller grass which is mown every two months or only once at the end of the summer include:

- Ox-eye Daisy (*Leucanthemum vulgare*)
- Cuckooflower (*Cardamine pratense*)
- Cat's ear hawkbit (*Hypochoeris radiata*)
- Cow parsley (*Anthriscus sylvestris*)
- Meadow buttercup (*Ranunculus acris*)
- Yellow Rattle (*Rhinanthus minor*)
- Meadow Crane's-bill (*Geranium pratense*)
- Musk mallow (*Malva moschata*)
- Common Knapweed (*Centaurea nigra*)
- Red clover (*Trifolium pratense*)
- Tufted vetch (*Vicia cracca*)
- Wild carrot (*Daucus carota*)

The results of a survey by Plantlife in 2019 gave the following results.

Flower and nectar sugar production was highest on lawns cut once every four weeks. This gives 'short-grass' plants, like daisies, selfheal and white clover, a chance to flower in profusion, boosting nectar production tenfold.

But areas of longer unmown grass were more diverse in their flowers, with plants like oxeye daisy, field scabious and knapweed increasing the range of nectar sources for different pollinators and extending nectar availability into late summer.

These findings suggested keeping two different lengths of grass:

Short-grass: Keep the grass shorter by mowing once every month – 4 or 5 times a year – to a height of 1 or 2 inches (2.5 to 5 cm). Some flowers will be cut off when you mow but they will usually flower again quickly; you can even rotate patches around your garden so there are always some areas in flower.

Long-grass: Leave some patches completely unmown to let taller flowers bloom and set seed. Cut these areas at the end of summer or early autumn.

34. Trees and shrubs

34.1   Trees matter

It is a rare occasion when a discussion on climate change, fossil fuels and renewable doesn't involve trees. In turn such discussions increasingly raise the profile of deforestation, the loss of forest from human activities, either through the clearing of forests to release more land for agriculture or through extreme weather conditions. In the British Isles we are concerned about the poor management and destruction of woodland for example to make way for the HS2 railway and urban and commercial development. Some woods / forests are often of great antiquity and so easily destroyed or their special circumstances changed resulting in their loss or significant degradation and loss of associated plant, animal and microbial species diversity. We have also been made aware (if many of us were not already) of the health and mental benefits of both seeing and being close to trees and inside woodlands.

At the start of the 2020's this has led to the stated intention of politicians around the world to plant many millions, nay trillions of trees, a promise which falls easily from the politician's tongue. Quite how this will be achieved and whether there is the supply and skill base to cultivate and manage such an enterprise is another question. However, let us be positive and see how the interests of honey bees and their pollinating insects might benefit and how we can help shape policy and practices which will not only achieve the intended objective of locking up carbon but also to facilitate and promote a recovery in the biodiversity on our planet and on a more parochial basis the prospects for the British Isles. The British Isles currently has a low percentage of tree cover and such cover is not uniformly distributed throughout the country. The history of trees and woodlands in the British Isles is a fascinating one and those interested in reading more about it are directed to Oliver Rackham's book titled **Woodlands** and published in the New Naturalist series.

Parts of the country were planted with fast growing coniferous species and large areas of upland landscape were covered in the uniform characteristics of plantation-grown blocks of trees. The requirement for such a resource was huge following the experiences of the First

World War and the need for timber for the war effort. Having said that shortages of finding suitable sized and shaped oak trees for the British Navy in earlier times not only helped the rise of the iron ships but also demonstrated the strategic need for timber. Today's needs and future demands for wood- based resources are somewhat different and the relearning of old skills of water catchment management are being employed to help mitigate and manage the increasing risks of flooding and as importantly the means of removing carbon dioxide from the atmosphere.

Those involved in today's forestry and woodland management have as much an interest in the potential for creating biodiversity and the benefits of green and green 'spaces' to the mental health and well- being of the human population as well the production of the wood and wood- based products of the trees. Such expertise will be in huge demand as the desired uses, conception, design, planning and creation of the features into which the trees to be planted will be put together.

In 2020 the RSPB commissioned a review titled **Woodlands for climate change and nature** (Crane 2020) of woodland planting and management approaches in the UK for climate change mitigation and biodiversity conservation. The review concludes that there are many dilemmas to tackle when planting new woods and managing existing ones. There is little reference made to pollinator species, including honey bees.

Earlier in the book we referred to the hunting of honey bee nests for their honey and wax in forest and woodland in previous centuries and how honey bees were very much a woodland -inhabiting species. Today Tom Seeley's book (Seeley 2016) reminds us of the relationship of the honey bee with trees, a relationship which we as beekeepers often fail to recognise.

There is no doubt that honey bees kept in woodlands in the British Isles will thrive especially if they are sited in clearings within woodlands and are situated so that the sun can shine on the hives giving warmth and helping keep the hives dry and sheltered from wind and rain.

Log piles are being made to encourage invertebrates, lizards, newts, frogs and toads but there is still the drive to fell dead trees or trees which have hollow centres. If the leaving of such trees (assuming there were no safety issues) was encouraged this would benefit many species (birds and invertebrates) and would also provide potential sites where wild bees could exist. Later in the book the benefits of the co-existence of wild (native) honey bees and managed colonies are referred to in relation to the development and transfer of disease resisting genetic material from the wild bees developing as a result of natural selection and its transfer via genetic transfer through drones congregations into managed honey bees.

If the proposals to increase the tree cover in the British Isles goes ahead then there are many potential scenarios ranging from large landscape scale planting to the planting of one or two trees of a suitable size and shape for small spaces.

Many tree species are wind - pollinated and do not need an insect, bird or other animals to effect pollination so at first sight they may appear of little benefit to honey bees, however this is too simplistic a view to take. Some species, both hardwood (broad-leaved) and softwood (coniferous) species produce flowers which are visited by honey bees and the pollen is collected. Not all pollen has the same nutritive value as we shall see later, however these resources may still be of significance especially in locations where there is poor forage availability and diversity in the early part of the beekeeping season.

Thus far we have written about standing trees which on the one hand may be only a few metres high at maturity to those which have the capability to grow into large trees over many years.

There are many shrubby species which can be used to create understoreys under the standing trees, and which form an important part of the ecology and biodiversity which can be achieved through their use. Some of these shrubby species can if managed appropriately become small trees.

Traditional fruit orchards have considerable ecological value and the planting of non-cropping (in the sense of fruit) trees in orchards can provide many benefits to both the environment and fruit crop yields. Such trees, if selected carefully, can provide alternative pollen sources helping to maintain the pollinator populations. They can also lower wind speeds enabling the bees to fly and visit the flowers of the crop trees when the weather is windy.

34.2   Tree and shrub species visited by honey bees

The following table lists a number of tree and shrub species which are visited by the honey bee and hence of interest to the beekeeper.

The authors hope that this information will help decision-making in the selection and combinations of species to be used in schemes which involve planting trees.

The letter 'N' denotes whether or not the species is considered native to the British Isles. Included in the table is information on whether the species is used in hedges. For completeness some species are included whose flowers are not visited by honey bees for pollen or nectar, but may be attracted when honeydew is produced.

Common Name / Native	Botanical Name	Pollen	Nectar	Honey Crop	Good for Hedges	Notes
Alder (N)	Alnus glutinosa	+	−	-		Male catkins are an early source of fresh pollen for early brood rearing when they start to dehisce in February or March or even earlier in mild winters
Apple	Malus pumila	+	-	+		Excellent early source of pollen and nectar depending on the apple variety, location and season.
Beech (N)	Fagus sylvatica				H	Probably introduced by Stone Age humans. Honeydew much worked by honey bees
Wild Crab apple and other single flowered forms (N)	Malus sylvestris ssp. sylvestris	+	+	+	H	

Common Name / Native	Botanical Name	Pollen	Nectar	Honey Crop	Good for Hedges	Notes
Ash (N)	*Fraxinus excelsior*	+	-	-		Small inconspicuous male catkins are visited by bees for pollen
Aspen (N)	*Populus tremula*					
Birch Downy Birch (N) Silver Birch (N)	*Betula species* *Betula pubescens* *Betula pendula*	+	-	-		All birch species produce large quantities of pollen early in the year which bees will sometimes collect
Blackthorn (also known as the Sloe) (N)	*Prunus spinosa*	+	+	-	H	Usually found in hedges and in shrub forms and coppices but will grow into a tree. Seasonal weather conditions impact on the accessibility of bees to the flowers and the availability of nectar and pollen
Box	*Buxus* spp.	+	+			Early source of nectar and pollen
Bullace or Damson	*Prunus domestica* ssp. *insititia*	+	+	+		April to May
Wild plum, plums and gages	*Prunus domestica*	+	+	+		
Buckthorn (N)	*Rhamnus cathartica* (Purging Buckthorn); *Hippophae rhamnoides* (Sea Buckthorn)	+	+	-		Early summer

Common Name / Native	Botanical Name	Pollen	Nectar	Honey Crop	Good for Hedges	Notes
Alder buckthorn (N)	*Frangula alnus*	+	+	-		Early summer
Wild Cherry (Gean) (N)	*Prunus avium*	+	+	+		Common in woods on limestone soils. March to April / May
Bird cherry	*Prunus padus*	+	+	+		More commonly found in the north
Cherry laurel	*Prunus laurocerasus*	+	+	-		Has extra-floral nectaries on under-surfaces of leaves on young growth
Cherry plum	*Prunus cersifera*	+	+	-		Can be in flower very early in February
Dogwood (N)	*Cornus* spp. esp. *C. sanguinea,*				H	
Wych Elm (N) Smooth-leafed Elm	*Ulmus glabra, Ulmus U. minor*	+	+	-		Bees work flowers in February and early March and the abundant early source of pollen can be important. Nectar may be produced in a warm Spring. Can be a source of honeydew later in the year
False acacia	*Robinia pseudo-acacia*	+	+	-		Note this species is considered an invasive species and should not be planted in the wild. June
Guelder rose (N)	*Viburnum opulus*		+			May
Wayfaring Tree	*Viburnum lantana*		+			May

Common Name / Native	Botanical Name	Pollen	Nectar	Honey Crop	Good for Hedges	Notes
Hawthorn (also called May or Whitethorn) (N)	Crataegus species Crataegus monogyna	+	+	+	H	Most often grown in hedges but all hawthorns will make fine small trees if left to grow. There is the added benefit that the trees will not be mechanically flayed and the hawthorn berries lost to wildlife species that relish them in the late summer and autumn
Midland hawthorn (N)	Crataegus laevigata	+	+	+		Flowers usually open in the early part of May depending on the season.
Hazel (N)	Corylus avellana	+	-		H	No nectar is produced but the pollen is collected from late January until early March depending on the season
Holly (N)	Ilex aquilifolium	+	+	-	H	Flowering begins in May and nectar secretion can be significant.
Hornbean (N)	Carpinus betulus	+	-	-	H	Occasionally visited by honey bees for pollen only in April / May
Horse-Chestnut	Aesculus hippocastnum	+	+	+		Well worked by honeybees in mid-late April and May depending on the season. There are double-flowered cultivars which are not worked by honey bees
Red horse-chestnut	Aesculus carnea	+	+	+		Flowers slightly later than the horse-chestnut
Indian horse-chestnut	Aesculus indica	+	+	+		Source of nectar and pollen in the early and helps meet the June gap in the south where it is found to be more hardy

Common Name / Native	Botanical Name	Pollen	Nectar	Honey Crop	Good for Hedges	Notes
California buckeye	*Aesculus californica*	+	-	-		Flowers July to August but causes POISONING and PARALYSIS in both honey bees and bumblebees because of the toxin saponin called aesculin which is found in the pollen and the nectar of this species. Planting of this species is not recommended if intended to attract honey bees
Common Juniper (N)	*Juniperus communis*					
Indian bean tree	*Catalpa bignonioides*	+	+	-		Garden tree flowers July and August, and attractive to honey bees for the nectar secreted by extrafloral nectaries on the under-surface of the leaves in the axils of the primary veins
Lime	*Tilia* spp.	+	-	+		Lime trees are also a major source of honeydew which is collected by the honey bee
Common Lime	*Tilia x europaea*	+	+	+		Hybrid between the large and the small leaved limes. Grown in streets, parks and gardens
Large-leaved lime (N)	*Tilia platyphyllos*	+	+	+		Flowers June / July. Nectar secretion best when the water table is high and when at the time of flowering the weather is settled, heavy and humid
Small-leaved lime (N)	*Tilia cordata (T. parviflora)*	+	+	+		Woods containing small-leaved limes can be found in nature reserves in Lincolnshire. Mainly found as a wild tree and is not often planted in the UK in contrast to mainland Europe where it is often used as a street tree
Silver-lime	*Tilia tomentosa*	+	+			

Common Name / Native	Botanical Name	Pollen	Nectar	Honey Crop	Good for Hedges	Notes
Pendant silver-lime	*Tilia petiolaris*	+	+			
Other limes				+	+	Tilia x europaea / Tilia x euchlora (Caucasian lime) were thought to secrete nectar which has a high mannose sugar content and this could affect bumblebees and to some extent honeybees, depending on the dryness of the season, and cause them paralysis making them vulnerable to being eaten by bird such as blue tits. Further research has indicated that the nectar from these limes is not toxic to bees and that bees showing such signs of paralysis are in fact starving and have been trying to access any nectar available on the tree.
Maple	*Acer* spp.	+	+	-		Early producers of nectar and pollen (April to May). There are many sorts of maple grown as ornamental trees
Field maple (N)	*Acer campestre*	+	+	-	H	The only native maple species in UK. Often grown in hedges and is coppiced but if left will grow into fine trees. April- May
Mountain Ash (also known as Rowan)	*Sorbus aucuparia*	+	+	-		Flowers May / June
Pedunculate oak (N)	*Quercus robur*	+	-	-		Honeybees will work the male catkins for pollen in April / May and also collect honeydew
Sessile oak (N)	*Quercus petraea*	-	-			Honeybees will work the male catkins for pollen in April / May and also collect honeydew

Common Name / Native	Botanical Name	Pollen	Nectar	Honey Crop	Good for Hedges	Notes
Turkey Oak	*Quercus cerris*	+	-	-		Considered an invasive species and should not be planted in the wild
Evergreen oak	*Quercus ilex*	+	-	-		Considered an invasive species and should not be planted in the wild
Pear (N)	*Pyrus communis*	+	+	-		Produces nectar and pollen April – May depending on the variety of the pear and the season
Wild Pear	*Pyrus pyraster*					
Plum (domestic plum, wild plum, damson)	*Prunus domestica*	+	+	+		In flower early to mid-April to May depending on the variety and season
Poplar	*Populus species*	+	-	-		Worked for early (March to April) pollen. Poplar is also grown in short rotation coppice for the biomass and biofuel industries.
Balsam Poplar	*Populus trichocarpa* and *P. tacamahaca*					The buds are worked for the sticky resinous substances present on them and use this in the form of propolis which is an important material in the life of the colony because of its biologically active properties in controlling micro-organisms
Black Poplar (N)	*Populus nigra var. betulifolia*					
Aspen Poplar (N)	*Populus tremula*					

Common Name / Native	Botanical Name	Pollen	Nectar	Honey Crop	Good for Hedges	Notes
Sycamore	*Acer pseudo-platanus*	+	+	+		Flowers April to June depending on location and season. There are many ornamental forms. The tree is quite salt-tolerant and will withstand the salt and often windy weather conditions around the coast. Generally known in Scotland as the Plane Tree
Scots Pine (N)	*Pinus sylvestris*	+				Pollen sometimes collected by honey bees in the spring
Spindle (N)	*Euonymus europaeus*					
Strawberry Tree	*Arbutus unedo*					Native to Ireland only
Wild Service Tree (N)	*Sorbus torminalis*					
Common Whitebeam (N)	*Sorbus aria*	+	+			April - May
Willows	*Salix* spp. and hybrids	+	+	+	+	There are many hybrids and varieties of willows (Salix spp) and they can be short to tall shrubs as well as trees. Willows are often a key component of short rotation coppicing in the production of biomass and biofuels for power generation. Some of the varieties used can be important sources of early pollen and nectar accessible to honeybees and some are especially sugar rich. This sugar richness can be induced by cultivation techniques. Rothamsted Research is a source of more information

Common Name / Native	Botanical Name	Pollen	Nectar	Honey Crop	Good for Hedges	Notes
Crack Willow (N)	Salix fragilis					June
Grey Willow (N)	Salix cinerea					
White Willow (N)	Salix alba	+	+			Can be a significant early source of forage and nectar is eagerly sought out on warm spring days. May
Purple Willow (N)	Salix purpurea				H	
Goat willow	Salix caprea	+	+		H	Usually found as a large shrub. June
Common Osier (N)	Salix viminalis					Used for basket making and was the material used in wicker hives
Yew (N)	Taxus baccata	+	-	-		Flowers April to March and worked by honeybees when other pollen sources are unavailable

Today there are no tree species considered **endemic** to Britain and found nowhere else in the world. There are tree species which are considered to be **native**.

'N' in this table means the tree species is considered a true **Native** species to all or parts of the British Isles and colonised Britain during the time between the end of the Ice Age about 10,000 years ago and the formation of the English Channel by the gradual expansion of ancient rivers some thousands of years later. Trees that came after the Channel had formed are generally described as **Naturalised.**

'H' means the species can also be grown in hedges.

35. British floral sources considered to be of importance to the honey bee and their flowering periods

35.1 Early records

We have noted that honey bees have been kept for many years but it is only from the sixteenth century do we begin to see the formal recording and advice given to beekeepers as to the best sources of forage. This no doubt reflected the increasing interest in plants, their naming and descriptions and the search for order. These developments are well described in Anna Pavord's 2005 book, ***The Naming of Names- the search for Order in the World of Plants***.

It is interesting to have the opportunity to consider the plants regarded as important for bees and listed in three bee- related publications, namely:

- Charles Butler (1560-1647) -*The Feminine Monarchie or A Treatise Concerning Bees and the Due Ordering of Them*. Oxford (1609)
- John Worlidge (1640-1700) - *System Agriculturae* (1669)
- Richard Remnant - *A Discourse or Historie of Bees* (1637)

The following table lists the plants from the publications and the information has been abstracted a table in Walker and Crane (2001).

Column **B** refers to Butler

Column **R** refers to Remnant

Column **W** refers to John Worlidge

Column **Roman** refers to plants mentioned by Roman authors.

Botanical Name	English Name	B	R	W	Roman
Acer spp.	Maple	x	-	-	
Alnus glutinosa	Alder	-	-	x	
Anthemis cotula	Mayweed	x	--	-	
Bellis perennis	Daisy	x	-	x	
Berberis vulgaris	Barberry	x	-	-	
Borago officinalis	Borage	x	x	-	
Brassica oleracea	Seed cabbage	-	x	-	
Brassica rapa	Turnip	-	x	-	+
Buxus sempervirens	Box	x	x	-	
Calluna vulgaris	Heather	x	-	-	
Cannabis sativa	Hemp	x	-	-	
Carduus / Cirsium	Thistle	x	-	-	
Centaurea nigra	Knapweed	x	-	-	
Corydalis avellena	Hazel /filberd (filbert)	xs	-	x	
Crataegus monogyna	Hawthorn	x	-	-	
Cucumis melo	Melon	x	-	-	
Echium vulgare	Bugloss	x	x	-	
Erica cinerea	Heath	x	-	-	
Erysimum cheiri	Wallflower / gillyflower	x	x	-	+
Fagus sylvatica	Beech	x	-	-	
Geum urbanum	Benet /?herb bennet	x	-	-	
Hedera helix	Ivy	x	x	-	
Helleborus foetidus	Bear's foot	x	-	-	
Hysocyamus niger	Henbane	x	-	-	

Ilex aquifolium	Holly	x		-	
Isatis tinctoria	Woad	-	x	-	
Lamium spp.	Archangel	x	-	-	
Lavandjula angustifolia	Lavendar	x		-	
Malus pumila	Apple	xs		x	+
Malus sylvestris	Crab apple	x	-	-	
Malva spp.	Mallows	x	-	-	
Matthiola incana	Stock, gilliflower		x	-	
Melissa officinalis	Balm		x	-	
Narcissus pseudonarcissus	Daffodil, lide-flower	x	-	X	
Papaver rhoeas	Red-weed	x	-	-	
Phaseolus / Vicia	Bean	x	x	-	+
Phillyrea angustifolia / P. latifolia	Phyllirea		-	x	
Pinus sylvestris	Pine		-	s	+
Pisum sativum	Pea	x	x	-	+
Plantago lanceolata	Short plantain	x	-	-	
Polygonum fagopyrum	Buckwheat		-	-	
Prunus avium	Cherry	xs	x	x	
Prunus x domestica	Plum	s	x	x	+
Prunus x domestica spp. *instititia*	Bullace		-	x	
Prunus persica	Peach		-	x	+
Prunus spinosa	Blackthorn, sloe	x	-	x	
Pyrus communis	Wild pear	xs	x	x	+
Quercus robur	Oak	x	-	x	

Reseda luteola	Weld (yellow-weed)	x	-	-	
Ribes grossularioides	Gooseberry	x	-	x	
Rosmarinus officinalis	Rosemary	x	x	-	+
Rubus fruticosus agg.	Blackberry	x	--	-	
Salix viminalis	Willow,withy-palm, withy	x	x	x	+
Sambucus nigra	Elder	x	-	-	
Satureja spp	Savory	-	-	-	
Sinapsis arvensis	Charlock	x	-	-	
Taraxacum officinale	Dandelion	x	-	-	
Thymus vulgaris	Thyme	x	x	-	+
Trifolium pratense T. repens	Red & white honeysuckle (clover)	x	x	-	
Vicia sp.	Bean, vetch	x	x	-	
Viola sp.	Violet	x	x		+

X =Foraged by bees
S =For shelter or for swarms to settle on
-=Not cited
+= Referred to by Roman authors

35.2   The situation today

Today honey bee colonies are kept throughout the British Isles and the pattern and practices of beekeeping employed vary in relation to the forage available which in turn reflects the location and the weather experienced in the colony's locality. For example, for some beekeepers the main nectar yielding (and thence potential honey crops) crops may be oilseed rape and field beans grown in lowland England, whilst for others it may be heather, or clover found on upland moors and others the forage to be found in the urban and suburban garden.

For honey bee colonies to thrive they need a regular supply of nectar and, more importantly, continuity of pollen availability throughout the active season. Some beekeepers migrate their colonies to forage sources, especially high value honey crops such as ling heather or other specialist crops such as borage. A disadvantage can be that moving bees can stress them and increase their susceptibility to diseases such as Nosema (Nosema apis and N. ceranae) and its associated bee virus diseases.

Beekeepers should adapt their beekeeping practices to take advantage of these sources. Each beekeeper needs to ascertain whether there will be a continual supply of nectar and pollen throughout the brood-rearing season and to establish sufficient winter stores for their bees in the places where they intend to locate their hives.

The following table indicates the time of year when the potential sources are flowering and can be part of a foraging resource for honey bee colonies. Precise timing depends on the season and location and we are already seeing changes in the phenology of flowering in response to the changes in weather patterns which we now experience.

Most of the plants listed below are unlikely to grow in large quantities in any one location but it is the possibility of creating habitats, especially in gardens where plants can be grown which will contribute to continuing forage availability throughout the beekeeping season. The benefits of such an approach become even more enhanced when considered on a landscape scale, and more tangible the greater the number of gardens containing honey bee forage plant species in a local area.

Some of the species listed form part of the flora of uncultivated and semi-natural areas. Please refer to the end of the table to see the explanatory note on the Honey Potential Class.

Plant name	Feb / Mar	April	May	June	July	Aug	Sept / Oct	Honey Potential Class
Alder (P) (*Alnus glutinosa*)	•	•						-
Almond (*Prunus dulcis*)	•							1
Apple* (*Malus* sp. / *Malus domestica*)		•	•					1-2
Alyssum Golden (*Alyssum saxatile*)		•	•	•	•			-
Alkanets (*Anchusa officinalis*)			•	•	•	•		2-6
Arabis / Rock-cresses e.g., Hairy Rock-cress, (*Arabis hirsuta*)		•	•	•	•	•		-
Ash (P) (*Fraxinus excelsior*)		•	•					-
Aubretia (*Aubretia deltoides*)		•	•	•				-
Autumn Crocus (*Crocus nudiflorus*)						•	•	-
Balsam (*Impatiens glandifera*)					•	•	•	-

Plant name	Feb / Mar	April	May	June	July	Aug	Sept / Oct	Honey Potential Class
Beech (P) (*Fagus sylvatica*)		•	•					-
Bell heather* (*Erica cinereal*)					•	•	•	-
Berberis / Barberry (e.g., *Berberis darwinii, B. vulgaris*)		•	•	•				-
Bilberry (*Vaccinium myrtillus*)		•	•	•				-
Bindweed Large (*Calystegia sylvatica*)				•	•	•	•	-
Birch (P) (*Betula pubescens* and *B. pendula*)			•	•				-
Bird's-foot-trefoil (*Lotus corniculatus*)				•	•	•	•	2-4
Blackberry* / Bramble (*Rubus fruticosus*)			•	•	•	•	•	-
Blackthorn (*Prunus spinosa*)	•	•	•					1
Bluebell (*Hyacinthoides non-scriptus*)		•	•	•				-
Borage* (*Borago officinalis*)					•	•	•	4
Brassicas (Cabbages) (*Brassica oleracea*)			•	•	•	•	•	3
Broom (*Cytisus scoparius*) (P)		•	•	•				-
Buckwheat (*Fagopyrum esculentum*)				•	•	•	•	3-5
Buttercup Meadow (*Ranunculus acris*) Note POISONOUS due to pollen containing protoanemonin		•	•	•	•	•	•	

Plant name	Feb / Mar	April	May	June	July	Aug	Sept / Oct	Honey Potential Class
Campanula / Bellflowers (Various including Canterbury-bells *Campanula medium*)			•	•	•	•		-
Catmint (*Nepeta gigantea*)				•	•	•	•	4-5
Ceanothus / Californian lilac (*Ceanothus thrysiflorus*)		•	•	•	•	•	•	-
Lesser Celandine (*Ranunculus ficaria*)	•	•	•					-
Cherry* Wild (*Prunus avium*)		•	•					2
Cornflower (*Centaurea cyanus*, also Perennial Cornflower) *Centaurea montana*)				•	•	•		3
Crab apple (*Malus sylvestris*)		•	•	•				2
Crane's-Bill (Various spp. including Bloody Crane's Bill, *Geranium sanguineum*)				•	•	•	•	3
Crocus Spring (*Crocus vernus*)	•	•	•	•				-
Currants (Redcurrant, Ribes rubrum; Blackcurrrant, *Ribes nigrum*; Flowering Currant, *Ribes sanguineum*)		•	•					2-3
Damson* (*Prunus domestica* ssp. *insititia*)		•	•					2
Dandelion* (*Taraxacum officinale*)	•	•	•	•	•	•	•	4

Plant name	Feb / Mar	April	May	June	July	Aug	Sept / Oct	Honey Potential Class
Deadnettle 3 (White Deadnettle *Lamium alba*; Purple Deadnettle *Lamium purpurea*). Major bumble bee plants. Honey bees may gain access to nectar through holes made in flower base by bumble bees		•	•	•	•	•	•	4-6
Doronicum Leopard's Bane (*Doronicum caucasicum; Doronicum plantagineum*)		•	•	•	•			-
Dwarf gorse (*Ulex minor*)					•	•	•	-
Wych Elm (*Ulmus glabra*) English Elm (*Ulmus procera*)	•	•						-
Escallonia (*Escallonia macrantha*)			•	•	•	•	•	-
Fennel (*Foeniculum vulgare*)					•	•	•	-
Field bean* (*Vicia faba*)				•	•			3
Forget-me-not (Wood Forget-me-not, *Myosotis sylvatica*); Field Forget-me-not, *M. arvensis*)		•	•	•	•			-
Fuschia (*Fuschia magellanica*)					•	•	•	-
Goldenrod (*Solidago viraurea*)					•	•	•	5-6
Gooseberry (*Ribes uva-crispa*)	•	•						3
Gorse (*Ulex europaeus*)	•	•	•	•	•	•	•	-
Hawthorn* (*Crataegus monogyna*)			•	•				2
Heather (*Calluna vulgaris*)					•	•	•	4

Plant name	Feb / Mar	April	May	June	July	Aug	Sept / Oct	Honey Potential Class
Hazel (P) (*Corylus avellana*)	•							-
Hogweed / Common Hogweed / Cow Parsley *Heracleum sphondylium*		•	•	•	•	•	•	4
Holly (*Ilex aquifolium*)		•	•	•				-
Hollyhock (*Alcea rosea*)					•	•	•	4
Horse-Chestnut (*Aesculus hippocastanum*) and Red Horse-Chestnut (*A. carnea*)			•	•				4
Ivy*2 (*Hedera helix*)							•	5
Jacob's-ladder (*Polemonium caeruleum*)			•	•	•	•		3-4
Knapweed Greater (*Centaurea scabiosa*)				•	•	•	•	2-3
Bay (*Laurus nobilis*)		•	•	•				-
Lavender (*Lavandula x intermedia, Lavandula angustfolium*)					•	•		4
Lime* or Linden (*Tilia* spp.)				•	•			4-6
Ling heather* (*Calluna vulgaris*)						•	•	4
Lucerne (*Medicago sativa*)				•	•			5
Mallow (*Malva alcea*; Musk Mallow *Malva moschata*); Common Mallow (*M.sylvestris*)				•	•	•	•	3-4 2
Maple* (*Acer campestre*)		•	•					6
Marjoram (*Origanum vulgare*)						•		4

Plant name	Feb / Mar	April	May	June	July	Aug	Sept / Oct	Honey Potential Class
Meadowsweet (P) (*Filipendula ulmaria*)				•	•	•	•	-
Michaelmas daisy (*Aster novi-belgii*)						•	•	2
Mint (*Mentha* spp.)					•	•	•	5
Mullein (e.g., Great Mullein, *Verbascum thapsis*)				•	•	•		-
Hedge Mustard (*Sisymbrium officinale*)			•	•	•	•	•	-
Oak (*Quercus robur, Quercus petraea*)			•	•				-
Oil-seed-rape- autumn sown* (*Brassica napus* ssp. *oleifera*)		•	•					4-5
Oil-seed-rape- spring sown* (*Brassica napus* spp. *oleifera*)				•	•			4-5
Onion (*Allium cepa*)				•	•	•	•	3
Pear* (*Pyrus communis var sativa*)		•	•					1
Poached Egg Plant (*Limnanthes douglasii*)								-
Phacelia* (*Phacelia tanacetafolia*)				•	•	•		5-6
Plantain Ribwort (*Plantago lanceolata*)		•	•	•	•	•	•	-
Plum* (*Prunus domestica*)		•	•					2
Common Poppy (P) (*Papaver rhoeas*)				•	•	•	•	-
Privet (*Ligustrum ovalifolium and L. vulgare*)			•	•				-

Plant name	Feb / Mar	April	May	June	July	Aug	Sept / Oct	Honey Potential Class
Purple loosestrife (*Lythrum salicaria*)				•	•	•		5
Firethorn (*Pyracantha coccinea*)				•				-
Ragwort (*Senecio jacobaea*)				•	•	•	•	-
Raspberry* (*Rubus idaeus*)			•	•	•	•		3
Red clover* (*Trifolium pratense*)			•	•	•	•	•	5
False-Acacia (*Robinia pseudoacacia*)				•	•			6
Rockrose (P) (*Helianthemum mummularium*)				•	•	•	•	-
Rosemary (*Rosmarinus officinalis*)		•	•	•				4
Sage (e.g., *Salvia officinalis; Clary Salvia sclarea*				•	•	•		5-6
Scabious Field, (*Scabiosa arvensis*)					•	•	•	4-5
Sea Lavender Common (*Limonium vulgare*)					•	•	•	2
Stonecrops Sedum (*Sedum spectabile* and spp.)				•	•	•		-
Snowdrop (*Galanthus nivalis*)	•							-
Sunflower (*Helianthus annuus*)				•	•	•	•	2
Sweet Chestnut (*Castanea sativa*)					•			2
Sycamore* (*Acer pseudoplatanus*)		•	•					4-6

Plant name	Feb / Mar	April	May	June	July	Aug	Sept / Oct	Honey Potential Class
Thistle Dwarf (*Cirsium acaule*)				•	•	•	•	3
Thrift / Sea Pink (*Armeria maritima*)		•	•	•	•	•		-
Thyme (Various *Thymus* spp. including Garden Thyme, (*Thymus vulgaris*); Wild Thyme, (*Thymus polytrichus*)			•	•	•	•	•	6
Toadflax Common (*Linaria vulgaris*)					•	•	•	-
Traveller's-joy (*Clematis vitalba*)					•	•	•	-
Veronica, Speedwells (various *Veronica* spp.; Germander Speedwell, (*Veronica chamaedrys*)	•	•	•	•	•			-
Vetch Common (*Vicia sativa*)		•	•	•	•	•	•	2
Violet Sweet (*Viola odorata*)	•	•	•					-
Viper's- bugloss (*Echium vulgare*)				•	•	•	•	6
Virginia-creeper (*Parthenocissus quinquefolia*)				•	•			-
Wallflower (*Chieranthus cheiri*)	•	•	•	•				-
White bryony (*Bryonia dioica*)			•	•	•	•	•	-
White clover* (*Trifolium repens*)				•	•	•	•	3-4
Wild rose (P) Dog-rose (*Rosa canina*)				•	•			-
Willow Salix spp. Many species and types e.g., Goat Willow (*Salix caprea*)	•	•	•					4

Plant name	Feb / Mar	April	May	June	July	Aug	Sept / Oct	Honey Potential Class
Willowherb* e.g., Rosebay Willowherb (*Chamerion angustifolium*)				•	•	•	•	5-6
Winter aconite (*Eranthis hyemalis*)	•							-
Winter heath e.g. (*Erica carnea, E. herbacea*)	•	•	•	•				-
Wood anemone (*Anemone nemorosa*)	•	•	•					-
Wood sage (*Teucrium scorodonia*)					•	•	•	3
Ribbed / Common Melilot (*Melilotus officinalis*)				•	•	•	•	4

Notes

(P) – Pollen source only, otherwise plant produces both nectar and pollen

 * - Major bee plant, widespread or locally significant.

- In coastal regions sea lavender will yield good nectar producing a light- coloured honey.

– Ivy can produce a useful supply of nectar from October until the first frosts.

– White Dead-nettle is perennial and regularly flowers throughout the winter months whilst Red Dead-nettle is annual.

The last column (Honey Potential Class) in the table contains information taken from Table 1.3/1 in Crane (1979) grouping into classes the potential honey yields estimated as the maximum quantity (expressed in kilograms) of honey that could be obtained in the course of a season from one hectare of land of the plant species (in Imperial measure terms approximately pounds per acre). Optimal growing conditions were assumed as well as there being an adequate foraging force to collect all the nectar.

Class	Kg honey/ ha	Class	Kg honey / ha
1	0-25	4	101-200
2	26-50	5	201-500
3	51-100	6	> 500
' –' represents no data given in Crane (1979) or that the plant species does not produce nectar			

36. Some likely effects of climate change on species change and flowering times in plants visited by honey bees

It is probably true to say that the species mix and habitats throughout the British Isles have continuously changed and that a changing climate is nothing new in a geological time perspective. Anthropogenic factors such as the development of agriculture, population growth and the relatively more recent industrial revolution, with its dramatic increase in the 'greenhouse gas' carbon dioxide, have influenced many of the weather- related processes and cycles which the biological world experiences. It is believed that the activities of man are changing the trajectory of the world's climate and weather patterns. In talking about such matters, we roll them up into phrases such as 'climate change' and 'global warming', largely as a means of trying to communicate something simply but which is far from simple and not, as our politicians seem to believe something which can be micro-managed in a political timeframe.

We have briefly referred to evidence of changing flowering times already experienced and now we will consider what might be the consequences of climate change for some of the plant species visited by the honey bee in the British Isles.

Plant species which are adapted to growing and flowering in northern and montane locations will become more restricted and localised and this has been recorded comparing current distribution maps with those made by botanical recorders over many decades.

The surviving calcareous grasslands not put to the plough have shown decreases in species such as:

- Dwarf thistle (*Cirsium acaule*)
- Greater Knapweed (*Centaurea scabiosa*)
- Wild parsnip (*Pastinaca sativa*)
- Bloody crane's-bill (*Geranium sanguineum*) has shown significant increases. As an aside as well as being well-visited by the honey bee this species is also very attractive to bullfinches who relish the seeds.

Plants associated with broad-leaved ancient woodlands have marginally decreased:

- Wood forget-me-not (*Myosotis sylvatica*)
- Ramsons (*Allium ursinum*)

Currently predominately southern located species would be expected to spread northwards and this has already occurred with the Wild teasel (Dipsacus fullonum).

A number of plant species introduced into the British Isles have been introduced from warmer places, such as southern Europe and these have become escapes from garden or landed estates. Today we often think of them as native species and having naturalised.

These include:

- Red Valerian (*Centranthus ruber*)
- Purple toadflax (*Linaria purpurea*)
- Cherry laurel (*Prunus laurocerasus*)
- Silver ragwort (*Senecio cineraria*)
- Summer lilac (*Buddleja davidii*)
- Rock Cotoneaster (*Cotoneaster horizontalis*)
- Orange balsam (*Impatiens capensis*)
- Mexican fleabane (*Erigeron karvinskianus*)
- Evergreen Oak (*Quercus cerris*)

37. Archaeotypes, neophytes and invasive non-native species

37.1   Some plant introductions into the British Isles

The transboundary movement of plants (and animal) species from one country to another has gone on for many years in the form of introductions into the British Isles. These may have been plant species (flowers, trees and shrubs), herbs and medicinal plants, fruits, vegetable and crop plants and the plants we see in our fields, the wider landscape, in roadside verges, in our gardens and other cultivated and 'wild' places. Such introductions may have been intentional through collectors, international trade or accidental, through human collecting and cultivation activities or species that have arrived with other materials or mixed in with other plants.

Campbell-Culver's book **The Origin of Plants, the People and the Plants** published in 2001 describes the people and the plants that have shaped Britain's Garden History since the year 1000 AD and lists the dates of the introductions of many plant species into the British Isles since this date.

The following table contains a non-exhaustive list of examples abstracted from the book detailing the dates of plant introductions from before 1000 AD until 2000 AD and which are visited by honey bees.

Common Name	Botanical Name	Year of Introduction
Before AD 1000		
Sycamore	*Acer pseudoplatanus*	NSD
Borage	*Borago officinales*	
Carrot	*Daucus carota*	
Mint	*Mentha* spp.	
Broad Bean / |Horse Bean	*Vicia faba*	
Christmas Rose	*Helleborus niger*	
Sweet Chestnut	*Castanea sativa*	
Peach	*Prunus persica*	
White Mustard	*Sinapsis alba*	
Period AD 1000 to 1099		
Wallflower	*Erysium cheiri*	c 1066
Period AD 1100 to 1199		
Hyssop	*Hyssopus officinalis*	c 1100
Period AD 1200 to 1299		
Sweet Marjoram	*Origanum majorana*	c 1200
Garden Thyme	*Thymus vulgaris*	c 1200
Lavender	*Lavandula angustifolia*	1265
Hollyhock	*Alcea rosa*	1260
Period AD 1300 to 1399		
Rosemary	*Rosmarinus officinalis*	c 1340
Almond	*Prunus dulcis*	NSD
Lupin	*Lupinus albus*	NSD
Yellow Lupin	*Lupinus luteus*	NSD
Sweet Basil	*Ocimum basilicum*	c 1340

Common Name	Botanical Name	Year of Introduction
Period AD 1400 TO 1499		
Autumn Crocus	*Crocus nudiflorus*	c 1400
Period AD 1500 TO 1599		
Snowdrop	*Galanthus nivalis*	NSD
Jacob's Ladder	*Polemonium caeruleum*	1510
Globe Thistle	*Echinops sphaerocephalus*	1542
Apricot	*Prunus armeniaca*	1542
Rock Rose	*Cistus salvifolius*	1548
Period AD 1600-1699 **Many species introduced during this period, not many visited by honey bees**		
Bluebell	*Hyacinthoides italica*	1605
Horse Chestnut	*Aesculus hippocastanum*	c 1616
Rose of Sharon	*Hypericum calycinum*	c 1676
Spanish Bluebell	*Hyacinthoides hispanica*	1683
Period AD 1700-1799 **Many species introduced during this period, not many are visited by honey bees**		
Michaelmas Daisy	*Aster novi-belgii*	1710
Verbena	*Verbena bonariensis*	1726
Woolly Mullein	*Verbascum phlomoides*	1739
Barberry	*Berberis canadensis*	1759
Rhododendron	*Rhododendron ponticum*	1763
Period AD 1800-1899 **Many species introduced during this period, not many visited by honey bees**		
Catmint	*Nepeta racemosa*	1804
Period AD 1900-2000 **Many species introduced during this period, not many visited by honey bees**		
	Nothing particular to note	

NSD refers to 'no set date'.

Plants introduced to our flora before AD 1500 are called archaeophytes and those introduced after AD 1500 are called neophytes.

37.2   Archeophytes

Many archaeophytes have become considered as weeds and become associated with arable / corn fields. They include:

- Corn buttercup *(Ranunculus arvensis)*
- Pheasant's eye *(Adonis annua)*
- Charlock *(Sinapsis arvensis)*
- Common poppy *(Papaver rhoeas)*
- Horse radish *(Armorica rusticana)*
- Ground-elder *(Aegopodium podagraria)*

37.3   Neophytes

Neophytes include the following species:

- Buddleja *(Buddleja davidii)*
- Monkey flower *(Mimulus guttatus)*
- Orange balsam *(Impatiens capensis)*
- Japanese rose *(Rosa rugosa)*
- Evening primrose *(Oenothera biennis and O. glazioviana)*
- Oxford ragwort *(Senecio squalidus)*
- Red valerian *(Centranthus ruber)*
- Russian comfrey *(Symphytum x uplandicum)*
- Japanese knotweed *(Reynotria japonica)*
- Wild Turkish rhododendron *(Rhododendron ponticum)*
- Himalayan balsam *(Impatiens glandulifera)*
- Giant hogweed *(Hieracleum mantegazzianum)*
- Spanish bluebell *(Hyacinthoides hispanica)*
- Turkey oak *(Quercus cerris)* – introduced in the 18th century.
- Sycamore *(Acer pseudoplatanus)* – *introduced into British Isles in the Tudor period by 1500 but is believed to have been present for many years having been introduced by the Romans.*
- Horse-chestnut *(Aesculus hippocastanum)* -*introduced into British Isles from Turkey in late 16th century.*
- London plane *(Platanus x hispanica)* – *a hybrid created and planted in British Isles in the mid-17th century.*
- Robinia *(Robinia pseudoacacia)* -*introduced into Britain in 1636.*
- Sweet-Chestnut *(Castanea sativa)* introduced into the British Isles by the Romans.

37.4   Invasive non-native species

There has been a long history of plants and animals being introduced into the British Isles from all around the world. Some have been introduced deliberately (legally and illegally) and some have arrived by chance or through accident. Some of these species become considered to be a problem either because they compete with and displace the native species in the same ecological niche or because they cause damage and economic loss. Plant and animal health inspection regimes are in place throughout the administrations in the British Isles and this aspect of biosecurity is co-ordinated by the GB Non-Native Species Secretariat (NNSS) and the British Irish Council. At the time of writing the following plant species are listed and considered Non-Native and there are measures in place for their control / eradication.

Common Name	Botanical Name
Cotoneaster	*Cotoneaster* spp.
Curly Waterweed	*Lagarosiphon major*
Evergreen Oak / Holm Oak	*Quercus ilex*
False Acacia	*Robinia pseudoacacia*
Garlic	*Allium* spp.
Giant Hogweed	*Heracleum mantegazzianum*
Giant Rhubarb	*Gunnera* spp.
Giant Salvinia	*Salvinia molesta*
Himalayan Balsam	*Impatiens glanduliflora*
Hottentot Fig	*Carpebrotus edulis*
Japanese Knotweed	*Fallopia japonica*
Japanese Rose	*Rosa rugosa*
Montbretia	*Crocosmia x crocosmiiflora*
Rhodedendron	*Rhodedendron ponticum*
Russian Vine	*Fallopia baldschuanica*
Turkey Oak	*Quercus cerris*

To date, many conservationists have been very zealous in the approaches they take to the presence of some of the neophyte species, however such attitudes will have to change as the impact of climate change, changing weather patterns will impact on our 'native' flora. No doubt there will be the need to introduce more drought-tolerant species in some areas and in others grow species which are tolerant to inundation by flooding. This in turn will influence the insect pollinator species in our fauna.

Reference sources used include:
Mabey, R. *Flora Britannica.* (1996). 480pp, Chatto and Windus
Stace, C., *New Flora of the British Isles 2nd ed.* Cambridge University Press
Blamey, M., and Grey-Wilson, *Wild Flowers of Britain and Northern Europe.* (2003). Cassell

38. Land management and its impact on forage availability for honey bees throughout the year

38.1   Current situation

In spite of the ever-increasing human population in the British Isles, by far the greatest proportion of the surface area of the country is comprised of non-urban or non-suburban development; principally agriculture, farming and horticulture as well as large areas of landscape, e.g., moorland whose existence is perpetuated by the land management practices of the landowner.

There are probably no areas of the land mass of the British Isles which have not been affected by human activities ever since the ice retreated at the end of the last glaciation.

The disciplines of plant and animal ecology developed the concept of communities of plant and animal species which co-existed as assemblages which could be defined and described, the term 'habitat' is used. It is common also refer to natural, semi-natural and native vegetation. More recently the term 'biodiversity' has entered our lexicon referring to the variety and variation of life on Earth, as well as the number and types of plants and animals that exist in a particular area. This is commonly measured and expressed as 'species richness'.

The State of Nature Reports (2019) and other recent reports demonstrates the progressive loss of biodiversity throughout the British Isles, but it is particularly evident in the practices employed in agriculture which have actively sought to eradicate non crop plants both in the crop and in the wider agricultural landscape.

This loss of plant diversity impacts all pollinator species. In the case of the honey bee it is now well documented that this loss of biodiversity not only impacts on the availability and variation in forage for protein (pollen) and nectar (sugars) but that this affects the bioequivalence of other micro-nutrients present in pollen and nectar such as flavonoids, steryl glucosides, and alkaloids which are necessary to enable the synthesis of immune peptides and other substances crucial for honey bee metabolism and enable the overwintering bee and the colony survive the winter.

Loss of the plant diversity and the lack of uncultivated land (defined as not growing or being prepared for a crop) in the farmed landscape results in a restriction of foraging options throughout the active year during periods of critical reproduction and development with consequent negative health effects which impact on the ability of the individual bees to meet their potential and for the colony to function and thrive. The abundance and diversity of the plant species providing pollen and worked by honey bees, especially those which are part of the 'natural' flora (including weed species) play a crucial role in keeping fit and healthy honey bees (Di Pasquale et al. 2016).

Intensive agricultural systems often expose honey bees to large temporary variations in the availability, quantity, quality and diversity of nutritional resources.

No single pollen source provides all their nutritional need, so honey bees must have a number of pollens available to them from different plant species to remain healthy and to produce the jelly required to feed the queen and rear brood.

Whilst pollen may be freely and widely available during the mass flowering of some crops, e.g., maize and oilseed rape, the pollen from some crop species can fail to provide bees with an adequate diet for their development and longevity.

In recent years it has often been said anecdotally that bees kept in an urban environment in the British Isles are more prosperous than those kept in an agricultural environment. Conversely food accumulation and wax production in honey bee colonies have been found in the United States of America to be positively correlated with agricultural crop and negatively correlated to the urban environment.

Other significant impacts humans make to the landscape over a significant area of the British Isles include monoculture grassland for livestock grazing and the planting of coniferous forest. This has been found to affect the diversity of bacteria associated with the honey bee food supply. Bee bread sampled from hives kept close to agriculturally improved grassland made up of a single grass variety and from hives sampled near coniferous woodland contained lower bacterial species variation than from hives sampled near habitats with a richer variety of plant species such as broadleaf woodland, rough grassland and coastal landscapes (DeGrandi-Hoffman and Hagler (2000).

Earlier in the book we wrote that the Western honey bee (*Apis mellifera L.*) is a species which was found in open forest habitats (principally with broadleaved trees) with a wide range of ages in the trees which would mean there were suitable cavities in old trees in which colonies could be established and the development of glades and open clearings caused by the falling and decay of old trees thus creating a habit for flowering plant species.

Dense shrub layers of native or non-native species however reduce plant cover and forage diversity and availability.

38.2   Agro-forestry

Agro-forestry is described as the cultivation and use of trees and shrubs with crops and livestock in agricultural systems in some form of spatial arrangement or temporal sequence.

Looking further ahead the adoption of 'agro'-forestry practices into our land management as well as the increased growing of trees in our urban and suburban places and on a smaller local, but nationally significant, scale in our gardens should increase the forage available for the honey bee and other pollinator species.

Continuing, maintaining and improving or even the creation of such habitats will bring many benefits:

- For the managed honey bee helping to satisfy bees needs such as forage (nectar / pollen / propolis), shade and shelter for apiary sites.
- Helping in the management of carbon dioxide and other greenhouse gas emissions.
- The provision of green spaces for human mental and spiritual needs.
- Fruits, seeds, nuts, and berries supporting food webs.
- Use in flooding prevention and water management.

Climate change will determine the changes needed to be made in the use and management of our land to enable humans to continue to live in the British Isles together with the many other plant and animal species. Not the least is the growing recognition of the need to cherish and manage our key resource, **soil**, upon which we all ultimately depend.

Policy makers and land managers should act to:

- Ensure that habitats for pollinators should be protected and that management plans should consider practices to help support pollinator.
- Ensure they have Agri-environment schemes in place to incentivise management. practices which provide wildflower field margins and hedgerow management to help pollinator populations, including honey bees.
- Place less reliance on the use of chemical controls and promote the use of integrated pest management.
- Engage in greater interaction between beekeeper, farming and land management organisations and their members.
- Ensure that Local and national planning authorities are provided with clear guidance on how to implement Green Infrastructure to enhance the quality of the built landscape for pollinators.

38.3   Managing farmed landscapes

On-line resources exist to enable farmers and land managers to assess the provision of food and shelter for pollinators, including honey bees on their land. A whole farm approach is encouraged including non-cropped areas such as trackways, around yards, field margins and corners.

Bees operate and draw their needs from an extensive landscape area and so farmers and land managers are encouraged to co-operate and consider low level changes which could be implemented across a wide area as such actions will have a bigger cumulative beneficial stimulus to pollinators, including the honey bee.

There are various schemes running in parts of the British Isles to encourage farmers and land managers to introduce or extend management schemes which will involve improvements to biodiversity.

These are continually evolving and any detailed descriptions and information we might write into this book would soon become outdated. The use of the internet using search terms such as 'agri-environment schemes' should provide signposts to up to date information as well as government agency websites and those of organisations representing the interests of farming, horticulture and landowners and not forgetting the farming press.

The changing relationship of the British Isles with the European Union will mean a significant change of approach to the current subsidy support to farmers as part pf the Common Agricultural Policy (CAP).

The CAP currently costs taxpayers more than £50 billion per year of which £3.5 billion goes to UK farmers. The CAP system currently subsidises EU farmers (there are some other beneficiaries) for owning or using farmland. It is designed to guarantee higher yields through increasingly intensive agriculture. This has led to a perverse incentive to clear wildlife habitats even in areas which cannot be farmed to produce empty ground which then qualifies for subsidy support.

This will change because of the implementation of the Agriculture Bill the provisions of which mainly apply to England, but there will be equivalent changes enacted into legislation in Scotland, Northern Ireland and Wales. The provisions of the Bill will be phased in over a period of seven years. Ireland of course remains in the EU.

Assuming the changes do proceed, in the future agri-environment scheme payments for farmers and land managers could include incentives and payments aimed at improving habitat heterogeneity and connectivity of semi-natural habitats near arable land. Such linking of payments for ecosystem services to the value of the public goods provided by farmers rather than to the size of the cultivated area would help to target pollinator needs, including those of the honey bee. The current plan envisages paying farmers about £3 billion a year to protect wildlife, ecosystems and for other measures to do things for the public good.

Further incentives could be given at the farm level to encourage mixed cropping and the management of arable land to improve pollinator diversity (such as reduced cutting of hedgerows) would also benefit the honey bee.

Flowering crops such as apples or oil seed rape can provide abundant forage (nectar and pollen) for short periods, but it is essential to understand that the needs for the honey bee colony needs continue throughout the year.

Honey bees differ from other insects in that they are able to search for and communicate the location and forage of sites some distance from the hive and this means that they can ascertain the forage over a considerable area and then through the use of communication dance language and recruitment behaviours increase the capacity of foragers to exploit the resource. A colony could theoretically search over a 16 square mile area but flying to such distances is energy consuming and in energy terms this means the foraging worker honey bee travels more distance for less returns of nectar and pollen. Ideally land management schemes to promote pollinators need to be done on a sufficiently large enough scale to be effective.

 A 2% flower-rich habitat (equal to 2 hectares) and up to 1 kilometre of flowering hedgerow in every 100 hectares is the kind of size of planting required to support wild bee species and would make a significant contribution to the forage available to honey bees. But in the case of bumble bees there is a need for these patches of flowers to be at least 0.25 hectares in size and no more than 250 metres apart. Bumble bees do not have the same colony foraging behaviour and store hoarding as is found in the honey bee and so fragmented habitats with inadequately sized areas of forage make them very susceptible to land management changes.

The wider foraging and communication behaviour of the honey bee means it is less vulnerable to the need to have such a small scale mosaic of flowering plants in the landscape.

In summary, good practice would be for all landowners, managers, farmers and estate managers to:

- Sow a mix of summer-flowering legumes.
- Sow diverse mixes of perennial wildflowers alongside.
- Managed hedgerows cut on a three - year rotation and plant trees (e.g., willow) to provide spring forage.
- Leave verges or ivy uncut to provide late summer flowers, and then cut.
- Vary the grassland sward height.
- Manage any scrub habitat sympathetically but aim to increase the amount of light reaching the ground surface to encourage herbaceous species.

38.4   Managing urban and suburban landscape areas

Urban areas make up 9% of the land area in the British Isles and these areas continue to grow apace as the demand for housing, commercial and industrial development as well as infrastructure is forecast to continue to grow.

Wild insect and other pollinator populations continue to decline.

There has however been an increasing interest in food growing in urban areas and for local food production with more fruit and vegetables being grown in gardens, allotment and small holdings and many of these will benefit from insect pollinators. Creating circumstances which will encourage the insect pollinator population will be an important part of ensuring the success of the growing of local food.

Wildflower species also grow in urban environments and require pollinators to enable them to set seed and ensure survival through successive generations.

Colonies of honey bees are successfully kept by beekeepers in many urban and suburban locations.

Even today there are many urban and suburban areas where there is extensive forage available through gardens, park and green spaces and these can be havens and pro-actively managed for insect pollinator species.

Urban and suburban gardeners and land managers can help honey bees (and other insect pollinators) by:

- Selecting a range of plants which are promoted and labelled as 'Good for pollinators' or 'Perfect for Pollinators' and can provide good sources of pollen and nectar to ensure a continual availability of forage from early spring until autumn.
- Such plants will have single rather than double flowers which are often found in sterile cultivars or where there is little, or no pollen or nectar produced.
- Mowing the grass in parks, road verges and other amenity grassland less frequently and allow flowers chance to bloom and to set seed.
- Reduce the use of pesticides wherever possible.
- Be more relaxed about an obsession with tidiness and neatness and extensive hard surfaces.

Many local authority and national policy makers should be expected to be able to and actually access help and advice in what to do in the promotion of development and in the case of infrastructure- related land assets to make and manage these very extensive areas of land in a way that they can support thriving populations of insect pollinators, including honey bees.

38.5   Managing the Semi-natural landscape

There remain significant areas of the British Isles which have been given a designation and protection of varying kinds including Ramsar sites, National Parks, Nature Reserves of all sizes (National and Local), Country Parks, Areas of Outstanding Natural Beauty and so on as well as privately and publicly- owned estates within which public access has been agreed and places and spaces managed by charities and local government. Within these areas there is a wide range of land uses including agriculture, both arable and pasture, livestock, forestry and woodland management and leisure. Very often these areas have become designated because it is the patterns of land management which have produced the landscape we see today.

It is hoped that those responsible for the management of such places recognise the role and importance of insect pollinators and seek to, as a minimum protect them, but ideally to promote and encourage them to enhance biodiversity. Honey bees would benefit from such activities.

39. The feeding of honey bee colonies – options available

39.1 Initial consideration

Understanding the relationship of the needs of insect pollinators and matching them with ways to achieve significant resources to help recover their populations and help sustain them is becoming a key activity in the British Isles and internationally.

Part of the answer as to how to help pollinators survive and thrive is more available forage throughout the year and places and spaces where the species can live and reproduce.

Whilst repeated reviews, evidence gathering and all the other mechanisms which can be used and misused to scope the problem, progress is painfully slow at building and communicating a national and local consensus on the benefit of these insect pollinators and the recognition of the vulnerability of the human species without them. And more importantly by taking some action.

In the time being beekeepers in the British Isles continue (with minimal public and state understanding and appreciation) to sustain a significant population of a pollinating insect, the honey bee, which in part have had to substitute for and take on the role of many pollinating insect species as their populations and distribution diminish. Those species which continue to decline probably do so because of varying combinations of factors including lack of forage and habitat destruction of their living spaces.

Beekeepers (often personally heavily subsidising their craft) are doing their bit, however as we have noted during the course of this book consistency of forage availability, variations and other aspects of forage are a challenge to sustaining honey bee colonies in the British Isles.

39.1.1 Why feed?

Ideally honey bee colonies should be self-sufficient and not require support in terms of additional nutrition supplied by the beekeeper.

But it is with increasing frequency that alerts are formally issued and received making beekeepers aware of colony starvation alerts and the need to check and, if necessary, to feed their bees.

These warnings may be issued in the spring as the winter stores become depleted but are becoming more frequent during the summer months. The lack of forage may be due to no suitable plants available, or that the flowers have become damaged through drought plants wilting or dying off or there being no nectar secretion, or the nectar drying out. The weather may also have been problematic for the bees to forage for any resources that might be available.

Because of the increasing frequency of whole or parts of beekeeping seasons in which the colonies become short of forage beekeepers increasingly turn to supplementation of the forage through feeding carbohydrates or through using pollen and protein supplements.

It is still unclear as to the precise requirements of the bees' nutritional needs. It is clear however that the requirements for essential micronutrients and the role of microbiota in the bee gut and in the preparation of bee bread must be considered in the development of supplemental food in the future.

These requirements may change depending on the colony, its life stage, and the season.

The continual need for supplemental feeding or keeping honey bee colonies in a perpetual stressed condition is morally, ethically wrong and bad beekeeping. It is not a sustainable or economical solution to keeping honey bees in an environment where the changing land use is resulting in the elimination at worst and reduction at least in the quantity, quality and availability of adequate bee forage. Having said that we are often faced in today's beekeeping with having to give supplemental feeding to colonies.

39.1.2   Terms used

Terms are used in the feeding of bees and it is important to appreciate their differences.

'Supplement' implies something *additional* required to be added to naturally occurring and foraged for by the honey bee (pollen and / or nectar), but has one or more deficiencies in their nutritional component profiles as well as the need to ensure the colonies receive the required amount of food.

A 'pollen supplement' is a supplement that contains natural pollen in an amount not less than 10% on a dry weight basis.

'Substitute' implies that either pollen or nectar are completely deficient in the field.

Substitutes for nectar are usually other carbohydrate supplemental foods. Substitutes for pollen mean the replacement of pollen with mixtures of proteinaceous and some amino-acid substances.

39.1.3 HMF (5-Hydroxymethylfurfural)

This organic compound is also known as 5-furfural, a furan.

HMF (hydroxymethylfurfural) is toxic to bees and is formed from the acid- catalyzed dehydration of hexose sugars, especially fructose and formed in honey because of heat treatment or storage. Associated with the production of HMF is a change in the diastase (amylase) activity.

Fresh natural unprocessed honey contains little (<1ppm or 1 mg/kg) or no HMF. This can increase with temperature and time and in well- stored, fresh, natural honey the HMF content is on average 12.5 mg/1kg.

High HMF levels are also considered a risk in the feeding of invert sugars or HFCS (High Fructose Corn Syrup).

HMF levels in honey increase with time and so there is the risk of toxic effects occurring in bees if they are fed back old honey as honey stored for 2 years and fed may contain >40 mg/kg and long- term storage may result in HMF levels of up to 200mg/kg and more.

Little is known about the lethal or sublethal effects or mechanisms of HMF or toxic sugars on larval development.

HMF has been shown to half the life of a bee if fed with honey containing more than 125ppm HMF. Currently an HMF content in sugar syrup of up to 30 mg/kg (equivalent to 30ppm) is considered safe for adult bees and larvae and an HMF content of >30mg/kg is considered toxic to bees.

39.2 Honey as a feed

Earlier in the book mention was made of the increasing use of honey in medicinal products and the properties that honey has and exploited in these products. These are very relevant to the storage, use and benefits of honey in the honey bee colony.

Honey's antimicrobial properties include:

- A high sugar content (and consequently a low water content) which creates an unsuitable osmotic environment for most microbial species and extracts the water from the microbe and drying it out and death. Note sugar solutions given as feed as susceptible to microbial attack because of the higher water content compared to capped honey.
- The pH of honey is low (acidic) in the range 3.2-4.5 and this acidic environment is too low for most microbes to grow in.
- Honey bees add an enzyme, glucose oxidase, during the processing of nectar into honey and subsequently when the honey is diluted with water (then more prone to microbial attack) the enzyme generate hydrogen peroxide killing the microbes. The enzyme secreted by the hypopharyngeal gland facilitates the conversion of glucose to gluconic acid, the primary acid in honey and hydrogen peroxide. The gluconic acid is responsible the pH of the honey (ranging from 3.2-4.5) and this suppresses bacterial growth and helps to preserve honey quality.
- Honeys vary in their antimicrobial properties and this may be due to the inclusion of addition components which were present in the nectar gathered from the floral sources.

Such an example is methylglyoxal (MGO) and dihydroxyacetone found in the nectars of some *Leptospermum* sp. (Tea Tree) which are contained in manuka honey.

Honey seems to upregulate (increase in the regulation of gene expression) genes associated with protein metabolism and reduced oxidation contributing to honey bee health through immunity stimulatory effects in colonies which are subject to stressors including environmental toxic substances, pests and diseases.

There is evidence that constituents of natural honey which are absent from sucrose and HFCS (High Fructose Corn Syrup) are important in the honey bee's ability to detoxify xenobiotics. Furthermore, it seems that honey, sucrose and HFCS may affect honey bee physiology and health differently (Wheeler and Robinson 2014).

Honey can be fed to bees, but great care should be taken on its provenance and its condition before making it available to the colony to be fed. Ideally it should not have been heated and be free from pesticide contamination. Pesticides and their interactions in honey bee colonies may be a factor in stress-induced colony losses.

The combs or jars of honey to be given as a feed should not show any signs of fermentation (e.g., a yeasty smell, or comb cappings exuding liquid, or a very shiny surface on the faces of the comb or the top of the honey in the jar). There should be no signs of bee faecal matter (brown streaky material) on the comb faces or across the frames. The origin of the combs should be known and if they have come from a dead colony the reason for the colony's death should be known and a decision made as to whether the comb is likely to be a source of a pathogen.

Honey from unknown sources should **never** be fed to bees unless it has been irradiated as it may contain pathogens such as American Foulbrood.

Combs of honey, and tubs of honey should ideally be kept in a dry, dark place at 11^0C and ideally the honey should have a moisture content of $<17.1^0$C. The honey is then not likely to ferment.

Alternatively, the combs of sealed honey can be bagged in freezer bags and placed in a freezer where it can be stored until used as an emergency feed source placing the unfrozen frames close to the bees having scored the surface of the wax capping to reveal the honey.

Or the frames and the comb can be brought up to room temperature the combs cut out, warmed and the honey can be separated by passing the wax / honey through a fine sieve and then prepared for consumption by the bees.

Unprocessed honey stored in the temperature range $0-27^0$C starts to form crystals and granulate. Glucose in the honey crystallizes because it is less soluble in water than fructose. When the temperature of the honey in the comb or storage place is lowered to the point of supersaturation it granulates. This can be observed in the honey in combs which are at the edges of the hive or are not covered by the over winter cluster. If the combs of honey comprise oil seed rape honey this will granulate quickly because it has a high glucose to fructose ratio.

39.3   Honeydew honey – as a feed

Feeding honey to honey bees containing honey dew in preparation for overwintering can cause harmful effects to the bees. The slow decomposition of melezitose, raffinose, and melibiose present in honeydew containing honeys to simple sugars causing the rectum to become overloaded and then initiate diarrhoea. Diarrhoea stops the absorption of simple sugars such as fructose and glucose in the digestive tract, weakening the bees, causing starvation, shortening their lives and this can lead to the death of whole colonies. Honeydew honeys are also poor in their protein content and rich in minerals and sugars thus making the overwintering colonies on honeydew honey more difficult.

Crane, E., ***Honey. A Comprehensive Survey***. (1979). *Heinemann, London*
Crane, E., ***Honey. A Comprehensive Survey***. (2020). IBRA and Northern Bee Books

39.4   Carbohydrates - the honey bees' energy source

Carbohydrates are the basis of the source of energy to enable growth, body temperature, flying and reproduction. When used as a feeding stimulant at high concentrations they also provide chemical building blocks for the growth and development.

Colonies are often fed carbohydrates as part of routine beekeeping practices such as when honey stores in colonies are low and there is a lack of forage after honey has been removed (harvested) from colonies, such that they do not have adequate stores to deal with dearth periods of forage or over winter.

Having considered honey as a supplementary feed if there is insufficient in store colony carbohydrate stores need to be augmented by supplementary feeding.

The next part of the book considers the feeding of honey bees with carbohydrates.

Sucrose (not contaminated with other sugars) is the best form of carbohydrate substitute for nectar to feed bees because it is the most attractive and has a higher nutritional value to them.

The order of preference has been found to be sucrose>glucose>maltose>fructose.

Liquid carbohydrate feeds of sucrose solution, invert sugars, HFCS or various fruit sugars may be given to the bees inside the hive and presented in feeders to enable the bees to access the liquid and remove it for further processing without drowning in it.

Colonies being managed in the British Isles may have to be fed syrup in the autumn to achieve at least 20-25kg of carbohydrate stores to last between November and April.

Bees that ingest and process sugar solutions being fed inside the hive are mainly the same age as the nectar receivers in a colony.

Feeding sugar to provide carbohydrates which can be processed and laid down as winter stores or used in colony management and beekeeping techniques is not a simple issue.

The beekeeper needs to consider the following aspects of feeding:

- Why feed sugar?
- When to feed the sugar, and in the particular circumstances which make supplementary feeding of sugar necessary?
- What form of sugar should the feed be in?
- How is the sugar to be fed to the bees?
- What is the target uptake of feed?
- When should the supplementary feeding begin and end?

39.4.1   Why feed sugar as a supplementary feed?

White sugar (sucrose) is the usual form of carbohydrate given to honey bee colonies and is considered by beekeepers to replace the sucrose which is found in naturally- occurring nectar and collected by foraging bees. Pure white granulated sugar is very clean and pure and is totally digested by the honey bee with no residues being left in the gut. If the bee needs to take a cleansing flight, there will be no residue from the bee having ingested sucrose.

Although the nectar collected in the field principally contains sucrose this is subsequently broken down into fructose and dextrose (glucose) which are much less attractive to foraging bees.

If left to their natural instinct colonies of honey bees will develop, reproduce (raise new queens and swarm to produce daughter colonies) and establish winter stores in relation to the amount of natural forage available. It is likely that colonies which have access to adequate forage all the active seasons and which do not have a disease or pest burden and hence, for example, in the case of *Varroa* and its associated viruses, where beekeeper intervention is still required to manage the pest) will be able to survive without addition of sugar as a supplement feed.

39.4.2   Main reasons for feeding sugar

- To enable the colony to have sufficient stored food to provide energy during periods when there is a dearth of forage.
- To stimulate the colony into rearing brood.
- To be used to provide energy in emergency situations.

39.4.3   General Precautions in feeding sugar to colonies

- Autumn feeding should be completed before the weather turns cold to give the colonies enough time to take down the syrup and reduce the water content to a safe level thereby avoiding fermentation or mould growth in the syrup which has not been taken down by the bees and processed.
- Avoid feeding fermented syrup and honey.
- If making one's own syrup use only refined sucrose or white table sugar. Do not feed the bees with brown sugar.
- Alternatively feed with a proprietary ready- made syrup such as Ambrosia, Apisuc or equivalent available from beekeeping equipment suppliers and often sold by beekeeping associations to their members.
- Regularly replace the combs in a hive; preferably only use comb that is less than three years old, not misshapen or brown in colour in your colonies.

- Do not use comb from infected colonies. Use new comb or foundation and *in extremis* old comb can be used provided it has been fumigated with acetic acid. Care should be taken when handling the acetic acid because the liquid and fumes are hazardous and will also attack and corrode concrete floors.

39.4.4   When to feed the sugar?

Timing is critical to the success of feeding sugar as a supplement.

If the ambient temperature is <7-8⁰C, bees are more reluctant to move the syrup from the feeder.

It is often the beekeeper's interference with the natural processes and the relationship that the colony has with the environment that causes situations to arise where supplementary feeding is necessary for colony survival.

This interference may be through the stimulation to produce colonies in a situation where there is inadequate quality and quantity of forage. In this case nectar.

The removal of honey from the colony may be the stores the colony was laying down for a dearth period and could not be made up in the rest of the season.

The beekeeper may be manipulating colonies to rear brood early in the season to catch early flowering crops such as oilseed rape, to produce drones, to encourage starter and finishing colonies and to help the development of nuclei colonies.

Feeding sugar syrup to bees for winter storage really should start in the mid- summer when the higher air temperature helps the bees to process the sucrose and reducing the amount of their body protein reserves which are required to elaborate the enzymes to process it.

We have already referred to the fact that in recent years, the National Bee Unit through its Bee Base communication service has been sending out an increasing number of alerts to beekeepers advising them to check their colonies for food reserves, and this has taken place not just in the late summer as colonies prepare for autumn but also during the summer months when the expectation would be that colonies would have stores. This they may have but the earlier injudicious harvesting of honey and a subsequent lack of forage enabling the colony the potential to rebuild its stores coupled with the fact that the beekeeper may be unaware of the stores status of their colonies can spell disaster for the colony.

Beekeepers when considering carbohydrate supplementation to their colonies should bear the following principles in mind.

- The nutritional requirements of honey bees depend on the activity of the colony which itself is affected by the ambient air temperature and this in turn is closely related to the seasons and the weather.
- All nectar / pollen which is collected are not equal in nutritive value for bees and a few are toxic to bees.
- Highly concentrated sucrose solutions are not necessarily attractive to foragers because the forager may have problems handling the syrup with its mouthparts due to the viscosity of the syrup.

39.4.5 What form should the sugar be in?

Sugar syrup is a suitable food for bees except in very cold weather when the bees are tightly clustered. Honey bees in this situation cannot easily remove excess water from the sugar syrup and depending on the climate and the time of year this will dictate the form of the sugar syrup given to the honey bees.

It is estimated that bees use the energy generated from 100 g of sugar to remove every 0.5 litres of water from the syrup.

Sugar (sucrose) can be fed to bees in the following form-

* Dry sugar (sucrose) – this does not cause any stimulating effect on the colony to any significant extent. It does not induce robbing and can be fed to nuclei at the end of the summer.
* Powdered sucrose (icing sugar or confectioner's sugar).
* Thick syrup (67%) – is less stimulative to colonies than thin syrup.
* Thin syrup (50%) which is very stimulative and can be directly utilised by the bees without further need for processing.

Note that Dry unrefined brown sugar in the form of a cake is not recommended as the bees need to frequently defecate and opportunities for bees to do this over winter or in cold periods can be limited. Molasses (not fully refined beet and cane sugar) are harmful to honey bees.

39.4.6 Dry Sugar

White granulated sugar packaged in paper bags can be fed to colonies.

It is preferable to make a few pin pricks in the bag of sugar and immersing it in water for a few minutes. The bags should be removed from the water then allowed to harden, split open along their length and then given directly to the bees by placing the bag on the top of the crown board.

Alternatively, 200 ml of water / kg bag of sugar can be poured into a small hole in the bag producing some liquified sugar.

Bees are unlikely to continue to feed on the granulated sugar once there is a suitable nectar flow and at this time the bag of sugar should be removed from the colonies marking each bag with the hive number (identifier) in case it is needed again. This will help reduce any potential disease transfer from one colony to another.

The bees need to be able to obtain enough water to liquify the sugar crystals so weak colonies should not be fed with granulated sugar.

Even other colonies which have problems locating enough water may start to utilise their own body water.

39.4.7   Sugar syrup

Sucrose syrup is a good substitute for nectar and honey especially when fed to bees that have or are provided with bee bread.

Whether the syrup is fed in from a bought ready-made supply or made from granulated sugar into a sugar solution it has a limited shelf life because of the growth of yeast in the syrup which then cause fermentation with increasing speed the warmer the temperature.

Syrup prepared from dry sugar should be fed to the bees within three days of preparation. It is possible to prolong this period by boiling the sugar syrup for 10-15 minutes before storing it in sterile, clean containers. Best practice is to only mix the amount of syrup required for immediate needs and to only use clean containers which do not contain any residues of previous mixes.

Quantities of sugar syrup can be made and stored in clean plastic jerry can style containers which have handles and are thus easier to pick up and pour in a controlled way avoiding spillage of the syrup which is not only costly but may set off robbing in the apiary. The containers should be labelled with an adhesive or have a swing tag type label with the contents, its strength and date of preparation.

Salt (sodium chloride) should not be added to the syrup because it causes dysentery and can increase mortality.

Syrup which has been prepared and stored for a long time can develop a black mould which often grows as a scum on the surface of the syrup. This can be prevented by the addition of a preparation of thymol solubilised in surgical spirit (thymol is not water soluble) and then added to the syrup. See later section for more details on preventing the fermentation of sugar syrup.

Older beekeeping books refer to the addition of 'Fumidil B' to the syrup to control *Nosema* spp.

Fumidil B is no longer an authorised product, nor is it permitted to be used in beekeeping.

Bees prefer a warm syrup when the weather is cold and a cooler syrup when the weather is hot.

The essential equipment needed to prepare syrup includes equipment for measuring liquids, weighing solids and a mixing vessel which ideally has a heating element with a thermostatic control together with a container which can be easily cleaned and can be closed to the atmosphere.

In the British Isles syrup can be bought premixed with a formulation of around 67% sucrose to around 33% water. This has a longer shelf life because it has been sterilised during the manufacturing process.

The sugar content is referred to in **Brix**. The term Degrees Brix (symbol °Bx) is the sugar (sucrose) content of an aqueous solution.

Degrees Brix is the sugar content of an aqueous solution. One degree Brix is 1 gram of sucrose in 100 grams of solution and represent the strength of the solution as a percentage by mass.

If the need is for only relatively small quantities of syrup to be made at a time the sugar syrup can be made in a clean plastic or stainless -steel bucket. The sugar is emptied into the bucket and the bucket lightly bumped on the ground to level the sugar. The level of sugar can be marked with a felt tip pen or sticker on the side of the bucket and clean water (hot if syrup is needed in a hurry, or cold if prepared to wait for it to dissolve) added until its level reaches the mark.

39.4.8   Syrup for stimulating a colony

This will also promote more flights to collect pollen.

One litre (1kg) of water mixed with 1 kg of dry sugar provides a 50% sucrose solution.

This can be utilised straightaway by the honey bees, without further dilution or concentration.

One litre of water plus two kg of dry sugar provides a 67% sucrose solution.

To prepare a thin syrup (50%) mixing equal parts of water and sugar:

Water (litres)	Sugar (kg)	Syrup (litres)
1	1	1.575
10	10	15.75
100	100	157.5
2	2	3.13
20	20	31.5
200	200	315

When considering making a specified quantity of thin syrup at 50% use the following equation

Volume of thin syrup required divided by 1.575 will give the quantity of sugar and the volume of water required.

e.g., for 200 litres required amount of syrup mix 127 litres of water and 127 kilograms of sugar (200 divided by 1.575 equals 127)

e.g., for 25 litres of thin syrup mix 16 (15.87) litres of water with 16 kilograms of sugar

The following table shows the mixtures required for increasing volumes of a sugar syrup solution at 50% brix.

Sugar syrup (litres)	Water (litres)	Sugar (Kg)
10	6.35	6.35
20	12.7	12.7
50	31.7	31.7
100	63.5	63.5
200	127	127
1000	635	635

In Imperial measures 1lbs of sugar mixed with 2 pint of water gives a 28.0 % sugar concentration syrup solution.

39.4.9 Thick syrup suitable for processing by bees into overwinter stores

A 67% sucrose solution is a supersaturated solution, and it is not possible to dissolve any more sugar into this solution.

To prepare a thick syrup solution (e.g., 67% brix) the ratio of one litre of water added to 2 kg dry sucrose sugar. This produces 2.26 litres of syrup.

The following table details the quantities of ingredients for producing a 67% sucrose sugar syrup solution.

Water (litres)	Sugar (kg)	Syrup (litres)
1	2	2.26
10	20	22.6
100	200	226
2	4	4.52
20	40	45.2
200	400	452

The proportions of water and sugar to produce various quantities of a 67% brix sugar syrup solution can be calculated from the following equation.

Volume of syrup required divided by 2.26 to obtain the amount of water to be used and then double this value to calculate the amount of sugar needed.

The following table details the quantities of ingredients for producing a 67% sucrose sugar syrup solution.

To produce Sugar syrup (litre)	Water (litres)	Sugar (kg)
10	4.46	8.9
20	8.85	17.7
50	22.1	44.2
100	44.2	88.4
200	88.4	176.8
200	442	884

In Imperial measures 2lb of sugar mixed with 1 pint of water yields a 61.5% sugar concentration syrup solution.

Syrup can also be purchased ready formulated with a low HMF (hydroxymethylfurfural) content and the pH adjusted. An example is Ambrosia syrup. Crystallisation of the syrup in the feeder and subsequently in the comb may be prevented by formulations with a higher fructose concentration.

39.4.10 Emergency feeding to recover starving bees

If on inspection a colony is discovered starving and close to being lost through death, the bees can be sprayed (using a clean hand-held spray bottle) containing a thin lukewarm sugar syrup solution). To further aid their recovery adjacent empty combs should be filled with the syrup and then if successfully recovered the bees can be fed with a thin syrup using a rapid feeder or an inverted honey jar with a screw on lid in which a fine mesh has been inserted placed directly above the bees. These modified lids are available from bee equipment suppliers. An eke will be necessary to accommodate the height of the jar.

Starving colonies can be fed with moistened granulated sugar in bags but should first be fed with a thin liquid syrup as described above.

39.4.11 Preventing the Fermentation of sugar syrup in feeders

Some beekeepers add tiny amounts of thymol to the syrup to help stop fermentation occurring and help protect the bees against *Nosema* spp.

For example:

- Mix 4 g or 0.14 oz of thymol crystals dissolved in 12oz of surgical spirit,
- A level teaspoon of this mixture can be added to 2 pints of syrup.

Alternatively:

- Obtain small sealable glass bottle.
- Fill to one third with thymol crystals obtainable from beekeeping equipment suppliers.
- Top up the bottle with surgical spirit obtainable from pharmacies.
- Seal and label the bottle **'Thymol in surgical spirit'** and add the date.
- Either add 2.5 ml of this preparation to 4.5 l. of syrup and mix well or
- Add half a teaspoon of the preparation to one gallon of syrup and mix well.

39.4.12 Feeding in the comb

Strong colonies can be selected whose purpose is to draw a full deep box of foundation placed over the queen excluder and fed with sugar syrup. The foundation is well drawn into straight combs and when capped the combs can be cleared, removed from the hive and stored in brood boxes in a cool dry, vermin -free place until needed. Combs can be removed and given to colonies running short of stores and when the stores have been removed the bees will use the comb and these combs can be used to remove old comb as part of the beekeeper's brood comb replacement programme.

It should always be remembered that there are legal limits on the amount of sucrose contained in honey which is offered for sale and so care must be taken to reduce the risk of honey which is removed from a colony which has been fed in the recent past does not contain any sucrose derived from the feeding of sugar in whatever form.

40. Types of sugar feeders

Beekeepers are good at adapting and using materials and this is certainly the case with devices devised and used to feed honey bees. Nevertheless, beekeepers are well provided for in the varying types of feeders to be found on offer in the catalogues of beekeeping equipment suppliers.

Where honey bees should have easy access to the syrup easily but care is needed to ensure and there are no places whether they can climb out of the feeder and fall into the syrup where they will readily drown. Unless attention is paid to this possibility large numbers of bees can be lost. Such situations may also make opportunities for robbing and attract the attention of wasps and ants.

40.1   Small quantity feeders

40.1.1   Jar Attachment feeders

A plastic lid into which is inserted a phosphor bronze mesh / gauze is screwed onto the top of the glass honey jar, filled with syrup and inverted before placing it on the tops of the frames so the mesh is situated between the top bars and the bees can access the syrup. Alternatively, the jar can be inverted on top of the crown board over a ventilation hole. The jar is best inverted firstly over a container before the partial vacuum is created so that the syrup which initially flows from the jar can be collected, and not become a trigger to start robbing in the apiary. Typically fitted to 1Ib (454g) honey jars.

40.1.2   Wooden entrance feeders

Here the jar attachment feeder which is inserted into the top of a wooden block which is hollowed out to provide the syrup at the hive entrance. The proximity of access to the feeder to the entrance can increase the risk of robbing, especially if the colony is weak or there is a persistent threat from wasps.

40.1.3   Plastic entrance feeders

In this instance a plastic reservoir (400 or 800 ml) is placed on a horizontal lighting board of the hive and a feed channel is positioned in the hive entrance. The same caution applies as for wooden entrance feeders.

40.1.4   Frame feeders (also called division board feeders)

Feeders placed within the body of the hive and take the place of one or more of the frames, usually the brood frames, and placed to one side of the colony. They may be made of wood (plywood) or plastic. This type of feeder typically holds around 3½ pints of syrup (1.989 L)

Frame feeders are accessed by bees and filled with fresh syrup from the hole in the top bar of the frame feeder.

Bees can easily be drowned unless there is a piece of wood easily able to float on top of the syrup inserted into the feeder and the beekeeper checks to ensure it does not become stuck but moves with the level of the liquid in the frame feeder.

Because of the proximity of the frame feeder and the syrup to the brood or cluster area the syrup is slightly warmed and is also close to the bees to access and hence readily consumed.

Frame feeders are an excellent way to ensure nuclei have access to syrup.

40.1.5   Tin feeders

Tin cans with perforated removable lids can be used by inverting the filled can over a container to allow the partial vacuum to form before placing it over the frames or on top of the crown board over one of the ventilation holes. An empty super placed on the crown board acts as an eke to accommodate the height of the filled can on top of which the roof can be placed.

40.2   Larger quantity feeders

40.2.1   'English feeder' / tray feeder

This is a so-called 'rapid' type of feeder is a shallow tray with covers the whole cross section of the hive and has a central opening over which is a hollow central cone. A cap or cup over the hole allows the syrup to access the base of the cone but prevents the bees from falling into the syrup and drowning. The feeder may or may not have a lid which fits over the feeder and prevents robbing or access to the syrup by bees from another colony.

The feeder is placed on top of the crown board and the bees access the central cone from the holes in the crown board.

The syrup is easily replenished by simply pouring fresh syrup into the feeder removing the lid if one is present.

The capacity is around 6 litres of syrup, but this is dependent on whether the hive is level.

40.2.2   Rapid feeder

This type of feeder works on the same principles as above but is a smaller cake tin style and size. It also has a central cone cover and a lid.

This type of feeder requires a bespoke eke or an empty super or brood box to be added over the crown board with the roof on top to enclose the feeder and so it can be kept inside the hive and the hive remains protected from the elements and robber bees and wasps.

The feeder can be easily replenished by removing the lid and pouring in fresh syrup.

The capacity of the feeder is around 4 pints of syrup.

40.2.3   Ashworth feeder

This type of feeder is a shallow tray which covers the top of the hive but instead of having a centrally placed hole and cone the bees access the syrup one side of the tray climbing over a weir to gain access to the syrup.

In all of these types of feeder a small quantity of the syrup should be trickled down the inside surface of the opening using your hive tool to create a syrup trail which the bees will detect, follow and then make contact with the bulk of the syrup feed.

The feeders which cover the whole cross section of the hive are ideal for feeding large quantities of syrup and are best used for the feeding of syrup in preparation for winter. In the late summer and early autumn there is still an amount of heat being generated by the colony which warms the syrup slightly and aids its uptake by the bees.

40.2.4   Contact feeders

These are food grade plastic buckets with snap on lids into the lids of which are set phosphor bronze gauze. The containers are filled with the syrup and upended over a bucket until a partial vacuum is formed before being placed over one of the holes in the crown board. An empty brood box or super is used as an eke and the roof placed on top.

Typical capacities are ¼, ½ and one gallon. A ½ gallon feeder will usually fit inside an empty super. These are also available in metric equivalents.

These types of feeders are ideal for emergency and spring feeding when the close contact of the bees with the syrup through the mesh insert is easier to achieve.

40.3   Good apiary hygiene when feeding syrup

Ideally syrup should be given in the early evening or on a day when the weather conditions will not encourage the scout and foraging bees out from the hive.

Feeding during the day encourages the foraging bees to leave the hive and investigate where the new source of feed has come from and this can promote robbing by other bees or wasps and ants.

It is essential not to spill syrup in the apiary or allow puddles of it to be left on the tops of hives. Do not leave matters to chance but use a handful of grass / vegetation to wipe away the syrup and take the removed syrup / vegetation back to your store / bee shed and dispose of the vegetation in a manner where other bees or wasps cannot access it.

Always use an entrance block with an aperture sufficient for a couple of bees to pass side by side in and out of the hive so the guard bees can prevent robber bees and deter wasps, who will harry a colony to access the hive for the syrup.

Take care to ensure that bees cannot get under the roof or through spaces between the hive bodies creating opportunity for bees to gain access by- passing the guard bees.

Ensure everything is square, especially if using an eke. They are easily dislodged allowing robbing bees and wasps a means of access. Put a weight (e.g., a brick) on top of the roof to help prevent the eke being moved or dislodged.

Robber bees and wasps can be deterred by propping a small piece of glass or preferably Perspex at an angle over the width of the hive entrance. This allows the bees to access their own hive but deters the robbers.

Wasp and hornet traps should also be in place. In the apiary.

41. Sugars in the form of fondant and candy

41.1   Fondant

Fondant may be a ready -made paste consisting of refined sugar sucrose, fructose and glucose. Some fondant products contain thymol which is added to disrupt Varroa development. Note thymol can taint honey that is already in the colony.

Fondant can also be purchased as a ready-made product. An example is Ambrosia fondant.

It can be used for emergency use in winter or feeding nuclei and mating colonies throughout the year. It is softer, more pliable and it cuts easier than harder candy.

Split the plastic sleeve and slicing the patty with a sharp serrated knife and place the patty on top of the brood frames directly above the brood nest. Accommodate the depth of the patty by using a super as an eke.

41.2   Candy

Candy can be made from honey to which is added icing sugar until the consistency of a stiff paste is made.

41.2.1   Hard Candy

Hard Candy is used as an emergency feed for bees in late winter and early spring. A typical recipe is 5 parts by volume of sugar added to one part of water. The mixture is brought to the boil and allowed to bubble for two minutes at 121^0C. The container and syrup is then put into a sink full of cold water, allowed to cool and stirred from time to time. When the syrup begins to turn cloudy then the mixture should be stirred vigorously and as it thickens emptied onto a candy board or into suitable containers.

41.2.2   Queen Candy

This candy can be used for providing food for feeding the queen during queen rearing and queen transport purposes.

Mix equal quantities by weight of starch-free icing sugar with invert sugar, preferably fructose, as it reams softer and easier for the bee to consume and then add a few drops of water. Do not use honey because of the risk of American Foul Brood (AFB) and European Foulbrood (EFB). Heat the invert sugar to 50 – 60^{0}C adding a few drops of water at a time. Then wearing clean household gloves knead the mixture into a ball and leave it to stand overnight. If the ball loses shape overnight, then add more icing sugar and re-knead. Store the candy in a plastic bag in a labelled airtight plastic container to retain the moisture in the candy.

41.3   Syrup

41.3.1   Invert sugar (acid-inverted sugar) syrup

Invert sugar or invert sucrose is a product in the form of a syrup where the aqueous sucrose has been chemically hydrolysed through using hydrochloric acid, citric acid or invertase.

Supplementary cakes based on invert sugar syrup are suitable for feeding bees in the spring and at other times. The shelf life of partly inverted sucrose syrup stored in a room temperature of 15-25^{0}C is 6 months.

However, it can contain toxins that aggravate the accumulation of water in the bee gut and can significantly increase mortality through high levels of HMF (around 150ppm).

41.3.2   HFCS (High Fructose Corn Syrup) - HFS (High Fructose Syrup) also called Isoglucose

HFCS is used by some commercial beekeepers and fed to bees to promote brood production in the spring in readiness for commercial pollination contracts. The usual form of HFCS used is HFCS-55.

HCFS can be successfully fed to colonies as part of a mixed diet of free forage and syrup feeding. Deleterious effects have been found when the obtain bees their carbohydrates almost exclusively from HCFS when fed before going into winter.

In the production of HFCS starch is first separated from the sweet corn kernel and the starch suspension is then liquified and is initially converted by acid or enzymes into a low solution strength glucose. Further conversion of the starch takes place, and the process can be stopped depending on the specification of the end-product which are given identifier numbers, e.g.,42 or 55.

A common mixture is 55% fructose and 32% glucose. HMF begins to form at 45^{0}C (no HMF is produced during the mixing process) and this temperature can be reached if the large metal tankers which are used the transport the HFCS is moved during hot weather and this heating may continue when the syrup is offloaded from the tanker into smaller containers which the retailers of the syrup may use.

Many corn syrups have high starch contents and are toxic to bees.

Major problems have occurred using corn syrup not only from a toxicity point of view because of the residues of starch, acids, enzymes and the presence of higher saccharides in the product, but also granulation in the comb.

Austrian beekeepers report the sucrose / invert sugar or starch syrup fed to bees did not affect the overwintering mortality of colonies when the feeding was done in the period between July and October.

Honey bees can be successfully fed iso-glucose provided the beekeeper:

- Adheres strictly to the manufacturer's requirements concerning the storage conditions in terms of temperature and hygiene to avoid producing conditions which are suitable for microbial growth in the syrup as it has no antibiotic properties, unlike honey.
- Ensure there is a current six -month validity certificate of quality.
- Colonies to be fed isoglucose must be healthy and strong with good body protein content and supplies of bee bread especially before winter.
- Crystallization of the stored syrup in the comb may occur where the bees cannot cover the combs because they are too weak because of poor pollen foraging, Varroa and virus infections.

41.4　Other potential sources of sugars

Honey bees are sometimes intentionally fed with waste sugar residues from the manufacture of confectionary and sweets and in some case the beekeeper only becomes aware of the fact their bees have found such materials when they notice the colour of the honey in the comb is blue / green or red from the dyestuffs added during the manufacturing process and the honey itself may also have a smell, for example of mint.

Such waste sugar sources should be carefully assessed before using them. If the source and composition are known the producer of the residues should be required to store them in a bio-secure way to prevent honey contamination and prevent the potential spread of foulbrood diseases. Protection from flies and wasps is also necessary.

41.5　Sugars poisonous to honey bees

The following sugars are poisonous to honey bees even at very low levels:

- Galactose – occurs in mammalian milk, milk products and is only slowly digested by honey bees. If it is the only sugar feed given to the bees it will cause starvation.
- Glucuronic acid
- Galacturonic acid and polygalacturonic acid
- Arabinose
- Xylose
- Melibiose
- Mannose – nectar from silver lime (*Tilia tomentosa*) and to some extent Broad-leaved lime (*Tilia platyphyllos*).

- Raffinose – occurs in some soybean products.
- Strachyose
- Lactose- occurs in milk, milk products.
- Cellobiose
- Mannitol
- Sorbitol
- Ramnose

42. Protein - through the feeding of pollen supplements or substitutes

42.1 The need to Feed Protein

Colonies require about 22 kg pollen per year which has to be available throughout the brood-rearing period in quantities sufficient to meet the requirements for colony development.

There are two main reasons why beekeepers may need to feed protein supplement,

- To stimulate colonies to produce brood at certain times during the year.
- To provide a source of protein when there is insufficient pollen forage available.

The use of pollen- supplemented foods requires some experience and judgement to estimate the available protein stocks but also to understand the potential effects, beneficial and non-beneficial, on the colony.

Beekeepers need to be aware of the nutritional status of their colonies, whether or not they wish to take measures which will help the colony continue brood rearing.

Such awareness is required throughout the whole season and if the weather is expected to worsen for an extended period or there is a wish to promote the rate of increase in the colony build up.

These include:

- In the spring when there is no pollen available for 6-8 weeks or at least 41 days before the first flowers are open and there is pollen available.
- When there is a shortage or lack of good pollen, e.g., when bees are working a single crop or there is poor diversity of forage.
- At least 6 weeks mainly in the late summer and not later than early September.
- The movement of colonies to a particular crop, including prior to, during and after the pollination of the crop.

- Spring and summer divisions of colonies.
- In queen rearing operations.
- Package bee production.
- Overwintering.
- Building up of colonies after damage by pests and diseases and pesticides including colony confinement.
- The application of bee medicines.
- Helping prevent or cure brood diseases.

42.2 Whether to feed

Times when beekeepers should carefully consider whether the feeding of pollen supplements or substitutes is necessary. This can occur throughout the year and include:

42.2.1 Colony size and its nutritional status

Is the colony too small and are there signs of starvation, e.g., larvae being discarded sucked dry? Does the colony have any stores of bee bread?

42.2.2 When conditions are unsuitable for bees to fly and forage

For example, during periods of cool and / or windy weather. Reduction in pollen availability may result in reduced brood rearing e.g., drones.

42.2.3 Weak colonies

Weak colonies especially early in the season when a colony may be devoting a lot of its adult bee resources to the activities of brood rearing and nurse bee activities.

42.2.4 Timing and availability of surrounding forage

Extended periods of inclement weather in the spring may mean a delay in flowering or conditions unsuitable for the foraging bees to fly. Supplementary feeding may be necessary to maintain a strong colony and prevent decline or starvation. Warm weather may mean good flowering and no need for extra feeding.

42.2.5 Pollen-deficient 'nectar' flows

Pollen-deficient nectar flows – situations where there is a considerable honey flow, but the pollen produced by these crop plants is insufficient in quantity or quality to maintain healthy and strong colonies.

42.2.6 Pollen sources of doubtful quality

On examination of the combs determine whether there are cells containing old pollen and which may have pollen mites? Is the pollen mouldy and are there fungal sporulating bodies visible? When handling mouldy comb, the beekeeper should be aware that there is the potential for the fungal spores produced from the moulds to present a health risk to them if they inhale the spores. Wearing an appropriate face mask is desirable.

42.2.7   Pollen clogged combs

Sometimes when there is an abundance of pollen and the colony continues to collect more than it requires one or more combs will become full of pollen. These combs can be sealed and kept frozen and then fed back to the colonies.

If left in the hive and the bees do not utilise it in humid conditions the pollen become vulnerable to microbial degradation by the fungal mould species *Bettsia alvei* which turns the pollen into hard lumps which are of no nutritive value to the bees.

Sometimes pollen clogged combs become infested with tiny pollen mites which both consume the pollen and turn any residues into a yellow powder.

In some years stimulative feeding with protein supplement may not be necessary or of any benefit and there is little to be gained feeding such supplements when there are adequate amounts and varieties of pollen with good nutritional characteristics.

42.3   Quality of the protein supplement to be fed

It is unclear in the British Isles whether the need for stimulative feeding with a protein supplement widely exists largely because of the relatively small scale of the landscape, its diversity, the type of beekeeping carried out in the British Isles and the locations where beekeeping tends to be practised.

If the carbohydrate stores available in a colony are low and there is no nectar flow the bees will sometimes obtain the sugar from the feed supplement and leave the rest. Likewise, if there is bee bread still present the bees may remove the sugar content and leave the rest.

Feeding sugar syrup to a colony when there is no pollen available may cause the bees to have to utilise their body protein vitellogenin reserves resulting in a protein stress in the colony.

A lack of interest in the pollen substitute or supplement may also be a sign that the queen is weak.

There are numerous recipes for pollen substitutes and supplements involving combinations of a wide variety of substances. They include soya flour, canola flour, linseed flour, sunflower flour, torula yeast, Brewer's yeast, vitamin and mineral supplements, fish meal, peanut flour, skimmed milk powder, powdered casein, sodium caseinate, lactalbumin and pollard. Some are attractive to honey bees, others have high oil contents and others have inadequate essential amino acid levels.

42.4   Factors affecting the consumption of pollen substitutes and supplements

When natural pollen is available, and the bees can forage for it a colony is reluctant to consume the pollen substitute or the supplement or other forms of pollen provided to them.

If no pollen is being brought into the colony these bees are reluctant to use the substitutes or supplements but will utilise them if there is any natural pollen being foraged for and brought into the colony.

When bees are fed with pollen substitutes, they only digest 25% of the protein in the diet, but this may be due the almost ubiquitous use of soy flour as the protein source. Such colonies are also found to have higher levels of infection with *Nosema* (DeGrandi-Hoffman et al. 2016).

In general, artificial diets do not mean indefinite brood rearing. Prolonged feeding will result in a decrease in the colony population because of the production of workers which have shorter lives.

Spring pollen supplementation can be helpful especially when the weather is bad and feeding will help in a faster spring build-up and consequently a potentially higher honey yield. It can also reduce the effects of parasitism and nosema infection.

The feeding of supplements should begin around 6 weeks before the expected natural pollen will become available. If the honey stores are low, then thin sugar syrup should be fed. Pollen substitutes should not be fed to starving colonies.

If feeding the protein supplement is planned in early spring the bees should have adequate honey stores and if they don't have enough food stores they should be fed a sugar syrup the same time as the protein supplement.

42.5   Supplementation by British Isles beekeepers

A majority of beekeepers in the British Isles have 5 or less colonies and so it is unlikely that they will want to develop the skills to produce their own pollen substitutes or supplements but are content to buy them in.

Pollen can be trapped at the entrances to hives using various devices to strip / dislodge the pollen loads into a receptacle and then stored so that it can be fed back to the colony at a time when conditions require supplemental feeding with pollen.

Ready-made pollen traps can be purchased from beekeeping equipment suppliers.

Pollen trapping should only be carried out when the colony is strong and there is an abundance of good quality pollen sources available for the honey bees to forage for.

Trapped pollen should be frequently collected from the traps and quickly dried otherwise mould will cause the pollen to deteriorate and lose its nutritive value and, in any event, it should be utilised before two years elapses unless it is irradiated otherwise it is of little use to feed the brood.

Trapped pollen can also be frozen, preferably at -15°C, but even under these conditions its nutritive value is reduced. Some of the vitamins will deteriorate and some fats will oxidise.

Whichever way the pollen is stored when required it can be mixed with sugar and fed to the bees.

Soya flour is often use as a substitute for pollen, however it contains too little riboflavin (B2) and much less niacin (B3) than dried baker's yeast. The soya flour must be expeller processed to remove the high oil content which is 15%. Whilst the protein levels in soya can be as much as 50% it is deficient in the essential amino acid tryptophan

Other substances which should not be included (see above for list of toxic sugars) in pollen substitutes in high concentrations (when intended to be given to younger bees) are pectins, starches, cellulose, glucans and arabinoxylans.

Lipinski (2018) gives a much fuller description of the materials which can be used to prepare pollen substitutes and protein supplements and we makes no attempt to cover these details.

Suffice it to say the growth and well-being of a colony depends on the quality of its diet and in particular the amounts of protein and lipids in the form of fats and oils.

A typical pollen substitute should include the following ingredients:

- Protein
- Lipid
- Water / syrup
- Pollen

43. Pollen / protein supplements with / without fondant

Protein and pollen supplements can be prepared and fed to honey bee colonies, but it has to be recognised that as of today there is no true substitute available for pollen.

43.1 Physical forms in which protein supplements can be given to honey bee colonies

Pollen and protein supplements can be fed to honey bees in a variety of physical forms and with or without the use of fondant

43.1.1 Pollen / Bee bread

- Freshly collected pollen should be kept in a freezer or cold room and although it can be fed directly to the bees in a powdered form without any additives it is usually mixed (minimum of 5% dry weight of the total mix) with soya flour or Torula yeast or a mix of both.
- If you are buying pollen, ensure the pollen has been sterilized by gamma irradiation.

43.1.2 Dry Feed

- One form of this type of feed is corn gluten and can be fed using a frame feeder or by being placed on top of the crown board under the roof.

43.1.3 Patties

- Bees utilise patties best if the sugar content is at least 50% by dry weight.
- The larger the surface area of the patty exposed to enable more bees to access and feed and therefore consumption is increased.
- There are different types of protein supplements put up in the form of patties and which are given to colonies usually around the end of February / early March.

- The patties can be pre-cut and part-wrapped in cling film or wax paper before placing them near the holes on the crown board or on top of the frames directly over the cluster, opening up the wrapping sufficiently to allow bees access to the patty.

43.1.4   Recipe for a pollen supplement (bulk)

- Pollen   1.50 kg
- Lactalbumin or caseinates   3.28 kg
- Brewer's Yeast   6.56 kg
- Sugar   18.67 kg
- Water   4.5 litres

Use a clean concrete mixer to mix the ingredients and them form the mixture into 500g patties, 10 mm thick which can be wrapped in cling film or plastic bags and kept in an airtight container.

43.1.5   Recipe for pollen substitute (smaller quantities)

- lactalbumin or caseinates   120g
- Deactivated Brewer's yeast   230g
- White sugar   650g
- Water to half the weight of sugar   320ml

Make into 500g patties (10mm thick), feeding one 500g patty per hive every 6 days for 4 weeks.

43.1.6   Recipe for high protein and high carbohydrate pollen substitute

- 3 parts soy flour made from ground soybeans
- 1 part brewer's (or other) yeast
- 3 parts sugar

This can be mixed and fed dry inside the hive or mixed with a little water into a paste.

43.2   Commercial pollen substitute

Products can also be purchased such as NEKTAPOLL- a ready to feed pollen substitute / fructose syrup patty. Use from early March until late summer if there is little pollen for the bees to forage for. Feed between 250 and 500g per colony by splitting the plastic sleeve and slicing the patty with a sharp serrated knife. Then place the patty on top of the brood frames directly above the brood nest. Accommodate the depth of the patty by using an eke, or a super.

43.3   Candy Board Cake

Protein supplement is mixed with sugar, water, corn syrup and powdered pollen substitute (protein). After cooking the resultant cake is placed on the top bars of the hive in a special tray with a mesh in the bottom through which the bees access the cake.

43.4   Liquid Amino Acid and Protein Supplements

These are a protein supplement food in a thick liquid form which can be fed to the colony in a frame feeder. They may be further enriched with vitamins and minerals.

Examples include:

- Honey-B-Healthy
- BeeStrong

Some of these products can also be mixed in a sugar syrup into pollen patties or sugar dough.

The least attractive pollen supplement to honey bees is one that contains no pollen.

DeGrandi-Hoffman et al. 2010 suggest that there is a connection between diet, protein levels and immune response and that colony losses might be reduced by alleviating protein stress through supplemental feeding.

However, it must be understood that the addition of protein supplements to colonies in the spring does not always guarantee desired effect, for example extending the longevity of the bees. The pollen supplements may be consumed by the bees for their sugar content and not for the protein and other components.

The concentration of linoleic acid is a plant oil found in pollen. The amount varies between plant species and influences honey bee feed intake and resulting longevity.

Linoleic acid is also good for

- Maintaining the highest reproductive performance.
- The digestion and absorption of fatty acids during the colony expansion in the spring.

The effectiveness (as measured by bee development and amount of brood reared) are improved with the addition of no more than 5-8%) of cholesterol, fatty acids, gibberellic acid, plant growth hormones and certain trace elements.

Consumption is a good indication of the acceptability of the diet supplement. The phagostimulants present in pollen mean that lipid or pollen containing supplements are consumed more readily.

44. Other Supplementation

44.1 Vitamin supplementation

Honey bees require relatively high levels of vitamins, including those in the Vitamin B complex, some of which are likely to be provided by the activities of the bee's gut bacteria.

Vitamin supplements formulated for the honey bee can be purchased to supplement sugar syrup and different kinds of cakes and patties. They include vitamins B1, B2, B3, B5, B6, and C as well as essential amino acids.

They have been added to formulations at a rate of 1-3%.

It should be remembered that vitamins are toxic to honey bees at higher concentrations.

44.2 Propolis supplementation

Lipinski (2018) refers to an old Polish beekeeping tradition where small amounts of a wax free alcohol extract of propolis is added to the sugar syrup when feeding the colonies to prepare them for winter.

The raw propolis is kept in dark containers and protected from light and heat. It can also be freeze-dried. The propolis is dissolved in 70% ethyl alcohol. This solution is then freeze dried. Wax and other impurities then settle in the bottle and a clean solution can then be removed (e.g., by using a small syringe body) from the top.

44.3 No additional feeding

Ideally no additional feeding should be required, however in most years supplementary feeding is now a regular practical necessity and insurance policy.

Beekeepers should always endeavour to help the bees acquire and accumulate sufficient materials to meet their needs through their own efforts if possible, but this should not encourage the beekeeper to avoid their responsibilities for their colonies' welfare.

45. Current supplementary feeding practices in England and Wales

45.1 Proportion (%) of beekeepers in England and Wales reporting feeding their honey bee colonies

The Annual Bee Husbandry Survey contains information on the proportion of respondents (England and Wales) to the survey on who reported feeding their honey bee colonies in the years 2009-2018 and these are reproduced in the table below.

Year	% Reported feeding their bees
2009	88.3
2010	78.2
2011	79.9
2012	87.8
2013	94.6
2014	88.6
2015	84.5
2016	87.4
2017	87.5
2018	89.8

45.2 Use of different types of bee feeds in England and Wales

The Annual Bee Husbandry Survey also contains figures on the proportion of respondents (England and Wales) to the survey on who used bee feeds in the years 2009-2018 and these are reproduced in the table below.

Type of Feed	2009	2010	2011	2012	2013	2014	2015	2016	2017	2018
Ambrosia	7.86	10.88	N/A	N/A	N/A	N/A	N/A	N/A	N/A	N/A
Candy / fondant	52.39	51.6	50.46	57.43	70.45	68.27	60.66	60.16	62.36	70.25
Feed supplement	3.89	2.54	N/A	N/A	N/A	N/A	N/A	N/A	N/A	N/A
HFC syrup	N/A	N/A	0.51	0.83	0.51	1.22	1.13	1.08	0.65	0.28
Honey	8.27	6.88	6.54	10.34	11.01	7.04	11.32	7.56	7.38	6.86
Inverted sugar syrup	N/A	N/A	7.15	7.49	10.5	12.35	12.64	12.56	12.15	17.05
Pollen / patty	3.48	5.74	N/A	N/A	N/A	N/A	N/A	N/A	N/A	N/A
Sugar syrup	65.44	72.63	62.72	63.85	53.68	75.95	66.98	73.8	71.69	71.08
Vitafeed	3.89	2.54	N/A	N/A	N/A	N/A	N/A	N/A	N/A	N/A
Other	1.71	1.74	3.27	4.76	6.94	3.3	4.06	4.81	4.56	4.26

In the 2019-2020 BBKA Overwinter Survival Survey respondents (BBKA Members) were asked to report on the type(s) of feed they gave to their colonies in preparation for and during the overwinter period. These data are contained in the following table:

Type of Feed	% of Respondents	
	Prior to 1/10/2019	1/10/2019 1/4/2020
Concentrate syrup prepared by beekeeper	50.75	13.43
Proprietary concentrated bee feed sugar syrup	16.53	5.28
Proprietary brands of bee fondant (candy)	14.62	54.54
Light sugar syrup	5.70	3.02
Homemade fondant (candy)	3.33	12.13
Pollen supplements	2.60	6.16
Pollen substitutes including with fructose and /or glucose syrup patty	2.53	7.16
Candy feeds which include polyphenols	1.34	3.41
I did not feed my bees because already had adequate amounts of stored honey	22.12	15.46
Not stated	2.45	2.83

46. Beekeeping strategy considerations for meeting colony nutritional needs throughout the year

46.1   The Seasons

The British Isles have a very wide range of geological and topographical variation which contribute to the very different climates, weather patterns and habitats to be found. Because of this there is no one size that fits all when it comes to timing and when to carry out beekeeping activities.

For the purposes of this book and the consideration of feeding strategies the meteorological seasons and their dates will be followed.Readers will have to adjust these for their own particular circumstances and experiences.

Season	Meteorological Season
Spring	1st March to 31st May
Summer	1st June to 31st August
Autumn	1st September to 30th November
Winter	1st December to 28th February (29th in a Leap Year)

The autumn is usually considered to be the beginning of the beekeeper's year and the nutritional status of the colony at this time of year is critical not only as to whether the colony survives the winter but the state it might be in at in the late winter / early spring when brood rearing will begin the start of an expanding colony with new bees to replace those that have overwintered and are dying off and importantly are in the right physiological condition to rear and feed the new brood.

46.2   Useful skills

Successful beekeepers are ones who understand honey bee biology and can synthesise scientific knowledge and blend it with the art of beekeeping and skilfully using the information gained from close observation of their honey bees and the surrounding environment to meet their bees' needs.

An essential beekeeper's skill is to be able to ascertain the nutritional (and health) status of a colony. Some key aspects of this can be done without having to take the colony apart and inspect the combs and the bees directly.

Storch in his book *At the Hive Entrance* (1985), published as a European Apicultural Edition, Brussels invites and encourages the beekeeper to spend time observing the entrance and the immediate area around the hive and watching the activities of their honey bees.

Much can be determined about the health and the nutritional status of the colony throughout the year. We have taken the guidance given in the book and applied it in each of the sections which describe strategies for nutrition throughout the year. The observations and things to look for are those with an immediate relationship with nutrition. Honey bee health and the nutritional status of a colony are inextricably linked and observing colonies which are in good shape is a very satisfying and therapeutic experience and at the same time being able to detect indications and awareness of situations where a colony may be struggling and taking remedial action can also be both satisfying as well as good bee husbandry.

Caution: Any moving in front of the hive or close head proximity to the hive entrance can prompt bees to fly out and investigate and may be motivated to sting the observer. Honey bees may react quickly to sensing the carbon dioxide or smell of alcohol in the human breath or of perfume. The colony temperament may be problematic because the colony has become queen-less, is being very defensive, there has been a sudden reduction in nectar- flow or the bees are responding to the weather conditions at the time. The beekeeper should bear this in mind and ideally always wear a veil as a piece of personal protective equipment to protect their face and head when in close to the hives or possibly in their flight lines. Wasps may also be in the hives robbing colonies.

46.3   Preparations for any Eventuality

There are potentially a number of occasions when there is a need to feed honey bee colonies either on a pro-active basis to encourage colony development, reproductive potential and overwinter survival or in emergency situations because of lack of forage or opportunities for foraging because of inclement weather. In these days of changing weather patterns together with the other pressures on the landscape shortage of forage can occur several times during a beekeeping season or at times when crucial events are happening within the colony's life cycle.

In anticipation of situations of a potential crisis it is reassuring to know that you have an adequate number of clean (best cleaned using washing soda solution and air-dried before putting into storage) feeders capable of delivering feed in the quantities required, ekes and the foodstuff (sugar crystals, fondant, ready prepared syrup) all stored in a clean dry place.

Keeping back combs of sealed liquid honey from extraction and stored in cool dry conditions and protected from attack by vermin, ants, wax moth and wasps as well as robber bees is a good strategy. Best practice is to label the combs so that where practicable they are only returned and used on the colonies from which they were taken.

Then using them when the bees are active (having scored the surface of the comb disrupting a small area of the comb cappings to expose the honey) and placing the frames as near to the cluster as possible).

Source an empty drawn comb together with a clean squeezy washing up bottle ready for emergency feeding.

Packets of proprietary made pollen supplements / substitutes in a fondant or syrup patty should also be available for use. Examples of such products available in the UK include Nektapoll and ApiCandy and there are others.

There is nothing worse than being faced with an emergency for which you are unprepared.

46.4   Assessing Stored Carbohydrates

It is hard to estimate the actual amount of honey stores a colony will require because of the year-on- year variation in weather conditions over the winter and as referred to earlier much will depend on the location of the hives and the strain of honey bee.

As a general rule of thumb, the following estimates of the amounts of honey that can be found in sealed combs of these types of hives. There is a lot of inconsistency in published information and these figures are for illustration purposes only.

Hive Type	Comb area both sides Sq"	Typical amounts of stores held in a comb (lb)	Typical amounts of stores held in a comb (kg)	Number of cells per box	Number of frames / combs per box	Typical amounts of stores held in a box (lb)	Typical amounts of stores held in a Box (kg)
Modified National Brood	199	4-6	2-3	57,100	11	44-66	22-33
Modified National Brood 14 x 8	292	6-9	3-4.5	78,550	11	66-99	30-45-
Modified National shallow	119	2-3	1-1.5	35,750	11	25-30	11-13
WBC Brood	199	4-6	2-3	52,000	10	44-66	20-30
WBC shallow	119	2-3	1-1.5		10	25-30	11-14
Smith Brood	199			57,100	12	-	-
Smith shallow	119	2-3	1-1.5	50,000	12	25-30	11-14
Langstroth Brood	272	7-9	3-4	72,500	10	70-90	30-40

Langstroth shallow	153	5-7	2-3.5		10	50-70	11.36
Modified Commercial Brood	275	7-9	3-4	74,250	11	70-90	35--45
Commercial shallow (when National Shallow is used)	119	2-3	1-1.5	35,750	11	25-30	11-14
Modified Dadant Brood	340	6-8	2.7-3.6	93,700	11	60-80-	27-36-
Modified Dadant shallow	173	3-4	1.3- 1.8	47,500	11	40	18

The term 'shallow' is used to define the size of the box whereas the term 'super' defines its use usually as the box containing frames/ comb in which the bees are encouraged to store honey, often placed over a queen excluder.

There are a number of factors which influence the type of hive selected by the beekeeper. Heath's book *A Case of Hives* and Kritzsky's book *The Quest for the Perfect Hive* describes many of the properties and characteristics of different hive types.

Beekeepers in the British Isles use a range of hive types and in the 2019-2020 BBKA Annual Honey Survey respondents indicated the following use pattern.

Hive types	% Respondents 2019	% Respondents 2020
Beehaus	0.2	0.7
Commercial	4.6	6.3
Dadant	0.5	0.5
Dartington Long Hive	0.1	0.3
Langstroth	4.3	5.3
National, Standard Pattern	68.4	67.6
National, with larger brood frames 14 x 12	18.6	20.0
Smith	1.7	0.8
WBC	10.6	9.0
Warré	0.8	0.7
Other	4.0	4.5

Readers will note we have not included the more unusual hive patterns, some are based on old designs and others more recent developments.

The frame size can also be assessed from the point of view of the maximum storage of food reserves which the colony could accumulate either for overwintering or the capacity to cope with periods of dearth. Larger frames sized hives are heavier to handle and manipulate and this factor should not be underestimated, especially of the beekeeper intends to periodically move colonies from one apiary to another to benefit from anticipated nectar flows and / or the availability of pollen.

Estimates vary from 20-40kg but there may be areas in the British Isles where because of a milder climate and the effects of changing weather patterns meaning the flowering periods of many plants are changing to adapt to these changes that the availability of nectar and honey is sufficient to require a lower quantity of stores, say 20kg / year. However continual fluctuations in the ambient temperature will accelerate food consumption and consequently more stores might be required compared to extended periods of cold when the colony clusters and is relatively inactive.

Beekeepers often end the season with a so-called 'brood and a half hive' configuration; a brood box and a super, usually separated by a queen excluder. If this is the case the beekeeper should pay particular attention to ensure the queen excluder is removed in mid- autumn and the super either left on the hive because it already contains stores or because a potential late flow of Ivy or Himalayan Balsam (ideally a mixture which gives a honey which does not granulate easily) will give extra stores for the colony. If the queen excluder is left in place and the cluster follows the stores up into the super over the winter period the queen may be left below the queen excluder and runs the risk of being lost due to becoming cold, immobile and then falling from the cluster. If this happens to a colony and you decide to remove the queen excluder turn the excluder over and carefully examine any bees which are adhering to the underside. One of them might just be the queen!

Surveys of colony losses overwinter in England carried out by the British Beekeepers Association and the reports from the from the regions of the Bee Inspectorate repeatedly report starvation as a cause of lost colonies.

In general winter starvation occurs because of one or more of the following reasons:

- Insufficient honey because of lack of stored honey.
- Poor honey availability because although honey is present the bees are unable to access and utilise it as it may be several frames away from the cluster or below it.
- The number of bees is too small and unable to warm the stores and utilise them.

Hive scales can also be inserted under the hives and a regular record of the change in weight throughout the year. The use of such equipment is good because a disruptive inspection of the combs to assess their weight is unnecessary. The status of the food reserves can be assessed by examining the trend of stores increase (hive weight gain) or decrease (consumption) in the case of the overwintering hive.

The development of hive monitoring systems, such as those developed by Arnia, enable data to be sent from electronic scales containing load cells placed under the hives to a monitor gateway which is connected to a data storage facility from which the user can access the data for their own hives. This can be done remotely without the beekeeper having to physically attend the hive and of course there is no physical disruption of the colony. Other information such as weather conditions can also be accessed.

46.5   Assessing protein / bee bread reserves

Beekeepers are usually taught to focus on the amount of carbohydrate (sugar/ honey -based stores), however the amount of pollen and bee bread available for consumption in the autumn in preparation for overwintering is equally important. In autumn the bees will process pollen if available and store it overwinter in the form of bee bread and any excess pollen placed into empty cells.

Ideally a colony going into winter should have 500-600 in^2 (3226-3871 cm^2) of bee bread stores ready for late winter / early spring brood rearing (Lipinski 2018). This is a substantial quantity to attain and it may well be that many colonies in the British Isles do not always succeed in having such quantities available.

An appreciation of the state of the colony at the beginning of winter and as you monitor its progress late winter / early spring will enable you to manage your expectations as to what the colony can do in the period before forage becomes available and then once available its diversity, quantity and likely availability throughout the coming season. This appreciation will help the beekeeper to plan a strategy for the colony in the coming season.

A quick inspection of the brood frames in which brood is developing can be made using an LED torch / magnifying lens to see if there is a pollen ring around the upper outer perimeter of the brood nest. If there is a ring of empty cells, often without nectar, this is the place where the bee bread (i.e., the protein reserves) should be. The slow build-up of a colony may be due to the lack of pollen / bee bread reserves and the depletion of the vitellogenin (storage protein) levels in the worker bees until the worker bee's body protein content reaches 20%.

The opening up and inspection of a hive, especially when combs are removed to be examined can provoke a stress reaction in the bees and even short periods (e.g., 60 seconds) when the colony is open, or the bees are being disturbed these stress effects can be seen in the colony for some time (an hour or so after) and even for several days.

46.6   Considering what resources will become available in the future

The nutritional status of a colony can be manipulated to encourage a greater ratio of foraging bees to collect pollen than nectar. This has implications not only for the health of the colony but also pollen- gathering bees are said to be better pollinators, especially in the pollination of crops.

In the case of field beans if there is no fresh nectar available in the field bees will often be reluctant to forage for pollen.

In the British Isles colonies of honey bees are able to survive the winter and other periods when there is a lack of nectar or honey by storing carbohydrates in the form of honey. If there are no honey stores and no nectar available and the beekeeper is unaware of the plight of the colony it will die in a very few days.

When there are nectar flows this stimulates the colony, encouraging the foraging for pollen and an expansion of the areas of brood within the colony. Conversely a lack of nectar will see a decline in the number of bees foraging for pollen.

Shortage of stored honey and a lack of nectar in the field can make colonies much more defensive of their hive and its contents.

The stimulus of providing sugar syrup or a good nectar flow keeps the brood area in the brood nest area open and encourages population replacement. If the stimulus is lost there is a reduction in area of brood being cared for by the colony.

The frequency of feeding sugar syrup depends on the reason why feeding is considered necessary.

Small amounts (1-2 litres) of a 50% solution provided once or twice a week in a rapid feeder will act as a stimulus whereas if say 10 litres of syrup are fed in one go to a colony this will have a less stimulatory effect.

If the colony is being fed to lay down the winter stores then feeding a good-sized colony a quantity of 10 litres of 67% solution is a good start, whereas only half this volume may be needed for a colony which only covers four or five frames.

In either of these feeding scenarios if the colonies do not consume the quantities of syrup fed to them within 5 days, the remaining syrup can be left on for a few more days and then removed from the colony. If left on the colony there is a possibility of developing black mould on the inner surfaces of the feeder container.

Inability to provide access to good sources of nectar, or feeding large amounts of sugar syrup to honey bee colonies before the colony has been able to collect a large amount of protein-rich pollen causes a rapid decline of a bees' physical condition and productivity.

Although we have previously stated the autumn to be the beginning of the beekeeper's year, we will move the clock on a little and start this section on strategies for the overwintering period, the time when the fate of the colony depends on factors which if addressed will mean the colony survives and is in a good state to start brood rearing in the new calendar year. Any factor which is inadequately addressed may put the colony at risk. Achieving a good position for the start of winter is very dependent on the skill, attentiveness of the beekeeper as well as the honey bees themselves.

Hopefully having survived the winter and the colony expansion has begun we can then plan the rest of the beekeeping season and how we will ensure are bees are in the best state and condition for the **next** winter as well as coping with the challenges to be faced during the current year.

46.7   The importance of Varroa management and awareness of Deformed Wing Virus (DWV)

The management and control of varroa and awareness of the presence of symptomatic DWV is important. Varroa management strategies should follow Integrated Pest Management principles and beekeepers are recommended to check the current advice given by national beekeeping authorities, beekeeping associations and their extension services.

Varroa management and control should be considered during all periods of the beekeeping season.

46.8   Acoustic Hive Monitoring

The development and use of acoustic hive monitoring systems by beekeepers and research scientists received a big impetus with the development of mobile phone and internet connectivity and opened up the possibility of being able to look at the current and historical patterns and behaviour for a number of key parameters which would inform the beekeeper / researcher about the status of the hive from a remote location. A leading developer of such systems was the company Arnia.

Technology and innovation continue apace but some of the features which such systems have include are listed in the following table.

Monitoring Feature	Benefits
Apiary temperature and rainfall	Current and historical
Hive Disturbance	Theft (removal) / Hive dislodged by animals or inclement weather
Temperature at hive entrance	Helps assess best location and aspect of the hive
Hive Activity	Foraging assessed by flight noise
	Fanning, nectar processing or hive ventilation
	Colony development
Brood Temperature	Brood state / Queen status
Hive Homeostasis	Humidity and colony temperature
Queen Status	Queen mating
Bees robbing honey from other hives / being robbed	Awareness raising
Swarm management	Indications of preparations for swarming
Hive weight and change in weight	Monitoring Stores increase / decrease depending on time of the season

In addition, correlations and overlaying weather data with hive monitoring provide powerful tools to help beekeepers in their practical colony management by being able to gains insights on what is happening inside the hive and colony without the need to disrupt it. Such systems also enable alerts to be set so that the beekeeper is alerted should a default threshold be triggered.

The utilisation of such technologies are undoubtedly useful skills which many beekeepers could embrace and make their beekeeping even more valuable, enjoyable and fascinating.

47. Overwinter colony survival

The overwinter nutritional requirements for a colony to survive depend on a range of factors, some of which may be more important than others, depending on the locality of the apiary.

- Colony size.
- Apiary location and aspect e.g., warm / cool / shaded / in a frost hollow / exposed to wind.
- Period of brood rearing, depending on the strain of the honey bee.
- Overwinter flowering species e.g., winter and spring bulbs available for early forage.
- Characteristics of the overwinter climate and weather.

Differences have also been reported on the relative amounts of stores consumed by different strains of bees. Honkó et al. (2009) cited in Lipińksi (2018) found in Finland that Carniolan and Central European Bees consumed less stores than the Italian and Buckfast strains studied.

Some beekeepers believe the age of the queen going into winter should be no more than 24 months old.

Beekeepers are often lulled into a false sense of security when they have strong colonies in the autumn ready for going into winter, but because of a lack of mainly carbohydrate (sugar) stores they become weakened for the time of spring brood rearing.

Strong well-fed (rich in sugars and bee bread) colonies start brood rearing in late winter and early spring leading to strong foraging colonies which are considered production colonies with at least 45-50,000 workers and 36-45,000 capped, older brood cells.

Weaker colonies which had insufficient winter stores consume them (sugars) and then, weather permitting seek to utilise nectar from willows, dandelions and other early flowering tree and plant species.

Malnutrition and even starvation itself become apparent at the end of the winter and into the early spring, especially after an extended period because of the inclement weather and lack of forage availability (both pollen and nectar). Often the dire state of the colony is recognised by the beekeeper when it is too late. Sadly, and too late the beekeeper opens a hive to find the bees dead with their heads embedded in bottoms of the comb cells.

This state of starvation may also have had its seeds sown in the previous year with the bees and the beekeeper not making enough stores provision perhaps because of the lack of forage availability or because the beekeeper fed insufficient or too late for the bees to take down and process the syrup into winter stores. The forage available may have yielded unsuitable honey, e.g., in the cases where the honey will granulate into large crystals in the comb which the bees cannot fully utilise; whereas access to balsam nectar at the same time that ivy is yielding will provide winter stores which do not granulate.

Beekeepers must appreciate the fact that it is starvation and not cold *per se.* either through running out of food reserves or even in a colony which has plenty of food stores which the bees have lost contact with are common causes of colonies deteriorating and then dying out.

Cluster size- there is an optimal cluster size of around 10-12,000 bees. The bigger the winter cluster the more honey will be consumed. If the cluster is too large it is hard for the bees to maintain a sufficiently high temperature for those bees on the outside of the cluster and even though the bees move through the cluster to access and warm up the honey stores, they can still become chilled and fall from the cluster to die on the floor of the hive and potentially interfering with the through ventilation of the hive (especially if an open mesh floor is used. This can lead to decaying bees and mouldy growths. The bees will always move upwards through the honey stores and always need to keep in contact with them to prevent isolation starvation. The beekeeper should make doubly sure that there is not a queen excluder in place which could trap the queen below it and leave her exposed as she follows the cluster as it progresses across the combs consuming its winter stores.

Lipinski (2018, p256) comments that at the end of November when brood rearing finishes a suitable overwintering sized colony should fill at least two standard 10 frame hive bodies from wall to wall and top to bottom at a clustering temperature of about 13.9⁰C. This would be a large colony in terms of colonies managed in the British Isles.

Normally the temperature in the centre of the brood rearing area is around 35ºC. but when no brood rearing is taking place, this may fall to 14ºC and as the ambient temperature outside the hive falls the temperature on the outside of the cluster of bees is maintained by the bees in the range 6-8⁰C. Energy requirements are lower at 10⁰C compared to 15⁰C.

Winter Cluster Size in relation to external temperature		
External temperature 0C	Cluster diameter cm	Cluster diameter in.
10	36	14
6.7	28	11
25.6	26	10
32.2	10	4

Adapted from Caron (1999).

In the winter cluster the bees in the inner cluster generate heat and maintain a temperature of 20-36^{0}C, whereas those in the outer insulation layers keep the temperature slightly above 10^{0}C.

Honey bees are said to utilise their winter carbohydrate stores most efficiently at 7^{0}C and cluster CO^2 production is minimal at 10^{0}C. Bees in the centre of the cluster use less food than those to the outside of the cluster. They do not have to flex muscles to generate heat as do those towards the outer edges of the cluster.

If a dwindling winter cluster becomes too small there may be honey and even bee bread close by on either side of the group of bees, but they will probably only move upwards following the small amount of heat the bees are generating and die when they reach the top of the combs as there is no food. The position of the cluster is important and if overwintering the colony in one brood box there is a high chance that the bees will reach the tops of the frames in the brood box and then have nowhere to go and may not move laterally soon enough and obtain before they starve.

There has always been a wide variation in the conditions which our honey bee colonies in the British Isles have to exist over the winter period. Variations in the climate from north to south and east to west as well as from low to elevated levels exist and weather patterns are changing so that our traditional four-season year is now not so easy to determine. Winters when there are long periods of cold weather with no milder interludes can result in colony clusters becoming isolated from their food stores because they have eaten all beneath the area of the cluster and at the same time it is too cold for the cluster to break move to a fresh area of comb with stores and regroup. Or the cluster may move in the wrong direction away from the areas of the combs where the food stores are located. Such a situation is called isolation starvation and for this reason the use of transparent coverboards or briefly removing the coverboard to determine the position of the cluster in relation to the stores is an essential part of overwinter management. It is insufficient to rely on only hefting (lifting) the hives and estimating the amount of stores present. Not only is the quantity of stores important but the relative position of the cluster to the food is vital to ensure the colony is not lost.

If the cluster is found to be detached or moving away from its food stores (potential for isolation starvation to occur) all is not lost and the application of a slab of fondant directly onto the tops of the frames above the cluster and / or the administration of ready for use sugar syrup can be given in a rapid feeder or an inverted honey jar with a screw on lid modified with a mesh insert in the lid to enable the bees to access the syrup can be placed directly on the frames over the cluster. An eke, or an empty super, can be used to accommodate the height of the feeders and allow the roof to be replaced securely on the top of the hive. There is a risk associated with feeding syrup over winter because such actions may stimulate an increase in levels of *Nosema* spp.

If, as is often the case, we have periods of cold interspersed with milder intervals then the clusters can move more readily from area to area and be less likely to suffer from isolation starvation, however these milder periods come at a price as milder conditions and more honey bee activity in the hive means an increased consumption in food stores.

In cold dry winters, the cluster is very tight, loses little heat and far less energy and some of the food reserves are conserved. In warm damp winters, or when the winter has frequent warm spells more stores will be consumed.

It is an important action that colonies are checked over the winter and that the beekeeper is prepared to take any remedial action necessary to help their colonies.

The opening up of colonies and the removal of comb for inspection or leaving the tops of colonies exposed to the elements is not good practice over the winter period. Small LED lights, especially those with the light at the end of a small flexible holder which can be lowered into the hive between the frames to illuminate the combs are especially useful and mean that a hive can be viewed even on dark cloudy days. Always remember as a minimum safety precaution to wear a veil and have a hand- held water spray to hand should you need to discourage the bees from flying from the cluster.

A starving colony must be fed with warm, concentrated sugar syrup at the earliest opportunity but do not give too much as the excess will be stored by the bees beyond the edge of the cluster and this may start to ferment. In addition, feed candy at the same time and repeat after one or two weeks.

Clustered bees produce carbon dioxide and water. The carbon dioxide will disperse readily if there is reasonable ventilation but the removal of excess water from the hive may be a problem if the bees have been unable to take cleansing flights and / or if the overwintering location is a damp place with poor natural air drainage away from the hives.

An optimum cluster should be able to overwinter quietly and calming without the need for many cleansing flights and taking advantage of periods of warmer weather to rearrange the position of individual bees within the cluster and on the frames.

Weaker colonies can be helped by filling excess space in the hives with insulation between the unused frames and the hive walls with expanded polystyrene, if other equipment options are not available which will keep them warmer (that is to say, reduce heat loss), resulting in them needing to consume less food and consequently producing less metabolic water which then has to be removed from inside the hive. Squares of radiator insulation foil can also be prepared and placed inside hive roof making sure that the ventilation holes in the roof are clear of spider debris and not covered by the insulation.

During very cold spells in winter when it is too cold even for the water foraging worker bees to fly from the hive the bees in the cluster may use moisture which has condensed on the inside surfaces of the hive.

Other aspects of successfully overwintering the colonies include some further practical measures.

These include the use of open mesh floors. In general colonies are very capable to survive winter with open mesh floors as there is good ventilation and little or no accumulation of the detritus on the hive floor reducing the potential for the accumulation of hive debris.

Beekeepers are advised to use mouse guards to keep mice out of the colonies. Unfortunately come the time when bees are beginning to forage for early pollen source the pollen collected is often lost to the colony as it is stripped from the workers pollen baskets as they attempt to regain access to their hives as they negotiate their way through the mouse guard. The pollen pellets either fall outside the hive onto the alighting board or on the floor just inside the entrance and are not retrieved or used by the bees.

Deterring mouse ingress into a hive can also be achieved using entrance blocks which have very much reduced entrances and though which mice would be unable to gain access thereby obviating the need for mouse guards and enabling colonies to benefit from all the pollen which has been collected, often during periods when there is considerable risk to the foraging bee.

48. During the winter until early spring

Assuming the colony has adequate carbohydrate- based food reserves a short- term dearth of pollen in a colony may not be too detrimental to the colony but a prolonged pollen dearth will result in the reduction or stopping of brood rearing with a consequent effect on colony development and future potential.

Colonies consume less stores in cold dry winters as the cluster loses little heat and so less energy is required. Warmer winters, or extended periods of warmer weather in winter can initiate brood rearing and the need to consume more food to maintain the brood nest temperature at $34\text{-}35^0C$.

It seems that honey bees are able to judge the weight of stores they have in their colony in response to the concentration of the smell of a specific aldehyde which is added when they process nectar or sugar syrup. They are also able to respond to the concentration of the smell in empty comb cells. Bees become stressed if they consume excessive amounts of sugar (Rinderer 1982).

A colony may consume honey in March at twice the rate it did in the previous December. Colonies may need supplemental carbohydrate feeding which can be given by placing fondant on top of the crown board or the directly over the cluster in the brood combs.

In good years it is possible to set aside whole combs full of stores for use in the winter if colonies require supplemental feeding. These combs should have a hole punched through the centre so that the bees can easily access both sides of the comb when the comb is laid horizontally across the top of the frames above the cluster. An eke may be necessary to accommodate the height of the comb placed on top of the brood frames.

At the Hive entrance

Look, listen, sniff	Signs to look for	Conclusion	Action needed
Listen at the Hive	A soft 'sh' or soft murmur / buzz might be detectable	Healthy colony	None
Shine light through entrance	A large number of dead bees visible on the floor	Colony not overwintering well	Ensure entrance is clear of dead bees
On the alighting board	Soiling with faeces	Sign of dysentery and /or queenlessness / unclean or fermented food stores	Clean faeces from alighting board / check guidance on management of Nosema / dysentery
Through the entrance	Heat and smell of fresh bread (yeasty)	Sign of dysentery caused by colony mainly living off honeydew stores. Can also be Nosema	Feed fondant or syrup and remove dead bees or replace floor and clean faeces from alighting board
Bees leaving and returning to the hive	Defaecation on the wing after reorientation flights facing the hive entrance in front of the hive	Cleansing flights	
Look at hive entrance	Bees do not appear to be making cleansing flights	If tapping side of the hive does not produce any noise or response noise to the beekeeper's tapping, then suspect the colony is dead and examine	Colony may have starved through no food or through isolation starvation or has already been robbed and weakened. Once cause of loss decided then close up the hive and dispose of bees and combs asap by burning in a pit
Look at hive entrance	Workers with swollen abdomens rush into the hive	These bees are water carriers and the need for water increases as the amount of brood increases	Ensure there is a source of water in the vicinity of the apiary and if not provide one.

49. The spring expansion

In most years, colonies begin brood rearing at the end of winter and will do so even if there is little or no fresh pollen available (no flowers or weather too inclement to forage in).

At this time, the bees will utilise their remaining bee bread stores and body proteins. This is a good time to stimulate or secure brood rearing with the provision of protein supplementation and sugars.

The queen has begun actively egg laying. Foraging bees returning to the hive with pollen loads are a good sign that brood is being reared.

The colonies are strong enough.

The nutritional status of the colony requires or can tolerate protein and carbohydrate feeding.

The status of the food reserves is not just relevant to the overwinter period when there is little or no nectar and / pollen available for forage or the weather conditions are not suitable for the bees to forage. Mention has already been made of the lack of distinctness in the seasons and spring in the many places in the British Isles can be little different to weather normally associated with winter.

As the days lengthen and the process of brood rearing gathers momentum colony food reserves, unless the beekeeper tries to intervene, will come under pressure. Not only is there the increasing amount of brood to rear (increased energy consumption to maintain the increasing brood area and temperature.

In the British Isles the northern adapted strains of *Apis mellifera* begin brood rearing when the daily maximum temperature averages only 4^0C and increases rapidly when the daily average temperatures are between 5^0C and 15^0C. This is often in midwinter. When there is no brood in the hive the temperature can be maintained at about 24^0C however once the queen is laying eggs and there is developing brood the temperature of the brood nest is increased to 34^{0}C. This is a big challenge for small colonies on cold nights as they need to expend much energy shivering their wings and muscles in their thorax to maintain the temperature of any brood in the brood-nest.

Colonies which start brood rearing in midwinter and swarm in late spring enables the colonies to achieve a maximum foraging force for the usually short summer season. Colonies where brood-rearing is delayed until late spring usually grow less strongly and have reduced swarming and may go on to starve during the winter.

In springtime the proper development of colonies requires between 5.0 and 7.5 kg of carbohydrate stores and bees are reluctant to rear brood if they do not have an adequate reserve of carbohydrates

It is not just the honey that the bees require to meet the demands of the expanding colony but also protein.

The late winter and early spring are often periods when the availability of pollen (and its quality as well) may be lacking. Colonies usually consume pollen one or two months after collection and during this time of year there may be periods of several weeks when pollen is in limited supply or the weather unsuitable for foraging bees to collect it.

 As already mentioned, there is little pollen stored and there may be relatively small amounts of pollen available to process into bee bread and feed the developing brood. The worker bees in the colonies utilise the storage proteins contained in their so-called 'fat bodies' and the importance of the preparation of colonies in the mid-late autumn for overwintering and encouraging them to continue to forage on pollen and convert the amino acids in the pollen proteins into storage cannot be over-emphasised.

Pollen supplement feeding in the British Isles to promote brood expansion are normally applied in late February or early March.

Very often it is the most populous and potentially valuable colonies that starve because of a lack of honey reserves in late March into early April. This is because of early extensive brood-rearing especially when there has been a mild winter. Early recognition of the colony's plight and prompt action by providing food (pollen or pollen substitutes) rich in proteins and carbohydrates in the form of sugars can help recover the colony.

Colonies which show early spring expansion can result in a large amount of brood to be fed and the brood nest kept warm, and the numbers of brood may exceed the number of adult bees which should there be a very cold spell leading to high losses of both adults and brood.

It is not good practice to open a colony for inspection in early spring during cool weather because if brood rearing has begun the colony cohesion is disrupted for a period of time.

If the volume of pollen is limited the area of brood being reared will reduce. If the quality of the pollen is poor the colony requires greater amounts of pollen to obtain adequate nutrition.

If poor quality pollen is available in limited quantities, the brood area will be reduced. Limited amounts of high -quality pollen will have less of an impact on the area of brood being tended.

When bees start to suffer increasing hunger, the queen will still lay eggs, the hatched larvae are thrown out of their cells so only eggs can be seen in the comb. Other signs that this is happening include finding the dead larvae, and shrivelled larvae with their haemolymph sucked out on the hive floor, under the combs and on the alighting board. Mid-aged so-called 'undertaker bees' which are no longer involved in the feeding of the brood remove the dead larvae.

When there has been more than 5 days of non-foraging the brood which is less than 3 days old is cannibalised to feed their proteins to other larvae

If the hunger still does not stop the bees not only eat the young larvae and will eat the eggs whilst the queen stops laying unfertilized eggs (i.e., the colony ceases to produce drone bees which consume a lot of protein in their development) and will eventually stop laying. The worker bees stop caring for the brood and the colony population weakens. The colony compensates for a pollen deficiency either by decreasing the quantity of brood reared by reducing oviposition or by cannibalism of young larvae thus enabling the older larvae to survive until the time when the pollen supply is deficient, and the bee body storage reserve is depleted. When this is happening inspection of the brood comb shows a 'scattered brood pattern' which may be confused with a brood disease.

The typical demographic of a normally functioning colony is for about 5% of the brood to be eggs, 10% are larvae, 21% pupae and the remainder adult (imago) bees.

All these points indicate the necessity for the beekeeper to consider the potential forage in the area in which the bees will utilise, the main species which will yield pollen and whether there is any information on the nutrient status of the pollen from the different plant species. These assessments are all part of good beekeeping practice and will help you to decide how many colonies you believe the area can support. Always remember to be aware of the presence of other beekeeper's bees and their potential foraging activities into your area. This is especially important in urban and suburban environments and in the wider landscape where the floral diversity and pollen and nectar availability may mean a marginal existence for the colonies.

Bees can be encouraged to use any capped honey in the hive if small areas of the capping's surface are scored with the hive tool to expose the honey.

If the colonies are short of carbohydrate stores (honey) then thick sugar syrup can be given to prevent starvation.

In strong, good, well fed overwintered colonies spring feeding is only advisable if there is a lack of food in the combs after winter and there has been dearth of nectar in the early spring flow and later.

Providing access to good sources of nectar, or feeding large amounts of sugar syrup to honey bee colonies before the colony has been able to collect a large amount of protein-rich pollen causes a rapid decline of a bees' physical condition and productivity. If a colony has access to nectar rich forage and there is a large proportion of uncapped brood in the colony which require a lot of protein to complete their development the bees divert / deplete their own body proteins when foraging. The nurse bees (young worker) supply the huge demand for protein in the form of a jelly to the developing brood in preference.

It is almost certainly the case that colonies will need attention and feeding when colonies going into winter have been large, when there has been a warm wet winter followed by a warm spring and then just as flowers start to bloom yielding nectar and pollen for the weather to turn cold or changeable.

Feeding bees with sugar syrup until the nectar flow restarts may mean the new food is placed in the brood box only to be moved again by an expanding brood nest and together with any existing food stores may be moved up into a honey super if one has already been placed on the hive. This will affect the quality of subsequent honey put into the honey super. When the nectar flow starts, for example in the case of oil seed rape then unwanted frames may need to be removed to prevent contamination of the honey being created from foraged nectar mixing with sugar syrup.

When a colony is seen to be producing drones (a check needs to be made to ensure that the queen is producing drones in which case feeding a protein supplement is unnecessary as drones are produced in protein rich colonies.

At the hive entrance

Look, listen, sniff	Signs to look for	Conclusion	Action needed
Listen at the Hive	A soft 'sh" or soft murmur / buzz might be detectable	Healthy colony	None
Shine light through entrance	A large number of dead bees visible on the floor	Colony may have died out	Ensure entrance is clear of dead bees if there are still live bees
On the alighting board	Soiling with faeces	Sign of dysentery and or queen-less / unclean or fermented food stores	Clean faeces from alighting board / check guidance on management of Nosema / dysentery
Through the entrance	Heat and smell of fresh bread (yeasty)	Sign of dysentery caused by colony mainly living off honeydew stores. Can also be Nosema	Feed fondant or syrup and remove dead bees or replace floor and clean faeces from alighting board
In front of the hive	Defaecation on the wing after reorientation flights facing the hive entrance in front of the hive	Cleansing flights	

Look, listen, sniff	Signs to look for	Conclusion	Action needed
On the alighting board and in the hive entrance	Dislodged pollen loads	Mouse guards, or the entrance is to narrow or other obstruction removing the pollen loads as bee tries to enter the hive	Identify cause and rectify
At the hive entrance	Returning foragers with shapeless green, grey or brownish material on hind legs	Workers have been foraging for propolis e.g. from chestnut, black poplar or cherry trees	None
At the hive entrance	Large numbers of small white crystals	Bees trying to utilise crystallized honey stores	Give a liquid sugar syrup feed immediately
On the alighting board by hive entrance	White or brownish pupae	Colony at the point of starvation	Give liquid sugar syrup feed and pollen/ candy supplement e.g. Nektapol directly above the cluster

50. Early and mid summer

The late spring and early summer show a wide variation in the growth and progress of flowering plants in the British Isles not only geographically but also with the natural variation in habitat and now the effects of changing weather patterns, a feature of climate change. On top of these factors can be overlaid the potential forage which may or may not be available because of land management practices in agriculture and horticulture and in the urban and suburban environments.

Administering a bag of granulated sugar which has been soaked in water and placed on the cover board using an eke to give headspace, for it is said to settle a colony when there has been a break of a few days in nectar availability. It also does not incite or encourage robbing behaviour.

If the break on the nectar flow becomes longer there is the potential for the brood rearing capabilities to be seriously set back and in these circumstances a light syrup can be fed

Cool, wet weather may mean there are breaks in the availability of nectar and at a time when colonies are or should be expanding rapidly such changes forage can cause the bees to become more excitable and agitated and have an increased tendency to swarm.

Oil seed rape (OSR) continues to be grown in many parts of the British Isles and depending on the soil condition when the seed was sown, the variety being cultivated as well as the weather factor this can mean OSR is flowering early or over an extended period. It usually also means there are few alternative forage sources to broaden the diet and nutritional value of the pollen and nectar available to the honey bee. Depending on the planting layout in landscape scale the colonies may have ready access to OSR within a short distance of their hives or they may have to fly some distance which is costly in energy consumption terms and the amount of flying and foraging capabilities they have remaining in their bodies.

Increasingly beekeepers and their honey bees are experiencing a more extended period of 'the June gap' a period when there are few flowers and thus limited amounts of nectar and pollen. Today the June Gap is likely to come earlier but has the same effect on the colony.

Colonies experiencing this dearth of forage may need to be supported because carbohydrate stores are being rapidly depleted as at this time the colonies should be reaching the point of strong build-up. It is best to feed a medium -light syrup because of the weather becomes hot and dry the syrup may crystallise if too strong and the bees have difficulty getting such crystals into solution.

If the bees begin to resume nectar collection and it is believed that this will be on-going then the feeders or bags of unused sugar should be removed from the colonies so that there is no mixing of the sugars with incoming nectar which will be processed into honey and hopefully yield a surplus for the beekeeper to harvest.

At the hive entrance

Look, listen, sniff	Signs to look for	Conclusion	Action needed
Listen at the Hive	A soft 'sh' or soft murmur / buzz might be detectable	Healthy colony	None
Shine light through entrance	A large number of dead bees visible on the floor	Colony may have perished	Ensure entrance is clear of dead bees if these are still live bees
On the alighting board	Soiling with faeces	Sign of dysentery and or queen-less / unclean or fermented food stores	Clean faeces from alighting board / check guidance on management of Nosema / dysentery
Through the entrance	Heat and smell of fresh bread (yeasty)	Sign of dysentery caused by colony mainly living off honeydew stores. Can also be Nosema	Feed fondant or thin sugar solution (1:1) and remove dead bees or replace floor and clean faeces from alighting board
In front of the hive	Defaecation on the wing after reorientation flights facing the hive entrance in front of the hive	Cleansing flights	
On the alighting board and in the hive entrance	Dislodged pollen loads from the pollen baskets	Mouse guards, or too narrow entrance or other obstruction roving the pollen loads as bee tries to enter the hive	Identify cause and rectify

51. Late summer / autumn

Protein intake

Providing the worker bees in a colony with a rich source of nectar or sugar supplement in the form of syrup before they have been able to collect a reasonable amount of pollen containing good quantities and variety of proteins, lipids, polyphenols and other macro and micro-nutrients, will lead to a rapid decline in the bees' physical condition. Such a shortage of pollen in the second half of the year can result in the protein content of the bee bodies containing <30%. This will result in the bees overwintering very badly, being very short-lived and vulnerable to diseases such as nosemosis. In the spring the colony is weak and not capable of rearing much brood. The lack of protein and essential amino acids for the synthesis of jelly, immune proteins, body tissues such as muscles and fat bodies affect both the developmental stages (larvae and pupae) and the mature stages of workers, the queen and drones. It will also affect sperm production in drones.

If sugar feeding is extended throughout the autumn and this extends brood rearing before winter actually 'sets in' it does not help autumn- reared bees. Colonies supplemented in the autumn with protein do not perform better in the following spring.

Nectar intake and processing

Strategically providing sugar syrup may, under the right conditions, induce the colony to collect greater volumes of pollen than they would without the intervention feeding of the beekeeper. Thick syrup can be added to provide food when the honey stores in the hive are low, but it should be noted that the thick syrup does not stimulate the production of brood and the collection of pollen.

In summary

Carry out an inspection to check and determine the following:

- The amount and position of the food stores.
- Whether the colony is still producing brood, its extent and pattern of brood area.
- Check and note whether there are any drones remaining because this could indicate future queen problems.
- Brood and adult bee health, especially varroa and deformed wing virus.
- The state of the combs and their correct positioning in the brood box.
- The weathertightness of the hive and vegetation cleared from under the hive to allow movement of air.
- Ideally the hive / hive stand should have an alighting board or ramp to help the bees land safely and be able to easily access the hive entrance. This will help bees flying in windy weather.
- If not already doing so, consider using a mesh (a varroa floor without the insert) rather than a solid floor. This will add ventilation and help reduce the build-up of detritus on the floor of the hive which will go mouldy over the winter period. Inserting an empty super or eke under the mesh floor as this will provide a boundary layer of air and help the colony maintain its cluster temperature by reducing the amount of draught the cluster experiences especially in windy and exposed apiary locations.

52. For colony consolidation before onset of winter

Bees need moisture to survive the overwinter period but too much moisture poses a threat to the colony.

- It cools the bees and the cluster.
- Where it condenses it will encourage the growth of moulds on combs and the floorboard.
- It can sour / ferment the bee bread.
- It can promote conditions suitable for the fermentation of the honey stores.

Beekeepers often believe that keeping a colony rearing brood late into the autumn and for as long as possible is desirable.

Observations made with 'native' or more locally adapted strains of bee which consistently overwinter well will terminate egg laying in response to the lack of forage pollen and nectar and in some years in the British Isles this may be as early as the end of August. These bees will not respond to stimulative feeding to keep the queen egg laying and any eggs that are laid are removed by the worker bees. The developing brood from eggs laid before this time still require substantial quantities of food in addition to the stores being laid down to overwinter. Insufficient forage or the removal of too much honey from the colony by the beekeeper can soon lead such a colony to starvation.

Any restrictions in brood rearing caused by malnutrition or starvation during the period before winter will significantly weaken the honey bee colony for the next beekeeping season. Any reduction of brood rearing because of nutritional factors together with the shorter lifespan of adult foraging bees will directly affect the size of the colonies.

Honey bees will, when able to, store large quantities of honey for overwintering reserves. In contrast however they only prepare and store small quantities of pollen in the form of bee bread such that the pollen stores can be fully consumed very quickly with raising the early brood before the spring forage opportunities become available, depending on the weather and the forage that is available for the bees to exploit.

For colonies to benefit from supplementary carbohydrate feeding and be able to properly process the sugar for storage and use over winter, the colony should begin to be fed late summer / early autumn (end of August onwards). This syrup should be the thicker and more concentrated mixture. It should be fed on a regular basis in quantities which the colonies can effectively process. The syrup should not be left for length of time (no more than a few days) as it will begin to ferment as the yeast population increases and this can be harmful to bees and promote the early death of adult bees.

Timing of feeding is important as too early feeding can result in the colonies turning this food into more brood, especially in Italian strain related bees with rather than laying it down for winter stores. It is necessary therefore always to check your colonies in early September to check the actual food reserves in the comb and if it is short to feed the colonies with additional thick syrup remembering not to leave the syrup on the hive for more than a few days.

Hygiene is of the utmost importance so containers of the syrup should be clean; the feeders thoroughly cleaned and washed between feeds of sugar syrup. Any fermented or sour tasting syrup should be discarded safely and in a way that it cannot be accessed by bees and any containers with a black mould present on their inner surfaces should be thoroughly cleaned in hot water.

Colonies may still be able to access late flows of nectar and pollen such as Ivy and Himalayan Balsam and in some years, they are flowering at the same time and the resultant honey from the mixed nectar source is slow to granulate. It not completely processed it may be left in the comb uncapped and if capped the moisture content of the honey may be high and the honey prone to fermentation. It may also contain indigestible material which the bee cannot void when restricted to the hive because the weather conditions outside the hive are not conducive to flying. This may be smeared over the frames as the bees become restless and fall from the cluster to die on the hive floor.

The nutritional status of the colony in the autumn has a direct effect on the colony's potential in the following year.

Long-lived winter bees (i.e., emerged in late summer) may be too weak to rear brood in the early spring as they change their physiological status from winter bees to those of the shorter-lived summer bees as they start brood-rearing, and the colony requires urgent foraging to obtain essential carbohydrate and pollen (for bee bread).

At the hive entrance

Look, listen, sniff	Signs to look for	Conclusion	Action needed
Observe bees	Pollen being brought in?	Brood being reared	
Wasps	Are wasps entering the colony?	Colony being robbed and at risk	Narrow the entrance. Use wasp traps. Place a piece of glass or perspex at an angle over the entrance
Drones	Drones dead on the alighting board?	Queen right colony	

53. When drawing foundation

Colonies require drawn comb to rear brood in and for the storage of honey, pollen, bee bread and propolis. Drawn combs are invaluable in helping young colonies with young queens and swarms to become quickly established and begin gathering a reserve of food and start rearing new bees as soon as possible.

Beekeepers are now encouraged to employ routine comb change management ideally on a three- year service life of the comb basis or replaced earlier if frames become deformed or is taken from a colony with diseases and conditions or dysentery.

As we have already noted the effort and energy expended as well as the nutrients consumed in the process should not be underestimated and are a significant drain on the colony's resources.

The effective drawing of foundation is only best attempted when the bees have access to abundant sources of nectar and pollen and a large colony population.

The foundation to be drawn should be placed in a super position, i.e., over the brood box where the heat generated from the brood rearing will help in the secretion and moulding of the beeswax into straight well- drawn combs and be well-attended to by the colony.

Make sure the same spacing is used in the foundation super and with the frames in the box below.

It is also practical to place a block of 2-3 frames in the centre of a box of already drawn foundation.

Note It is also possible to bottom super by placing the box with the foundation under the brood box.

If the foundation is going to be drawn into comb, then this will usually be completed in a week or so. Left any longer the bees will chew holes in the foundation, stain it walking across the foundation and if finally drawn the quality will be inferior and possibly a mixture of drone and worker cell-sized comb.

54. When queen rearing

The specialist indigenous UK industry which produces mated queens / colonies for the UK market is also relevant for enterprises wishing to export to or source and import mated queens and colonies into the UK.

The ideal conditions for rearing queens are a combination of a consistent but light nectar flow with pollen available from multi plant species. The aim is to keep the colony stimulated.

The cell building colony should be ready by the 2nd or 3rd week in May and this colony should have been built up through planned feeding support until it is well populated with adult bees and brood ready to receive the larvae to be grown into queens.

Likewise, colonies which will form queen mating nuclei should fed so there are good populations of bees in colonies which also have good stores and plenty of young bees.

If these conditions are not available (and they often are not) then a light sugar syrup can be fed to the drone mother colonies and this will further stimulate pollen collection.

Such syrup should also be fed to the queen cell starting and finishing colonies and nuclei with virgin queens. This will stimulate royal jelly production.

A sexually mature virgin queen takes 21 days to develop from the date the egg is laid in the queen cell.

The synchronisation of the drone honey bees reaching maturity is important and if properly nourished during their development they reach maturity and able to mate 9-12 days after emergence. In other words, the drone eggs need to have been laid about 15 days before the egg destined to be a queen is laid to ensure sexually mature drones are available to mate with the flying virgin queens on their mating flights.

If the availability of pollen in the producer's locality is scarce then the breeder colonies will cease to rear drone brood. Any drone eggs or larvae will be neglected and may by cannibalised by the nurse bees.

Such poorly- mated queens are superseded earlier than usual and the queen then fails without the colony replacing her leading to the death of the colony.

The lack of pollen can be due to:

- Too many colonies for the forage available.
- Land management practices and forage available.
- Changing flowering times, changing weather patterns, seasonal variation, or climate change.

Several factors contribute to the ideal conditions for rearing queens including a light nectar flow and pollen available from a number of plant species.

The colonies being used for queen rearing can be stimulated using a 50% sugar syrup fed in relatively small quantities (1-3 litres) twice a week.

Syrup should also be added to queen cell, starter and finishing colonies and the nucleus colonies created for the virgin queens and from which they fly to mate.

Feeding a 50% syrup solution to the drone mother colonies will encourage pollen collection which in turn will be used to stimulate drone rearing and feeding adults drones. If these colonies lack pollen the drone larvae will be neglected, drone eggs eaten by the workers and in extremis mature drones may be ejected from the hive.

If there is less than half a comb of stored pollen / bee bread in the colony then feeding protein in the form of patties is recommended. Pollen patties should be fed to the queen cell raising colonies by placing them **between** frames rather than on top of the crown / cover board or smearing the patty into empty cells. Placing on top of the frames means that the older bees (foragers) will consume it because they move away from the brood area as they age. Nurse bees consuming patty from between the frames will convert this to royal jelly and feed this to developing queen cells.

The queen will not be stimulated to lay drone eggs leading to an inadequate number of drones being produced and of the correct age to mate with the virgin queens when they are ready.

Timing the rearing of drones to maximise the number of flying sexually mature drones for when the virgin queens being their mating flights is vital for successful mating. Drone ability to mate is achieved 9-12 days after emergence from its cell and this can be delayed if there is a pollen shortage.

If, however you are a hobbyist beekeeper and does not want to use grafting techniques for queen rearing then the use of queens raised by utilising the natural swarming instinct or by using an artificial swarming technique can be carried out at a time when the biological rhythm of the colony is much more in tune with the status of the pollen and nectar resources available. Splitting colonies to make smaller nuclei may require supplementary feeding but judicious use of the combs (distribution of frames with honey and pollen stores), the queen cells present on them and the establishment of the nuclei may require supplementary feeding of 50% syrup. Great care must be taken not to start and encourage robbing of these small nuclei.

55. Preparing colonies for pollination services

If the pollination requirement is early in the year, e.g., stone fruits and apples then colonies may need early stimulation by feeding sugar syrup and supplemental pollen or protein supplement to enable the colonies to reach a large and expanding size in time to pollinate the crop. Foraging bees searching for pollen and nectar will promote pollination.

56. Feeding nuclei, small swarms, queen mating and small colonies

In these situations, the use of frame or dummy feeders placed in the hive box as close to the bees as possible so the bees can access the syrup is a good technique. Ensure the feeder has a float and that it does float on the syrup helping the bees access the syrup and not fall in and drown. Any loss of bees in a small colony is undesirable.

Recent years have seen an increase use of polystyrene hives and nucleus boxes. The nucleus boxes are often made with an integral feeder and provide ideal conditions for the nurturing of small colonies. Maintenance feeding of a syrup made from two parts sugar to one part water by weight is suitable and when being established these small colonies should be provided with at least one empty drawn comb to enable the bees to take down the syrup and store it in the brood area.

57. Emergency feeding

When the bees in a colony are active but are actually or close to starving then a piece of candy / fondant should be placed on the frames above the cluster and covered across the whole width of the box with plastic bubble wrap or the insulation which is used for the reflection of heat behind radiators. This helps the cluster retain some of its heat and energy enabling the bees to access the food. Woollen or fabric materials should not be used as insulation because they will absorb moisture which will encourage the growth of moulds and could help initiate fermentation of the feed.

Light syrup can be added to an empty comb held vertically using a thoroughly cleaned out container such as an empty squeezy washing up bottle or hand spray and placed as close to the bees as possible.

58. Hive hygiene

In researching material for this book, it became clear that honey bee colonies were well adapted to survive in the British Isles provided their fundamental needs could be met......namely food and shelter and freedom from disease.

Unsatisfactory internal hive conditions can lead to the harbouring of pests and disease and the extended use of old combs adds to the risk of re-infection.

Despite the economic costs involved in the ruthless removal and disposal by burning of combs and frames as part of a comb replacement programme, on balance the time spent trying to clean and sterilise used frames and dealing with the waste materials produced which have to be disposed of in a safe and responsible manner mean that disposal by burning and the preparation of new frames is the most cost effective and time efficient, biosecure and environmentally responsible way to achieve an effective comb replacement programme.

The National Bee Unit's guidance on hive cleaning and sterilisation refers to and promotes the use of the techniques set out in the following table:

Task	Technique
Cleaning and disinfecting wooden hives	Scorching with a blow lamp or hand-held electric paint stripper
	Sterilisation using Washing Soda crystals
	Chemical sterilisation with disinfectants such as sodium hypochlorite (bleach)
	Chemical sterilisation with acetic acid
	Boiling in caustic soda (Sodium hydroxide)
	Immersion in molten paraffin wax
	Irradiation using gamma rays from a radioisotope such as cobalt
Recycling wooden frames	Hot steam. Can also be used for wooden hive boxes
Cleaning and disinfecting plastic or polystyrene hive	Use sodium hypochlorite containing products or caustic soda
Queen excluders	Wire excluders scorched using a blow lamp
	Zinc slotted excluders scrubbing with solution of washing soda
Hive tools	Scorched with blow lamp or soaked in washing soda
Beekeeping clothing	Wash overalls with proprietary laundry product
	Soaking in washing soda can be used to remove propolis
Gloves	Leather gloves are best avoided with the preferable use of household washing up gloves or thin disposable nitrile gloves
Footwear	Scrubbed with washing soda solution

Several of the techniques involve potentially hazardous operations and chemicals. The use of appropriate equipment and correct personal protective equipment (gloves and safety glasses) are necessary to reduce risk of exposure to the chemicals or the processes involved. Beekeepers should not wear their veils when using a blow lamp or hot surface as the plastic mesh in the veil readily melts and is very combustible and if ignited could cause serious injury. In addition, several of the procedures will produce waste materials which will be required to be disposed of in a safe and environmentally satisfactory way.

The honey bee colony has elaborate mechanisms to help prevent and control diseases and conditions. Most of these mechanisms depend on the individual and the whole colony obtaining a diet rich in nutrients and other substances which will enable the colony to grow, reproduce, successfully survive dearth conditions and the changing climate conditions throughout the year and be able to deploy the metabolic, immunity and other mechanisms to prevent and control diseases.

It is clear that plant resins and propolis play a fundamental role in enabling the honey bee colony to do all these things and yet we as beekeepers perhaps have a negative view on the presence of propolis and how it can be a nuisance in our beekeeping practices. Many beekeepers spend much time removing propolis from the hive and frames and unless it is being harvested for human or other such therapeutic and nutritional purposes it is discarded. They mutter and curse when the hive parts are stuck together. Other beekeepers attempt to select and rear colonies which have a tendency for low propolis collecting and its use in the hive.

A better approach maybe a compromise where the beekeeper removes the propolis from key areas which do cause a problem and place the propolis collected on top of the crown board and see if the bees will utilise it.

We are also taught to scorch the inside of hive parts but the merits of doing this are worthy of consideration. Honey bees line all the internal surfaces with a material (propolis) which contains anti-fungal, anti-bacterial and anti-viral substances. Scorching the surfaces burns off this protective layer and provides a clean surface for mould spores and bacteria to recolonise if the humidity conditions are suitable for their growth. They pose a threat (especially to a weakened colony) until the colony has spent time effort and resources recoating the surface again. Scorching hive parts effectively and efficiently is not easy and there is the additional risk associated with the use of naked flames around combustible materials. If you must scorch in your workshop or honey house have a fire extinguisher or fire blanket close by and at minimum a metal bucket available into which you can place any parts which you have ignited. Even a bucket full of water will be a good safety measure.

Minimisation of undesirable spores and bacteria being present in the hive is probably best achieved by a good regime of comb changes and burning both the frames and the comb. The inner surfaces of the hive parts could be washed with a food grade surface spray or a laundry cleanser liquid and allowed to thoroughly dry before being used. Such practices are already being used with polystyrene hives.

The presence of excess moisture is to be avoided. Beekeepers who leave their open mesh floors on their hives without the inserts usually do not experience excessive condensation overwinter in their hives but if the floors are solid then it is important to ensure that the ventilation holes in the roof are not blocked with spider's webs or anything else preventing the vents operating properly. Think carefully before putting insulation material on top of the crown board. Will it impede the flow of air in the case of the use of solid floors?

59. So what is in store for our honey bees in the future?

There are no quick fixes which will enable insect pollinator population numbers to recover quickly. The requirements of the honey bee have also to be seen in a longer term view.

Initiatives which aim to address these initiatives must be capable of being envisaged outside and beyond the fixed period cycling of Parliament, the governing of the country, both national and local, above party politics and hopefully no reoccurrence of global pandemics..

In the last part of this book, we will draw the reader's attention to several developments which perhaps indicate that a corner may have been turned in the political thinking regarding the relationship of the human species with the planet Earth and the relationship of economics to nature.

Most recently we have seen the publication of the ***Independent Review on the Economics of Biodiversity*** by Professor Sir Partha Dasgupta in February 2021. This review was announced by the Chancellor of the Exchequer in March 2019 with the brief to

- Assess the economic benefits of biodiversity globally.
- Assess the economic costs and risks of biodiversity loss.
- Identify a range of actions that can simultaneously enhance biodiversity and deliver economic prosperity.

We will return to this review later, but in the meantime other important developments have been taking place which, if suitable and sufficient representational effort is made by beekeepers, could significantly improve the environment in which we expect our honey bees to, at least survive, but more importantly to thrive.

In 2018, the UK government (HMG) published its 25-year Environment Plan outlining a comprehensive and long-term approach to environmental matters. It was published as' A Green Future: Our 25 Year Plan to Improve the Environment' and is available on-line from www.gov.uk/government/publications. The plan relates to areas for which HMG is responsible and there are other documents covering the plans for the Devolved Administrations who all have the shared goal of protecting our natural heritage. The UK Government is responsible for a number of policies and programmes which affect sectors across the UK and internationally. Some aspects of the Plan will apply across the UK.

At the time of writing the Coronavirus (COVID 19) pandemic together with much work to be done post Brexit will no doubt hugely influence what can be done bearing in mind, in the social and economic situations in which the Plan is being implemented.

We have included this section in our book because the future of honey bees and their care of and management by beekeepers will be impacted by what happens next and it is incumbent on beekeepers and their representative organisations to actively participate in the policy and practical implementation of the Plan at all levels (Government, policy makers, the public and land managers). This takes time, effort and commitment and a strong sense of collective vision by beekeepers and those interested in the honey bee and its contribution to the environment (including man).

The Plan is set out into six key areas.

- Using and managing land sustainably.
- Recovering nature and enhancing the beauty of landscapes.
- Connecting people with the environment to improve health and wellbeing.
- Increasing resource efficiency and reducing pollution and waste.
- Securing clean, productive and biologically diverse seas and oceans.
- Protecting and improving the global environment.

With the possible exception of the seas and oceans, honey bees and beekeepers have a role to play in all of these key areas.

The Plan makes specific mention of managing pressure on the environment by:

- Mitigating and adapting to climate change.
- Minimising waste.
- Managing exposure to chemicals.
- Enhancing biosecurity.

All of which will have implications for honey bees and beekeepers.

The Plan lists the policies which will be formulated and implemented for each of these key areas and in these the implications and opportunities for honey bees and beekeepers become more tangible.

The Plan adopts the approach of 'Natural Capital' which it describes as the sum of our ecosystems, species, freshwater, land, soils, minerals, air and our seas. These are all considered elements of nature that either directly or indirectly bring value to people and the country at large. It is a tool, which can be used to make values and estimates and is one tool of many which will be used to formulate policy making.

Natural capital was defined by the English Natural Capital Committee in 2014 as 'elements of nature that directly or indirectly produce value or benefits, including ecosystems, species, freshwater, land, minerals, the air and oceans, as well as natural processes and function'.

Natural capital is an economic metaphor for nature; a concept that frames the world's resources including plants, animals, water, minerals as assets or stocks that yield a flow of benefits to people.

The Natural Capital approach involves measuring and valuing natural capital assets. Their values can reveal how natural capital is bringing important benefits to society and the economy.

These natural capital assessments can be used to support more sustainable decision-making.

Each of the devolved administrations in the UK evolved their own approaches to the subject of natural capital. In England it was the Natural Capital Committee, in Scotland the Scottish Forum on Nature, in Wales the Natural Capital Coalition and in Northern Ireland the Northern Ireland Environment link. In Ireland similar deliberations were made in the Irish Forum on Natural Capital.

Pollinators are mentioned in the UK Environment Plan, but their profile should be higher. Beekeepers should realise that for some, honey bees are considered livestock, a managed species and not part of the natural fauna and as was alluded to earlier in this book beekeepers have much to do in protecting the interests of honey bees in comparison with those promoting the benefits of other pollinator species, such as bumble bees. Beekeepers cannot just rest on their laurel. They must continuously demonstrate that their interests and the honey bee are an important part of the food webs and chains which exist on this planet and educate and inform the public and policy makers accordingly.

Large scale commercial beekeeping if not carried out taking into account other interests does attract criticism and reinforces the perception that the honey bee outcompetes other pollinator species, especially it seems rare species, for forage.

The long- term nature of the 25 Year Environment Plan transcends the national political cycle, and the same comment must be made for beekeeping associations, usually acting as charities, and having to conduct the management of their organisations within the legal constraints of charity legislation. They should gear themselves up for establishing an ongoing capability and capacity so that the corporate memory is maintained, and strategies decided and implemented over the longer term to ensure the interests and role of the honey bee and beekeepers are recognised and take an active part in policy making and the implementation of the Environment Plan.

We have referred throughout this book to the criticisms made of farming and other land management practices. The role of subsidies through mechanisms such as the EU Common Agricultural Policy in encouraging and directing them along particular pathways have resulted in environmental impacts, which in themselves are undesirable but also make significantly worse the effects of changing weather patterns, climate change and the pressures on producing enough food and accessing enough fresh water on a global basis. The British Isles are heavily dependent on a functioning global food market in supply, availability and price to sustain their populations with a diverse and nutritious choice of foodstuffs, fruits, salads and vegetables. More local production may become even more viable as well as the ever-increasing interest being shown in homegrown based fruits and vegetables and will probably become an environmental, an economic and political necessity.

The aim is said to be sustainable intensification – a decoupling of production from environmental impacts so that production can be increased whilst reducing the overall environmental footprint. (National Statistics).

Environmental Stewardship and other inducements have been used to encourage policies and practices in the farmed environment which are intended to bring benefits to wildlife and the environment. To date these have had differing degrees of success and of varying durability.

The basis of the support (subsidies) which farmers will be eligible for will radically change in the coming years as the Environmental Land Management (ELM) schemes are introduced. In essence farmers will be paid if they prevent floods, plants woods and help wildlife as well as the recognition of the need for fewer sheep and cows in the farming mix as we are encouraged to eat less red meat. The current area- based subsidies will be reduced by 50% by 2024 and abolished all together by 2028 with the cash allocated to the current subsidy system being transferred into the ELM system.

Farmers will be able to make claims for grants to:

- Protect heritage farm buildings and stone walls.
- Expand hedgerows.
- Capture carbon in soil and cutting the use of pesticides.
- Practice natural flood management including restoring river bends.
- Achieving landscape recovery, restoring peatlands and in planting new woods.
- Reducing the use of antibiotics.
- Improving animal health and welfare.

For the honey bee and beekeepers recognition of the need to expand hedgerows (and hopefully encourage sympathetic management to enable flowering and fruit and seed production) is welcome news.

Reduction in the use of pesticides, especially herbicides would benefit honey bees and other pollinating insects.

As a note of caution such schemes are only as good as the rigour of the assessment of the outcomes achieved by the applicant. Farmers and other land managers are exposed to the influences and influencers that the rest of the population experience and it is hoped that they are in tune with the increasing awareness and demand for positive and effective action being made by many areas of society including politicians.

60. Conservation agriculture

This is not to be confused with conservation in agriculture with reference to many aspects such as wildlife/habitat/hedgerows and woodlands.

At last, recognition of the primary importance of soil is to be welcomed and the critical relationships of soil health and food is now starting to attract attention, interest being accelerated because of the impacts of climate change and changing weather patterns. A new EU Mission for Soil Health and Food is currently being finalised. An assessment throughout the EU has shown that 60-70% of EU soils have suffered damage and their degradation. The reasons for this are many and varied and agriculture and the farming practices currently widely practiced pay lip service to soil health. Declining soil health and productivity is bolstered using fertilisers and other inputs, but soil microbial, invertebrate and plant life is damaged and even destroyed because of the use of insecticides, fungicides, molluscides and herbicides.

Conservation Agriculture is a set of soil management principles:

- Minimising Soil Disturbance
- Residue Cover
- Crop Rotation

It means the use of mini-tilling techniques create minimal disturbance to the soil and less disturbance means less chance of weed emergence.

Use of disc drills pushes the cover crop down as it is usually starting to die. The cover crop becomes a mulch on top of the new seed bed. It keeps the soil warmer and at a more constant temperature. The disc drill cuts minimally into the soil, places the seed alongside the disc and closes the slot; all achieved with little soil disturbance.

The cover crop harvests carbon dioxide and locks a lot into the roots, increasing the nutrient, water and the nitrogen- holding capacity of the soil. It solves many weed problems as the weed seeds and seedlings are not exposed to light. Oilseed, radish, sunflowers, spring oats, buckwheat, vetch and clover for example are used as cover crops.

Where grass leys end up back in combinable crops a white clover can be maintained as it can succeed against glyphosate and will kill any ryegrass. An example is to take out a grass ley put in spring beans which grows a crop with a mat of white clover growing underneath. When drilling rape one can also drill spring beans and thence have a mixture of plants from spring beans to buckwheat, berseem clover and vetch. This variety of plants confuses the flea beetle that lands on the rape, then finds the clover and vetch which it cannot feed from and so moves away.

In the UK, a new organisation, UK Soils (https://UKsoils.org), has been formed from a range of organisations contributing scientific, campaigning and awareness raising to promote the need to act and improve our nation's soil health.

The proposed UK Agriculture Bill will contain soil protection is a core issue.

These developments offer the opportunities for beekeepers to influence policy and practice and give guidance on the use of cultivation techniques and promote the selection and use of plant species which will benefit all pollinator species, including the honey bee.

61. Implications for decision makers

It should be remembered that **we** are **all** decision makers whether in our professional capacities, gardeners, members of society or enjoyers of the environment.

We need to secure the value of the honey bee to society and continue and re-affirm an age-old relationship.

Beekeepers and their representative associations and organisations should be much, much more proactive in ensuring that not only are the interests of the honey bees and beekeepers recognised but that public at large gain a greater appreciation and interest in the honey bee and its contribution to all our lives.

There is an increasing emphasis placed in the education system bringing a better understanding of the environment and life on planet Earth so the next generations can be more creative and develop ways of living which are more sustainable and ultimately more desirable if *Homo sapiens* is to survive and not just become another extinct species in the history of evolution on Planet Earth.

Discussions, decisions, publications, promotional activities need to be made and actions taken in several different policy areas, including:

- Agriculture.
- Biodiversity conservation.
- Land use planning.
- Pesticide regulation and use, including more detailed risk assessments on more subtle effects of pesticides e.g., behavioural changes.
- Bee health.
- Climate change.
- The creation of flower-rich and semi-wild habitat.

- The encouragement in the use of publicly (e.g., parks, roads, rail corridors) to manage the land and provide opportunities for creating habitats for pollinators and forage for honey bees.
- Forewarning, preparedness, ensuring effective biosecurity against potential or yet to be recognised exotic pests of honey bees.
- Identifying, managing and controlling the risks of introducing bee diseases and their associated pathogens in the importation of honey bees, whether as queens or colonies, into the British Isles.

These are all areas of activity which will impinge on and directly affect our honey bees.

In order to be effective and be carried through on a long- term basis they are but part of major changes which will need to take place in the global relationships of economics and biodiversity.

We briefly referred to the DasGupta Review and whilst it is a long read the findings of the review, if implemented, will see a paradigm shift in the relationship of economics and ecology. Ecology being defined by the British Ecological Society as the study of the distribution and abundance of organisms, the interactions between organisms and their environment, and structure and function of ecosystems. Ecology and ecosystems are two words whose original meanings are, like many other words in our language having their meanings changed and corrupted in today's society.

The DasGupta Review on the economics of biodiversity clearly identifies that the economy and economic and political actions must be **embedded** in nature. Humans are part of the natural world (not apart from it) and their future is utterly dependent on the functioning of the many ecosystems recognised around the world. They are not distant and remote and can no longer pursue activities and economic practices which continue to degrade and destroy these ecosystems with impunity. The recognition and acceptance of this reality will mean significant change.

The report refers to pollination as a key ecosystem service and to bees, but at first sight the <u>vital</u> importance of the pollinator and the key role of honey bees does not seem to be <u>fully</u> recognised and given due gravitas …..as yet. Beekeepers and those interested in honey bees clearly need to do more.

Recently we have endured the COVID 19 pandemic, but as Bill Gates has remarked solving COVID is simple in comparison to dealing with climate change whose increasing rate of change has been promoted by man's activities motivated by a world economic structure which is based on exploitation of the earth's resources for profit with little regard for the consequences.

The honey bee (and other pollinators) provide a key ecosystem service namely pollination and without it there may well have been no human species arising from evolution.

So, we come full circle but now instead of exploiting honey bees through honey hunting and robbing nests we need to husband and cherish our honey bees as they hold one of the keys to human survival……**pollination.**

62. In conclusion

This book is intended to be a small contribution towards the provision of information, guidance and suggestions in meeting the nutritional needs of the species *Apis mellifera* in the British Isles.

We also hope it will help in furthering the recognition of the importance of the honey bee in pollination and their contribution to biodiversity and providing foodstuffs for humans and domesticated animals, as well as other plant and animal species, all of whom form part of food chains and food webs of the ecosystems in the British Isles.

In doing so we hope that the reader reflects on the age-old relationship of humans and honey bees and the recognition of their importance which will be even more important in future years.

Above all we hope it will stimulate questions and encourage the reader to seek information and explanation and prompt them to action.

References

A Green Future: Our 25 Year Plan to Improve the Environment' and is available online (www.gov.uk/government/publications), published 2018

Alaux, C., Ducloz, F., Crauser, D., Le Conte, Y., (2010). Diet effects on honey bee immunocompetence. Biol. Lett. **6** (4): pp 562-565

https://doi.org/10.1098/rsbl.2009.0986

Alston, F. SKEPS. Their History, Making and Use. (1987). Northern Bee Books, Mythlmroyd, England

Amiri, E., Le, K., Melendez, C.V., Strand, M.K., Tarpy, D.R. Rueppell, O., (2020) Egg size plasticity in *Apis mellifera*: Honey bee queens alter egg size in response to both genetic and environmental factors. J. Evol. Biol. **33** (4) pp 534-543. https:/doi.org/10.1111./ jeb.13589

Annoscia, D., Zanni, V., Galbraith, D. et al. (2017) Elucidating the mechanisms underlying the beneficial health effects of dietary pollen on honey bees *(Apis mellifera)* infested by Varroa mite ectoparasites. Scientific Reports; **7** (1): 6258 https:/doi:10.1038/s41598-017-06488-2

ANON (1994) Beekeeping and Health Statistics. Ministry of Agriculture, Fisheries and Food; London, UK; 7pp

Antúnez, K., Harriet, J., Gende, L., Maggi, M. M., Eguaras, M., Zunino, P., (2008), Efficacy of natural propolis extract in the control of American foulbrood. Veterinary Microbiology **131**: pp 324-331

https://doi.org/10.10.1016/j.vetmic.2008.04.011

Arien, Y., Drag, A., Zarchin, S., Masci,T., Shafir,S., (2015) Omega-3 deficiency impairs honey bee learning. PNAS. **112**: pp 15761-66

https://doi.org/10.1073/pnas.1517375112

Aston, D. and Bucknall, S.A., *Keeping Healthy Bees*. (2010) Northern Bee Books, Mytholmroyd, England

Aston, D., and Bucknall, S.A., *Plants and Honey Bees – their relationships*. (2004). Northern Bee Books, Mytholmroyd, England

Bailey, L. and Ball, B.V., *Honey Bee Pathology*. (1991). Academic Press

Barker, R.J. *Poisoning by plants*. In Morse, R.A., *Honey bee pests, predators and diseases*. (1990). Cornell University Press pp 306-328

Barker, R.J., (1977). Some Carbohydrates found in pollen and Pollen substitutes are Toxic to Honey Bees. The Journal of Nutrition Vol **107** (10), pp 1859-1862

https://doi.org/10.1093/jn/107.10.1859

Barron, A., (2015). Death of the bee hive : understanding the failure of an insect society. Current Opinion in Insect Science **10:** pp45-50 http://dx.doi.org/10.1016/j.cois.2015.04.004

Basualdo, M., Barragan, M., Antunez, K., (2014). Bee bread increases honey bee haemolymph protein and promote(s) better survival despite of causing higher *Nosema ceranae* abundance in honey bees. Environmental Microbiology Reports **6** (4): pp 386-400.

https://doi.org/10.1111/1758-2229.12169

Basualdo, M., Barragán, S., Vanagas, L., Garcia, C., Solana, H., Rodriguez, E., Bedascarrasbure, E., (2013). Conversion of high and low pollen protein diets into protein in worker honey bees (Hymenoptera : Apidae). Journal of Economic Entomology **106** (4): pp 1553-1558

https://doi.org/10.1603/EC12466

Beaurepaire, A., et al., (2020). Diversity and Global Distribution of Viruses of the Western Honey Bee, *Apis mellifera*. Insects **11** (4) :25 pp .

https://doi:10.3390/insects11040239

Becher, M.A., Grimm, V., Thorbek, P., Horn, J., Kennedy, P., Osborne, J.L., (2014). BEEHAVE: a systems model of honeybee colony dynamics and foraging to explore multifactorial causes of colony failure. J. Appl. Ecol. **51**: pp 470-482.

https://doi:10.1111/1365-2664.12222

Beekman, M., and Ratnieks, F.L.W., (2000). Long range foraging by the honey *bee, Apis mellifera*. Funct. Ecol. **14** (4): pp 490-496 https://doi.org/10.1046/j.1365-2435.2000.00443.x

Blamey, M. and Grey-Wilson, C. *Wild Flowers of Britain and Northern Europe* (2003). Cassell

Bogdanov, S. (2014/2017). Pollen; production, nutrition and health: a review. Bee Product Science; see: www.bee-hexagon.net

Bogdanov, S., (2016). Pollen: Nutrition, Functional Properties, Health. In Chapter 2 *The Pollen Book* in Bee Product Science

Bonoan, R. E., O'Connor, L. D., Starks, P. T., (2017). Seasonality of honey bee (*Apis mellifera*) micronutrient supplementation and environmental limitations. J. Insect Physiol. **107**, pp 23-28.

 https://doi.org/10.1016/j.jinsphys.2018.02.2002

Bonoan, R.E. and Starks, P.T., *Western Honey Bee (Apis mellifera).* In C. Starr (ed) *Encyclopedia of Social Insects.* (2020). Springer Nature. Switzerland

Bortolotti, L. and Costa, C., (2014). *Chapter 5 Chemical Communication in the Honey Bee Society.* In Mucignat-Caretta C, editor. *Neurobiology of Chemical Communication*, Boca Raton (FL): CRC Press / Taylor & Francis

Brian, A.D., and Crane, E., (1959). Charles Darwin and Bees. Bee World Vol **40**, No 12, pp 297-303

Brillet, C., Robinson, G.G., Bues, R., and Le Conte, Y., (2002). Racial differences in division of labor in colonies of the honey bee (*Apis mellifera*). Ethology **108** (2): pp 115-126

https://doi.org/10.1046/j.1439-0310.2002.00760.x

Brodschneider, R. and Crailsheim, K., (2010). Nutrition and Health in honey bees. Apidologie **41**, pp 278-294

https://doi.org/10.1051/apido/2010012

Browne, K.E., Hassett, J., Geary, M., Moore, E., Henriques, D., Soland-Reckeweg., G., Ferrari, R., Loughlin, E.M., O' Brien, E., Driscoll, S., Young, P., Pinto, M.A. and MacCormack, G.P., (2020). Investigation of free-living honey bee colonies in Ireland. J. Apicultural Research

https://doi:10.1080/00218839.2020. 1837530

Cameron, J.J., Sai, Sree Uppala., Hannah, M., Lucas., Ramesh, R., Sagili, R.R., (2016). Effects of pollen dilution on infection of *Nosema ceranae* in honey bees.

J. Insect Physiol. **87:** pp 12-19

http://dx.doi.org/10.1016/jinsphys.2016.01.004

Campbell-Culver, M., *The Origin of Plants.* (2001) Headline Book Publishing, London

Carreck, N., (2008). Are honey bees (*Apis mellifera* L.) native to the British Isles? J. Apicultural Research and Bee World **47**(4) pp 318-322

Carreck, N., (1994). British Beekeeping Practice and Honey Sources in the Nineteen Eighties. Bee Craft May pp 151-155

Carreck, N., (1997). Changing Countryside- changing forage for bees. Part (June) and Part 2 (July) of Bee Craft

Caron, D. M. *Honey Bee Biology and Beekeeping.* (1999). Wicwas Press, Cheshire, Connecticut

Carpena, M., Nuñez-Estevez, B., Soria-Lopez, A., and Simal-Gandara, J. (2020). Bee Venom: A Updating Review of its Bioactive Molecules and Its Health Implications. Nutrients 12, 3360

Doi:10.3390/nu12113360

Carson, R., *Silent Spring* (1965). Penguin Books

Cervoni M.S., Hartfelder, K., (2019). Caste Differentiation: Honey bees. In Starr, C. Encyclopedia of Social Insects, Springer Nature Switzerland AG

Coggshall, W.L., and Morse, R.A., *Beeswax. Production, Harvesting, Processing and Products.* (1984). WICWAS Press, Ithaca, New York

Collins Beekeeper's Bible (2010). HarperCollins, London

Cooper, B.A., *The Honeybees of the British Isles.* (1986). British Isles Bee Breeders' Association

Copinger, I., *Heather Honey. An Anthology of Works* (2016). Northern Bee Books, Mytholmroyd, West Yorkshire

Corby-Harris, V., Snyder, L., Meador, C., Ayotte, T., (2018). Honey bee (*Apis mellifera*) nurses do not consume pollens based on their nutritional quality. PLoS ONE 13(1): e0191050.

 DOI:10.1371/journal.pone.0191050

Corona, M., Velarde, R.A., Remolina, S., Moran- Lauter, A., Ying Wang, Hughes, K.A. and Robinson, G.E., (2007). Vitellogenin, juvenile hormone, insulin signalling, and honey bee longevity: PNAS 2007; **104** (7) pp 7128-7133

https://doi.org.10.1073/pnas.070.1909104

Cowan, T.W., *Wax Craft. The History of Bees-wax and its Commercial Value.* (1908). Sampson Low, Marston & Co. London. Also available on the Internet Archive https://archive.org contributed by Cornell University Library

Crailsheim, K., (1990). The protein balance of the honey bee worker. *Apidologie* 21: pp 417-429

https://doi.org/10.1051/apido:19900504

Crane, E., (2020). Woodlands for climate and nature: A review of woodland planting and management approaches in the UK for climate change mitigation and biodiversity conservation. Report to the RSPB.

Crane, E., *The World History of BeeKeeping and Honey Hunting.*(1999) Duckworth

Crane, E. and Walker, P., (2000). Wall Recesses for Bee hives. Antiquity **74** (286): pp 805-811

Crane, E., (1951). Honey yields per acre of land. Bee World **32** (2):12-14

Crane, E., (*Honey. A Comprehensive Survey.*(1979). Heinemann. This book has recently (2020) been republished by Northern Bee Books and the International Bee Research Association (IBRA)

Crane, E., *Honeybees.* Chapter 65 In *Evolution of Domesticated Animals.* (1984) ed. I.L. Mason. Longman, London

Crane, E., and Walker, P., (1999). Early English Beekeeping: the evidence from Local Records up to the end of the Norman period. The Local Historian **29**(3): 130-151

Crozier, R.H. and Pamilo, P., *Evolution of Social Insect Colonies. Sex Allocation and Kin Selection.* (1996). Oxford University Press

De Grandi-Hoffman et al. (2018). Connecting the nutrient composition of seasonal pollens with changing nutritional needs of honey bee (*Apis mellifera* L.) colonies. J Insect Physiol. **109** pp 114-124.

https://doi.org/10.1016/j-jinsphys.2018.07.002

De Groot, A.P., (1953). Protein and amino acid requirements of the honeybee (Apis mellifica L.). Physiol. Comp. Oecol., **3** 197-285

Deans, A.S.C., (1957). Survey of British Honey Sources. Bee Research Association, London

Deans, A.S.C., (1958). The Pollen Analysis of some British Honeys. Thesis submitted to National Diploma of Beekeeping Board. Unpublished

Defra. PB 13981. (2013). Bees and Other Pollinators: their value and health in England. Review of Policy and Evidence. Available at www.gov.uk/government/publications

DeGrandi-Hoffman, G., Chen, Y.P., Rivera, R., Carroll, M., Chambers, M., Hidalgo, G., de Jong, E.W., (2016) Honey bee colonies provided with natural forage have lower pathogen loads and higher overwinter survival than those fed protein supplements Apidologie **47**: pp 186-96.

https://doi.org/10.1007/s13592-015-0386-6

DeGrandi-Hoffman, G., Chenb, Y., Huang, E., Huang, M.H. (2010). The effect of diet on protein concentration, hypopharyngeal gland development and virus load in worker honey bees (*Apis mellifera* L.). Journal of Insect Physiology **56** (9): pp 1184-1191 https://doi.org/10.1016/j-jinsphys.2010.03.07

DeGrandi-Hoffman, G. and Hagler, J., (2000). The flow of incoming nectar through a honey bee (*Apis mellifera* L.) colony as revealed by a protein marker. *Insectes Sociaux* **47** (4): pp 302-306

DOI:10.1007/PL00001720

DeGrandi-Hoffman, G., Yanping Chen., (2015). Nutrition, immunity and viral infections in honey bees. Current Opinion in Insect Science **10**: pp 170-176

http://dx.doi.org/10.1016/j.cois.2015.05.07

Delaplane, K.S., and Mayer D. E., *Crop Pollination by Bees.* (2000) CABI Publishing

Di Pasquale, G., Alaux, C., Le Conte, Y., Odoux, J., Pioz, M., Vaissière, B.E., Belzunces, L.P., Decourtye, A., (2016). Variations in the Availability of Pollen Resources Affect Honey Bee Health . PLoS ONE 11 (9): e0162818. https://doi:10.137/1journal.pone.0162818

Di Pasquale, G., Salignon, M., Le Conte, Y., Belzunces, L.P., Decourtye, A., Kretzschmar, A., Suchall, S., Brunet, J., Alaux, C., (2013). Influence of Pollen Nutrition on Honey Bee Health: Do Pollen Quality and Diversity Matter? PLoS ONE 8(8): e72016.

https://doi:10.1371/journal.pone.0072016

Döke, A.A., Frazier, M., Grozinger, C.M., (2015). Overwintering honey bees: biology and management. Current Opinion in Insect Science **10** : Supplement on Social Insects pp 185-193.

https://doi.org/10.1016/j.cois.2015.05.2014

Dolezal, A.G. and Toth, A.L., (2018). Feedbacks between nutrition and disease in honey bee health. Current Opinion In Insect Science **26**: pp 114-119

https://doi.org/10.10106/j.cois.2018.02.006

Doner, L., (1977). The sugars of honey- a review. J. Sci. Food Agric. **28**: pp 443-456

https://doi.org/10.1002/jsfa.27402.80508

Donkersley, P., Rhodes, G., Pickup, R.W., Jones, K.C., and Wilson, K., (2014). Honeybee nutrition is linked to landscape composition. Ecology and Evolution 4 (21): pp 4195- 4206.

DOI :10.1002/ece3.1293

Donkersley, P., Rhodes, G., Pickup, R.W., Jones, K.C., Power, E., Wright, G.A. and Wilson, K., (2017). Nutritional composition of honey bee food stores with floral composition. Oecologia **185**: pp 749-761.

DOI 10.1007/s00442-017-3968-3

Eisenhart, D., Guirfa, M., (eds)., *Honeybee Neurobiology and Behaviour* (2011), Springer, Dordrecht

Ellis, A., Ellis, J., O'Malley, M., Zettel Nalen, C., (2017). The Benefits of Pollen to Honey bees. University of Florida/IFAS Extension leaflet ENY 152; see http://edis.ifas.ufl.edu/in868

Engel, P., et al., The Bee Microbiome: Impact on Bee Health and Model for Evolution and Ecology of Host-Microbe Interactions. (2016) American Society for Microbiology March / April 2016 **7** (2): e2164-15, downloaded from http://mbio.asm.org/

DOI:10.11.28/mBio.02164-15

Erler, S., Denner, A., Bobis, O., Forsgren, E., Moritz, R.F. (2014). Diversity of honey stores and their impact on pathogenic bacteria of the honeybee, *Apis mellifera*. Ecol. Evol. **4**: pp 3960-3967.

Doi:10.1002/ece3.1252

Eyer, M., Neumann, P., Dietemann, V., (2016). A look into the cell: honey storage in honey bees, *Apis mellifera*. PLOS ONE 11(8):e0161059

Fearnley, J., *Bee propolis: Natural Healing from the Hive*. (2001). London: Souvenir Press

Ford, D.M., Hepburn, H.R., Moseley, F.B., and Rigby, R.J., (1981). Displacement sensors in the honeybee pollen basket. J. Insect Physiol **27**: pp 339-346.

https://doi.org/10.1016/0022-1910(81)90080-9

Forsgren, E., Olofsson, T.C., Váquez, A., and Fries, I. (2010). Novel lactic acid bacteria inhibiting *Paenibacillus larvae* in honey bee larvae. Apidologie **41**: pp 99-108.

DOI:10.1051/apido/2009065

Fraser, M.H. *History of Beekeeping in Britain*. (1958) Bee Research Association Limited, London

Free, J.B., *Insect Pollination of Crops*. Academic Press. (1970) (A Second edition was published in 1993)

Free, J.B., (1963). The flower constancy of honey bees. J. Anim. Ecol. **32**: pp 119-132

https://doi.org/10.2307/2521

Frias, B.E.D., Barbosa., C.D. and Lourenço, A.P., (2016). Pollen nutrition in honey bees(*Apis mellifera*); impact on adult health. Apidologie **47**: pp15-25. https/doi.org/10.1007/s13592-015-0373-y

García – García, M.C., Ortiz, P.L. and Dapena, J.D., (2004). Variations in the weights of pollen loads collected by *Apis mellifera* L. Grana **43** (3): pp183-192.

https://doi.org/10.1080/00173130410020350

Glover, B., *Understanding Flowers and Flowering an Integrated Approach*. (2007). Oxford University Press

Gmeinbauer, R., and Crailsheim, K., (1993). Glucose utilization during flight of honeybee *(Apis mellifera)* workers, drones and queens. J. Insect Physiol. **39** (11): pp 959-67

https://doi.org/10.1016/0022-1910(93)90005-C

Godfray, H.C.J., Blacquiere, T., Field, L.M., Hails, R.S., Petrokofsky, G., Potts, S.G., Raine, N.E., Vanbergen, A.J., McLean, A.R., (2014). A restatement of the natural science evidence base concerning neonicotinoid insecticides and insect pollinators. Proceedings of the Royal Society B-Biological Sciences **281**: Article ID 20140558.

https://doi.org/10.1098/rspb.2014.0558

Goodman, L., *Form and Function in the Honey Bee*. (2003). IBRA.

Grozinger, C.M., and Flenniken, M.L., (2019). Bee Viruses, Ecology, Pathogenicity, and Impacts. Ann. Review of Entomology **64**: pp 205-206.

https://doi.org/10.1146/annurev-ento-011118-111942

Hao Wei, Xu Jiang He, Chun Hua Liao, Xiao Bo Wu, Wu Jun Jiang, Bo Zhang, Lin Bin Zhou, Li Zhen Zhang, Andrew B Barron and Zhi Jiang Zeng., (2019). A Maternal Effect on Queen Production in Honeybees. Current Biology **29**, pp 2208-2213. https:doi.org/10.1016/j.cub.2019.05.059

Hanley, M.E., Franco, M., Pichon, S., Darvill, B., and Goulson, D., (2008). Breeding system, pollinator choice and variation in pollen quality in British herbaceous plants. Functional Ecology **22,** pp 592-598

doi:10.1111/j.1365-2435.2008.01415.x

Healthy Bees Plan Review (2020). Defra and Welsh Government. PB 14608 available at www.gov.uk/government/publications

Heath, L., ed., A Case of Hives. (1985). BBNO

Heil, M., (2011). Nectar: generation, regulation and ecological functions. Trends in Plant Science **16** (4) pp 191-200.

Doi:10.1016/j.tplants.2011.01.003

Hendriksma, H.P., Oxman, K.L., Shafir, S., (2014). Amino acid and carbohydrate trade-offs by honey bee nectar foragers and their implications for plant-pollinator interactions. J. Insect Physiol. **69**: pp 56-64

https://doi.org/10.1016/j.jinsphys.2014.05.025

Hendriksma, HP., Shafir, S., (2016). Honey bee foragers balance colony nutritional deficiencies. Behav. Ecol. Sociobiol. **70:** pp 509-517.

https://doi.org/10.1007/s00265-016-2067-5

Herrod-Hempsall, W., *Bee-keeping New and Old* Vol I. (1930). British Bee Journal, London

Herrod-Hempsall, W., *Bee-keeping New and Old*. Vol II. (1937). British Bee Journal, London

Hesse, M., (1981). Pollenkitt and viscin threads: their role in cementing pollen grains. Grana, **20** (3): pp 146-152,

DOI: 10.1080/00173138109427656

Hodges, D., *The Pollen Loads of the Honey Bee*. (1984). IBRA

Homitzky, M., (2004). How much canola pollen is in canola (*Brassica napus*) honey? Australian Government RIRDC Publication Number W04/189.

Hooper, T., and Taylor, M., *The Beekeeper's Garden*. (1988). Alphabooks

Horridge, A., *The Discovery of a Visual System. The Honeybee*. (2019). CABI

Hoskins, W.G. *The Making of the English Landscape*. (1955) Hodder and Stoughton

Howes, F.N., *Plants and Beekeeping*. (1945) and (1979). Faber and Faber, London

Hrassnigg, N., Crailsheim, K., (2005). Differences in drone and worker physiology in honeybees Apis mellifera. Apidologie **36**, pp 255-277. http:dx.doi.org/10.105/apido:2005015

Hung, Yu-Shan., and Ibbotson, M.R., (2014). Ocellar structure and neural innervation in the honeybee. Front. Neuroanat. **8**: Article 6, pp 1-11.

https://doi.org/103389/fnana.2014.00006

Hussein, A.E., El-Ansari, M.K. and Zahra, A.A., (2019). Effects of Honey bee hybrid and Geographic Region on Honey Bee Venom Production. J. Plant Prot. and Path. **10** (3) pp 171-176.

DOI:10.21608/JPPP.2019.40922

Johnson, B.R., (2010). Division of labor in honeybees: form, function, and proximate mechanisms. Behav. Ecol. Sociobiol. **64**: pp 305-316.

DOI 10.1007/s00265-009-0874-7

Johnson, B.R., (2008). Within-nest temporal polyethism in the honey bee. Behav. Ecol. Sociobiol. **62**: pp 777-784.

https://doi.org/10.1007/s00265-007-0503-2

Johnson, R.M., (2015). Honey bee toxicology. Ann. Rev. Entomol. **60**: pp 415-434

https://doi.org/10.1146/annurev-ento-011613-162005

Jones, L., Brennan, G.L., Lowe, A., Creer, S., Ford, C.R. and De Vere, N., (2021). Shifts in honeybee forage reveal historical changes in floral resources. Communications Biology (2021) 4:37

https://doi.org/10/1038/s42003-020-01562-4

Kastberger, G., (1990). The ocelli control the flight course in honeybees. Physiol. Entomol. **15** (3) pp 337-346.

https://doi.org10.1111/j.1365-3032.1990.tb00521x

Kelber, A., Somanathan, H., (2019). Spatial Vision and Visually Guided Behaviour in Apidae. Insects. November 22; **10** (12): 418, pp 17

DOI.10.3390/insects10120418

Kirk, W.D.J., and Howes, F.N., *Plants for Bees- A Guide to the Plants that Benefit the Bees of the British Isles.* (2012). IBRA

Koenig, P.A., Smith, M.L., Horowitz, L.H., Palmer, D.M., Petersen, K.H., (2020). Artificial shaking signals in honey bee colonies elicit natural responses. Scientific Reports Nature Research **10**:3746.

https://doi.org/10.1038/s41598-020-60421-8

Krell, R., *Value-added products from beekeeping*, (1996). FAO Agricultural Services Bulletin **124**, Rome.

Kritsky, G., *The Quest for the Perfect Hive.* (2010). Oxford University Press

Laidlaw, H.H., and Page, R.E., *Queen Rearing and Bee Breeding.* (1997). Wicwas Press, Cheshire

Langstroth, L.L., *On the hive and the honey-bee; a beekeeper's manual.* (1852). Hopkins, Bridgman and Co., Northampton, M.A.

Langstroth, L.L., *A practical treatise on the hive and the honey -bee.* (1859). A.P. Moore and Co. New York.

Lawes, G., *The Victorian Beekeeping Revolution.* (2011). Northern Bee Books

Lawton, J. et al.., (2010). *Making space for nature: A Review of England's Wildlife Sites and Ecological Network.* Report to Defra

Lawton, J.H. and Heads, P.A., (1984). Bracken, Ants and Extra-floral Nectaries. 1. The Components of the System. J. Anim. Ecol. **53** (3) pp 995-1014

https://doi.org/10.2307/4673

Li, A.Y., Cook, S.C., Sonenshine, D.E., Posada-Florez, F., Noble,I.I., Mowery, J., Gulbronson, C, J., Bauchan, G, R., (2019). Insights into the feeding behaviours and biomechanics of *Varroa destructor* mites on honey bee pupae using electropenetrography and histology. J. Insect Physiology **119** :1-8

https://doi.org/10.1016/j.insphys.2019.103950

Lilolios, V., Tananaki, C., Dimou, M., Kanelis, D., Goras., Karazafiris, E., Thrasyvoulou, A., (2016). Ranking pollen from bee plants according to their protein contribution to honey bees. J. Apicultural. Res. **54** (5): pp 582-592

https://doi.org/10.1080/00218839.2016.1173353

Linton, F., *The Observation Hive Handbook: Studying Honey Bees at Home.* (2017) Cornell University Press

Lipiński, Z., *Honey Bee Nutrition and Feeding. In Temperate / Continental Climate of the Northern Hemisphere.* (2018). Eva Crane Trust ISBN 978-83-939279-0-6

Loftus, J.C., Smith, M.I., Seeley, TD., (2016). How honey bee colonies survive in the wild: testing the importance of small nests and frequent swarming.

 PLoS ONE 11, e0150362.

Doi :10,1371/journal.pone.0150362

Mabey, R., *Flora Britannica* (1996), 480pp Chatto and Windus

Mabey, R., *WEEDS. How Vagabond Plants Gatecrashed Civilisation and Changed the Way We Think About Nature.* (2010). Profile Books

Maclean, N., *Silent Summer-The State of Wildlife in Britain and Ireland.* (2010). Cambridge University Press

Maclean, N., *A Less Green and Pleasant Land.* (2015). Cambridge University Press

Manley, R.O.B., *Honey Production in the British Isles*, (1936). Faber and Faber, London

Manning, R., (2001.) Fatty acids in pollen: a review of their importance for honey bees. Bee World **82**: pp 60-73

Mao, W., Schuler,M.A., and Berenbaum, M., R., (2015). A dietary phytochemical alters caste-associated gene expression in honey bees. Science Advances **28** August 2015 Vol 1, no. 7, e1500795

DOI:10.1126/sciadv.1500795

Martin- Hernandez, R., et al.., (2009). Effect of Temperature on the Biotic Potential of Honeybee Microsporidia. Applied and Environmental Microbiology **75** (8) pp 2554-2557

DOI:10.1128/AEM.02908-08

Muir, R. and Muir, N., *Hedgerows-Their History and Wildlife*. (1987) Michael Joseph, London

National Statistics (2020) *Agriculture in the United Kingdom in 2019*. Issued by Defra and the devolved administrations

Nguyen, V.N., (1999). Effect of protein nutrition and pollen supplementation of honey bee (*Apis mellifera L.*) colonies on characteristics of drones with particular reference to sexual maturity. Australasian Beekeeper **101**: pp 374-375, 419-425

Nicholls, E., Hempel de Ibarra, N., (2017). Assessment of pollen rewards by foraging bees. Funct. Ecol. **31**: pp 76-87

Doi.10.1111/1365-2435.12778

Nicholson, S.W., (2009). Water homeostasis in bees, with the emphasis on sociality, J. Exp. Biol. **212**, pp 429-434

doi:10.1242/jeb.022343

Nicodemo, D., Malheiros, E.B., De Jong, D., Couto, R.H.N., (2014), Increased brood viability and longer lifespan of honey bee selected for propolis production. Apidologie **45**: pp 269-275

https://doi.org/10.1007/s13592-013-0249y

Oliver, R., (2015). Understanding colony build up and decline. Part 3. The 'spring turnover'; see http://scientificbeekeeping.com/understanding-colony-build-up-and-decline-part-6

Online Atlas of the British and Irish Flora. https://www.brc.ac.uk/plant atlas

Park, S. and Thornberg, R.W., (2009). Biochemistry of Nectar Proteins. J. Plant Biol. **52**: pp 27-34.

DOI 10.1007/s12374-008-9007-5

Pavord, A., *The Naming of Names – The Search for Order in the World of Plants*. (2005). Bloomsbury

Percie du Sert, P., (2009). Les pollens apicoles. Phytotherapie **7**: pp 75-82

https://doi.org/10.1007/s10298-009-0375-x

Percival, M.S., *Floral Biology*. (1965). Pergamon Press

Percival, M.S., (1955), The Presentation of Pollen in Certain Angiosperms and its collection by *Apis mellifera*. New Phytol. **54** (3) pp 353-368

Pettis, J.S., Lichtenburg, E.M, Andree, M., Stitzinger, J., Rose R., van Engelsdorp, D., (2013). Crop pollination exposes honey bees to pesticides which alters their susceptibility to the gut pathogen *Nosema ceranae*.

PLoS ONE 2013: 8(7):e70182

https//doi.org/10.1371/journal.pone.0070182

Pirk, C.W.W., Crew, R., Moritz, F.A., (2017). Risks and benefits of the biological interface between managed and wild bee pollinators. Funct. Ecol. (31): pp 47-55.

Doi.10.1111/1365-2435.12768

Pollard, E., Hooper, M.D. and Moore. N.W., *Hedges*. (1974). New Naturalist Series Book **58**, Collins

Proctor, M. and Yeo, P., *The Pollination of Flowers*. (1979) The New Naturalist. Collins

Proctor, M., *Vegetation of Britain and Ireland*. (2013). New Naturalist Series. Collins

Proctor, M., Yeo, P. and Lack, A., *The Natural History of Pollination*. (1996) Timber Press, Oregon

Pulice, C.E. and Packer, A.A., (2008). Simulated herbivory induces extrafloral nectary production in *Prunus avium*. Funct. Ecol. 22, pp 801-807 https://doi.org/10.1111/j.1365-2435.2008.01440.x

Rackham, O., *The History of the Countryside*. (1986) J.M. Dent, London

Rackham, O., *Woodlands*. (2006) New Naturalist Series. Book **100**, Collins, London

Ramsey, S.D., van Engelsdorp, D., (2017). Varroa destructor feed primarily on honey bee fat body not haemolymph. Proceedings of the 2017 American Bee Research Conference, abstract 39

Ramsey, S.D., Ochoa, R., Bauchan, G., Gulbronson, C., Mowery, J.D., Cohen, A., Lim, D., Joklik, J., Cicero, J.M., Ellis, J.D., Hawthorne, D., van Engelsdorp, D., (2019) Varroa destructor feeds primarily on honey bee fat body tissue and not hemolymph

*PNAS Vol **116** no 5 1792-1801*

www.pnas.org/cgi/doi/10.1073/pnas.1818371116

Ransome, H.M., *The Sacred Bee in Ancient Times and Folklore* reprinted in 1986. Bee Books New and Old (BBNO), Burrowbridge, Bridgewater

Rasille, R. , Pirk, C.W.W. and Tautz, J.,(2008). Trophallactic activities in the honeybee brood nest -Heaters get supplied with high performance fuel. Zoology **111**: pp 433-41 https://doi.org/10.1016/j.zool.2007.11.002

Requier, F., Odoix, J-F., Tamic, T., Moreau,N., Henry, M., Decourtye, A. and Bretagnolle, V.,(2015). Honey bee diet in intensive farmland habitats reveals an unexpectedly high flower richness and a major role of weeds. Ecol. Appl. **25**: pp 881-890.

Doi:10.1890/14-1011.1

Rinderer, T.E., (1982). Volatiles from empty comb increase hoarding by the honey bee. Animal Behaviour **29**: pp 1275-1276

https://doi.org/10.1016/s0003-3472(81)80085-1

Roulston, T.A.H., Cane, J.H., Buchmann, S.L., (2000). What governs protein content of pollen: pollinator preferences, or phylogeny? Ecological Monographs **70**: pp 617-643

https://doi.org/10.1890/0012-9615(2000)070[0617:WGPCOP]2.0:2

Roulston, T.H. and Cane J.H., (2000). Pollen nutritional content and digestibility for animals. Plant Syst. Evol. **222**: pp 187-209

https://doi.org/10.12007/BF00984102

Roy, R., Schmitt, A, J., Thomas, J.B., & Carter, C.J., (2017). Nectar biology: From molecules to ecosystems. Plant Science, **262**: pp 148-164 https://doi.org/10.1016/j.plantsci.2017.04.012

Sammataro, D. and Yoder, C.R.C. Cellular response in Honey Bees to Non-Pathogenic effects of pesticides. Chapter 15 In *Honey Bee Colony Health Challenges and Sustainable Solutions.* (2012) ed Sammataro, D. and Yoder, C.R.C. Press,

Sampson, B., Noffsinger, S., Gupton, C., and Magee, J., (2001) Pollination Biology of the Muscadine Grape. HORTSCIENCE **36** (1): pp 120-124

https://doi.org/10.21273/HORTSCI.36.1.120

Schmidt, J.O., Buchmann, S.L., (1985) Pollen digestion and nitrogen utilization by *Apis mellifera L.* (Hymenoptera:Apidae). Comp. Biochem. Physiol. A Physiol 82 (3): pp 499-503https://doi.org/10.1016/0300-9629(85)90423-2

Schmidt, J.O., Johnson, B.E., (1984) Pollen feeding preference (Hymenoptera: Apidae), a polylectic bee. Southwestern Entomologist **9**: pp 41-47

Schneider, SS and LA Lewis, L.A., (2003. Honey bee communication: The "tremble dance", the "vibration signal" and the "migration dance". **Monographs in Honey Bee Biology, No. 1**. Northern Bee Books, West Yorks, Great Britain, 26pp

Scofield, H.N., Mattila, H.R., (2015) Honey bee workers that are pollen stressed as larvae become poor foragers and waggle dancers when adults.

PLoS ONE 10(4):e0121731

https://doi.org/10.10.1371/journal.pone.0121731

Seeley, T. D. *Following the Wild Bees. The Craft and Science of Bee Hunting.* (2016). Princeton University Press, Princeton and Oxford

Seeley, T.D. *Honeybee Ecology.* (1985). Princeton University Press

Seeley, T.D. (1986) Social foraging by honeybees – how colonies allocate foragers among patches of flowers. Behav. Ecol. Sociobiol. **19**: pp 343-354

https://doi.org/10.1007/BF00295707

Seeley, T.D. *The Wisdom of the Hive. The Social Physiology of Honey Bee Colonies.* (1995). Harvard University Press

Seeley, T.D. *Honeybee Democracy.* (2010) Princeton University Press

Seeley, T.D., *The Lives of Bees. The Untold Story of the Honey Bee in the Wild.* (2019). Princeton University Press

Showler, K. *The Observation Hive.* (1978). Bee Books New and Old.

Simone, M., Evans, J.D., and Spivak, M., (2009). Resin Collection and Social Immunity in Honey Bees. Evolution **63** (11): pp 3016-3022

Doi:10.1111/j.1558-5646.2009.0072.x

Simone-Finstrom, M.D. and Spivak, M., (2010). Propolis and bee health: the natural history and significance of resin use by honey bees. Apidologie **41**: pp 295-311

DOI:10.1051/apido/2010016

Simone-Finstrom, M.D. and Spivak, M., (2012). Increased resin collection after parasite challenge: a case of self-medication in honey bees?

PLOS ONE 7(3)e34601

https://doi.org/10.1371/journal.pone.0034601

Sims, D. *Sixty Years with Bees.* (1997). Northern Bee Books, Mytholmroyd, England

Smith, D.R. Biogeography of Honey Bees. In C. Starr (ed) Encyclopedia of Social Insects., (2020) Springer Nature, Switzerland

Somerville, D.C., (2000). Honey bee nutrition and supplementary feeding. Agnote ISSN 1034-6848, NSW Agriculture, DAI-178, 8 pages

Somerville, D.C., (2001). Nutritional value of bee collected pollens. A report for the Rural Industries Research and Development Corporation, RIRC Publication No. 01/047. RIRDC

Project No. DAN-134A, NSW Agriculture Australia

Somerville, D.C., (2005). Fat bees, skinny bees- a manual on honey bee nutrition for beekeepers. NSW Dept. of Primary Industries, Australia. RIDC Publication Number 05/54

Sparks, T., Whittle, L., Garforth, G., (2019). A comparison of Nature's Calendar with Gilbert White's phenology. British Wildlife **31** (4) April (2020). pp 271-275

Stace, C., *New Flora of the British Isles 2nd ed.* (1997). Cambridge University Press

Starks, P.T., Blackie, C.A., Seeley, T.D., (2000). Fever in honeybee colonies. Naturwissenschaften **87**: pp 229-231

https://doi.org/10.1007/s001140050709

Storch, H., *At the Hive Entrance* (1985) published as a European Apicultural Edition, Brussels

Tautz, J., *The Buzz about Bees. Biology of a Superorganism.* (2008). Springer

The State of Nature Reports (2019) available online from National Biodiversity Network. https://nbn.org.uk

Thomson, J., Draguleasa, M., Tan, M., (2015). Flowers with caffeinated nectar receive more pollination. Arthropod Plant Interact **9**: pp 1-7

https://doi.org/10.1007/s11829-014-9350-z

Tosi, S., Nieh, J.C., Sgolastra, F., Cabbri, R., Medrzcki, P., (2017). Neonicotinoid pesticides and nutritional stress synergistically reduce survival in honey bees. Proc. Biol. Soc. **284** (1869).

Doi:101098/rspb.2017.1711

Toth, A.L., and Robinson, G.E., (2005). Worker nutrition and division of labour in honeybees. Anim. Behav. **69** (2) : pp 427-435

https://doi.org/10.1016/j.anbehav.2004.03.017

van Doorn, W.G. and van Meeteren, U., (2003). Flower Opening and Closure: a review. Journal of Experimental Botany, **54**, Issue 389 pp 1801-1812

https://doi.org/10.1093/jxb/erg213

Visscher, P.K., and Seeley, T.D., (1982). Foraging strategy of honeybee colonies in a temperate deciduous forest. Ecology **63**: pp 1790-1801

https://doi.org/10.2307/1940121

von Frisch, K., *Dance Language and Orientation of Bees.* (1967) Harvard University Press, Cambridge, MA von Frisch., K., *The Dance Language and Orientation of Bees.* (2014) Harvard University Press, Cambridge, MA. With foreword by T.D. Seeley

Walker, P. and Crane, E., (2001). English beekeeping from c.1200-1850: evidence from Local Records. The Local Historian **31** (1): 3-30

Wang, T- H, Jian, C-H, Hsieh, Y-K, Wang, F-N, Wang, C-F., (2013). Spatial distributions of inorganic elements in honeybees *(Apis mellifera L.)* and possible relationships to dietary habits and surrounding environmental pollutants. J. Agric. Food Chem **61**: pp 5009-15

https://doi.org/10.1021/jf400695w

Weber, M.G., and Keeler, K.H., (2013). The phylogenetic distribution of extrafloral nectaries in plants. Annals of Botany, **111** (6): pp 1251-1261

https://doi.org/10.1093/aob/mcs225

Weber, M.G., Porturas, L.D., and Keeler, K.H., (2015). World list of plants with extrafloral nectaries.

www.extrafloralnectaries.org

Wheeler, M.M. and Robinson, G.E., (2014). Diet dependent gene expression in honey bees: honey vs. sucrose or high fructose corn syrup. Scientific Reports **4**: 5726.DOI:10.1038/srep05276

Whitaker, J.M., *The Ethics of Beekeeping.* (2018). Northern Bee Books, Mytholmroyd, England

Williams, I.H., Carreck, N., and Little, D.J., (1993). Nectar Sources for Honey Bees and the Movement of Honey Bee Colonies for Crop Pollination and Honey Production in England. Bee World, **74** (4): pp 160-175. https://DOI:10.1080/0005772X1993.11099182

Willmer, P., *Pollination and Floral Ecology.* (2011). Princeton University Press

Wilson, Bee., *The Hive. The Story of the Honeybee and Us.* (2004). John Murray

Winston, M.L., *The Biology of the Honey Bee.* (1987). Harvard University Press

Woodward, D., *Queen Bee: Biology, Rearing and Breeding.* (2007). Northern Bee Books

Wright, J., *A Natural History of the Hedgerow and ditches, dikes and dry stone walls.* (2017). Profile Books.

Wright, G.A., Nicolson, S.W., and Shafir, S., (2018). Nutritional Physiology and Ecology of Honey Bees. Ann. Rev. Entomol. 2018. **63**: pp 327-344

https://doi.org/10.1146annurev-ento-020117-043423

Annex I

Some families of plants used by honey bees, with notes on their nectaries

Aceracae (Acer family) e.g. Sycamore *(Acer pseudoplatanus),*
Nectar is secreted by thick fleshy central disc, freely exposed.

Apiaceae (Umbelliferae) (Carrot family) e.g. Carrot *(Daucus carota),*
Markedly protandrous, nectar secreted by epigynous disc easily accessible.

Arialaceae (Ivy family) e.g. Ivy *(Hedera helix)*

In ivy the nectar is secreted from yellowish-green disc surrounding the styles and is freely exposed.

Asteraceae (Daisy family) e.g. Daisy *(Bellis perennis);* Dandelion *(Taraxacum officinalis);* Thistles *(Carduus* spp.*);* Sunflower *(Helianthus annus)*

The corolla tube is usually short enough to enable the nectar secreted by a ring-like nectary at the base of the style to be reached. In longer-tubed flowers the corolla tubes are narrow and the nectar rises in them and is accessible.

Boraginaceae (Borage family) e.g. Myosotis spp. *(Forget-me- not); Borago officinalis* (Borage) Borago and *Symphytum officinale* (Comfrey) and *Pulmonaria officinalis* (Lungwort)

Pendulous flowers, nectar secreted by a nectar disc at the base of the ovary. In borage the insect touches the appendages of the stamens and disturbs the anthers and gets dusted with pollen.

Brassicaceae (Cruciferae) (Cabbage family) e.g.Aubretia *(Aubretia deltoidea); Oil Seed Rape (Brassica napus* subssp. *oleifera);* Turnip *(Brassica rapa);* Sweet Alison *(Lobularia maritima);*

Charlock *(Sinapsis arvensis)*; Thale Cress *(Arabidopsis thaliana)*; Candytuft *(Iberis* spp.); Wallflower *(Cheiranthus cheiri)*; Cabbage *(Brassica oleracea)*.

Nectaries are small green glands situated on the torus at the base of the short stamens. The nectar gathers in the pouches of lateral sepals. Each flower has four nectaries the two at the base of the short stamens and secrete much more nectar and of a higher sugar concentration than the two nectaries situated outside the ring of stamens.

Campanulaceae (Bellflower family) e.g. Giant Bellflower *(Campanula latifolia)*

Flowers are markedly protandrous. Nectar is secreted by a disc developed on top of the ovary and is protected by the triangular bases of the stamens.

Caprifoliaceae (Honeysuckle family) Honeysuckle *(Lonicera* spp.); American elderberry *(Sambucus canadensis)*

Fleshy nectar disc at the base of the corolla tube, some of the species with shorter corolla tubes visited by bees.

Caryophyllaceae (Pink family) e.g. White Campion *(Silene latifolia)*;Stitchwort *(Stellaria* spp.*)*

Nectar is secreted at the base of the stamens, often protandrous.

Dipsacaceae (Teasel family) e.g. Scabious *(Knautia arvensis)*; Teasel *(Dipsacus* spp.*)*

Nectar is secreted on the upper part of the ovary and protected from rain by hairs lining the corolla tube.

Ericaceae (Heather family) e.g. Ling Heather *(Calluna vulgaris)*; Bell Heather *(Erica cinerea)*;

Cross-Leaved Heath (Erica tetralix)

Flowers are often pendulous.Usually protandrous.Well developed nectar disc at the base of the ovary. 8 tiny swellings that alternate with the bases of the stamens. *E. cinerea* and *E. tetralix* – bees visit the pendulous flowers and first touch the projecting stigma and then shake the anthers by pushing past their appendages. *C. vulgaris*- often wind pollinated as the pollen tetrads are dry and readily blown out of the anthers.

Fabaceae (Leguminoseae) Pea family e.g. Red Clover *(Trifolium pratense)*;White Clover *(Trifolium repens)*; Sweet Pea *(Lathyrus odoratus)*; Broad Bean *(Vicia faba)*;

Gorse *(Ulex gallii)*.
In species where nectar is produced it is secreted between the base of the stamen tube and the ovary.

Geraniaceae (Crane's-bill family) e.g. Bloody Crane's-bill *(Geranium sanguineum)*; Dusky Crane's-bill *(Geranium phaeum)*.

Protandrous. 5 nectar glands representing a disc are found as little cushions just outside the bases of antisepalous stamens.

Grossulariaceae (Gooseberry family) e.g. Gooseberry (*Ribes uva-crispa*); Blackcurrant (*Ribes nigrum*).

Bell-shaped flower approx 5 mm deep bee extracts nectar from the open flower. Nectar is secreted at the base of the bell-shaped floral axis (receptacle) and is protected by stiff hairs projecting vertically from the style

Hydrophyllaceae (Phacelia family) e.g. Phacelia (*Phacelia tanacetifolia*)

The nectary is a disc at the base of the ovary protected by special appendages at the base of the stamens (no hindrance to honey bees).

Lamiaceae (Labiatae) (Dead-nettle family) e.g.Lavendar (*Lavandula x intermedia*); Catmint (*Nepeta cataria*); Rosemary (*Rosmarinus officinalis*); Thyme (*Thymus* spp.); Mint (*Mentha* spp.); Salvia (Claries)(*Salvia* spp.)

Flowers usually protandrous. Nectar disc is at the base of the ovary and is best developed on the anterior side.

Liliaceae (Lily family) e.g. Crocus (*Crocus* spp.); Lilies (*Lilium* spp.); Hyacinth (*Hyacynthus orientalis*); Tulip (*Tulipa gesneriana*); Onion (*Allium* spp.); Snowdrop (*Galanthus nivalis*).

In the crocus nectar is secreted by a nectary on top of the ovary. In the bluebell (*Hyacynthoides non-scripta*), nectar in glandulous tissue in partition between chambers of the ovary, whilst in the tulip there is no nectar and flowers are visited only for pollen. In the hyacinth nectar is secreted as large drops from each of three nectaries appearing as dots near the top of the ovary.

Malvaceae (Mallow family) e.g. Garden Tree Mallow (*Lavatera olbia*); Mallow (*Malva* spp.)

Hollyhock (*Alcea rosea*).

Nectar is secreted by the torus in five little pits lying between the bases of the petals and protected by hairs.

Onagraceae (Willowherb family) e.g. Willowherbs (*Epilobium* spp.); Rosebay willowherb (*Chamaenerion angustifolium*); Evening Primrose (*Oenothera biennis*)

Nectar disc is at the top of the ovary.

Primulaceae (Primrose family) Primrose

Homogamous. Nectar secreted around the base of the ovary

Ranunculaceae (Buttercup family) e.g. Bulbous Buttercup (*Ranunculus bulbosus*); Wood Anemone (*Anemone nemorosa*); Winter Aconite (*Eranthis hyemalis*); Traveller's Joy (*Clematis vitalbis*); Pasque Flower (*Anemone pulsitilla*); Marsh Marigold (*Caltha palustris*); Hellebore (*Helleborus* spp.).
Nectary at the base of the carpels In Winter Aconites the petals are reduced to hornlike structures with an oblique mouth in which nectar is secreted.

In Delphinium nectar is secreted at the bottom of a long spur and can only be reached by long-tongued bees. In Buttercups the nectary is at the base of the flower and is covered by a scale. In the Wood Anemone there are no nectaries, as also is the case in Traveller's-joy. In the Christmas Rose (Helleborus niger) there are numerous slipper shaped nectaries.

Rosaceae (Rose family) e.g. Rose (*Rosa* spp); Hawthorn (*Crataegus monogyna*); Raspberry (*Rubus idaeus*); Blackberry (*Rubus fruticosus*); Almond (*Prunus dulcis*); Plum (*Prunus domesticus*); Wild Cherry (*Prunus avium*); Cherry Laurel (*Prunus laurocerasus*); Peach (*Prunus persica*); Strawberry (*Fragaria* spp.); Dwarf Cherry (*Prunus cerasus*)

Flowers of the "open type" e.g.wild rose / strawberry / raspberry / apple / cherry / rowan and hawthorn: in most cases nectar is produced by the whole inner surface of the torus or there is a ring-like nectary around the torus mouth within the insertion of the stamens.

+ /- protogynous	apple, hawthorn, sloe
homogamous	common cherry
protandrous	roses

Self-pollination is apparently possible in all cases.

In the case of the Crab Apple (Malus pumila) nectar is secreted inside the calyx tube. In Raspberry it is secreted in a fleshy ring on the margin of the flower within the stamens. No nectar is secreted in the flowers of many Wild Roses (although they produce abundant pollen) with the exception of the Sweet Briar (Rosa rubigenosa) where a thin layer of nectar is secreted on the broad fleshy margin of the receptacle or calyx tube.

Salicaceae (Willow family) e.g. Poplar (*Populus* spp.); Willow (*Salix* spp.)

Both wind and insect pollinated. Goat Willow (Salix caprea)female flowers have a nectary. Other species have nectaries on both sexes of flowers

Scrophulariaceae (Figwort family) e.g. Foxglove (*Digitalis purpurea*); Common Figwort (*Scrophularia nodosa*); Mullein (*Verbascum* spp.)

Slightly protandrous and in most cases self pollination can occur

Nectar disc is at the base of the ovary / corolla

Thymelaeceae (Mezereon family) e.g. Daphne (*Daphne* spp.)

Here the nectar is secreted at the base of the ovary

Violaceae (Violet family) e.g. Violas (*Viola* spp.)

In the case of Heart's-ease (Viola tricolor) the lower petal becomes a spur into which nectar is secreted. The anther of each of the lower stamens has a nectar secreting area which projects into the spur of the lower petal.

The two antero-lateral stamens bear greenish horn-like appendages projecting into the spur of the anterior petal and functioning as nectaries.

Annex II

Recommended wildflower and grass seed mixtures to benefit bees on field margins or grassland sites of specific soil types

Cultivation notes

These mixtures can be sown in field margins or blocks. Ideally a few blocks of up to 1 ha scattered around the farm / land under management. The seed should be broadcast and rolled rather than drilled.

Glyphosate may be applied as an overall spray immediately before spring sowing, in order to help establish the crop. Otherwise, limit herbicide application to weed wiper or spot treatment, or cut occasionally in the first year, if necessary to prevent weeds dominating.

All of the sown areas should be cut after 15 September, at a height of 10-20 cm, ideally removing the cuttings using a forage harvester. Then, each year cut the same half of the area before the end of June to stimulate late flowering unless there are ground-nesting birds present and repeat the total cut in September.

The legumes in the following mixtures should last 3-5 years before re-sowing may be necessary.

Where appropriate, extensive autumn and winter grazing of cattle produces optimal conditions for bumblebee forage plants to flower in the following spring, by accelerating nutrient removal, and limiting the ability of competitive species dominance. Cattle are better than sheep in producing a more species diverse sward. More established, closer swards are even better for providing more suitable bee nest sites.

Neutral / loamy soils

Common Name	Botanical Name	Flowering period
Wild flowers		
Knapweed	*Centaurea nigra*	July-September
Meadow Geranium	*Geranium pratense*	April-October
Field Scabious	*Knautia arvensis*	June-September
Oxeye daisy	*Leucanthemum vulgare*	June-August
Bird's-Foot-Trefoil	*Lotus corniculatus*	June-August
Self-Heal	*Prunella vulgaris*	May-October
Meadow Buttercup	*Ranunculus acris*	May-July
Corn Rattle	*Rhinanthus minor*	May-August
Red clover	*Trifolium pratense*	May-October
Alsike clover	*Trifolium hybridum*	May-October
Tufted vetch	*Vicia cracca*	June-August
Grasses		
Common bent	*Agrostis capillaris*	
Sweet Vernal-grass	*Anthoxanthum odoratum*	
Crested dog's-tail	*Cynosurus cristatus*	
Sheep's fescue	*Festuca ovina*	
Red fescue	*Festuca rubra* **spp** *commutata*	
Yellow Oat-grass	*Trisetum flavescens*	

Calcareous soils

Common Name	Botanical Name	Flowering period
Wildflowers		
Kidney vetch	*Anthylis vulneraria*	June-September
Knapweed	*Centaurea nigra*	July-September
Greater Knapweed	*Centaurea scabiosa*	July-August
Wild Basil	*Clinopodium vulgare*	July-September
Field Scabious	*Knautia arvensis*	June-September
Oxeye daisy	*Leucanthemum vulgare*	June-August
Bird's- foot- trefoil	*Lotus corniculatus*	June-August
Musk Mallow	*Malva moschata*	July-September
Rest-harrow	*Ononis repens*	June-September
Marjoram	*Origanum vulgare*	July-October
Self-Heal	*Prunella vulgaris*	May-October
Meadow Crowfoot	*Ranunculus acris*	May-July
Common Corn Rattle	*Rhinanthus minor*	May-August
Red clover	*Trifolium pratense*	May-October
Tufted vetch	*Vicia cracca*	June-August
Grasses		
Common bent	*Agrostis capillaries*	
Crested dog's-tail	*Cynosurus cristatus*	
Sheep's fescue	*Festuca ovina*	
Chewings fescue	*Festuca rubra* **spp.** *commutata*	
Timothy grass	*Phleum bertolonii*	
Yellow Oat-grass	*Trisetum flavescens*	

Acidic / sandy soils

Common Name	Botanical Name	Flowering period
Wildflowers		
Knapweed	*Centaurea nigra*	July-September
Viper's Bugloss	*Echium vulgare*	June-September
Oxeye daisy	*Leucanthemum vulgare*	June-August
Common Toadflax	*Linaria vulgaris*	July-October
Bird's-Foot-Trefoil	*Lotus corniculatus*	June-August
Musk Mallow	*Malva moschata*	July-September
Self-Heal	*Prunella vulgaris*	May-October
Common Corn Rattle	*Rhinanthus minor*	May-August
Betony	*Stachys officinalis*	June-August
Red clover	*Trifolium pratense*	May-October
Common vetch	*Vicia sativa*	May-July
Grasses		
Common bent	*Agrostis capillaris*	
Sweet Vernal-grass	*Anthoxanthum odoratum*	
Crested dog's-tail	*Cynosurus cristatus*	
Wavy Hair-grass	*Deschampsia flexuosa*	
Sheep's fescue	*Festuca ovina*	
Red fescue	*Festuca rubra*	

Wet soils

Common Name	Botanical Name	Flowering period
Wildflowers		
Knapweed	*Centaurea nigra*	July-September
Meadowsweet	*Filipendula ulmaria*	June-September
Greater Bird's-foot-trefoil	*Lotus pedunculatus*	June-August
Ragged robin	*Lychnis flos-cucili*	May-August
Wood Betony	*Stachys officinalis*	June-August
Marsh Woundwort	*Stachys palustris*	June-August
Self-Heal	*Prunella vulgaris*	May-October
Devil's Bit scabious	*Succisa pratensis*	July-August
Red clover	*Trifolium pratense*	May-October
Grasses		
Meadow or Common Fox-tail	*Alopecurus pratensis*	
Sweet Vernal-grass	*Anthoxanthum odoratum*	
Tufted Hair-grass	*Deschampsia caespitosa*	
Red fescue	*Festuca rubra*	

Notes

Page No.

Notes

Page No.

Notes

Page No.

Notes

Page No.

www.ingramcontent.com/pod-product-compliance
Lightning Source LLC
Chambersburg PA
CBHW041801280326
41926CB00103B/4764